JOHANNES KEPLER • Mysterium Cosmographicum
The Secret of the Universe

TRANSLATION BY A.M. DUNCAN

INTRODUCTION AND COMMENTARY BY E.J. AITON

WITH A PREFACE BY I. BERNARD COHEN

ABARIS BOOKS • NEW YORK

Copyright © 1981 by Abaris Books, Inc.
International Standard Book Number 0-913870-64-1
Library of Congress Card Number 77-86245
First published 1981 by Abaris Books, Inc.
24 West 40th Street, New York, New York 10018
Printed in the United States of America
A William J. Prendergast Production

THE SECRET OF THE UNIVERSE

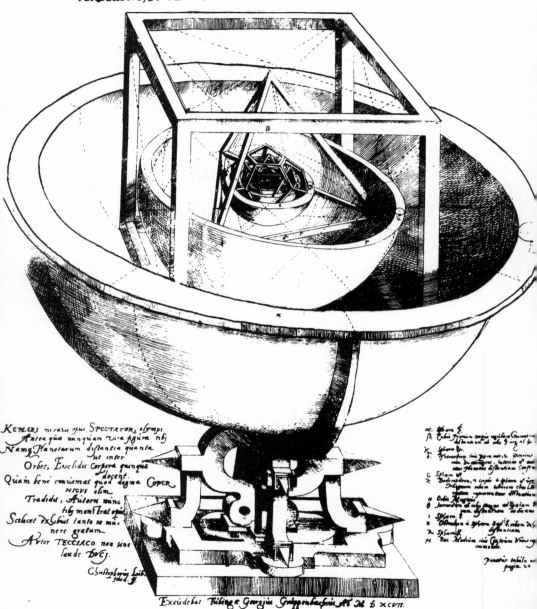

Johannes Kepler, *Mysterium cosmographicum*. Copper engraving
from the first edition (Tübingen, 1596).

CONTENTS

04387

PREFACE

Although Johannes Kepler is universally esteemed as one of the major scientists of the seventeenth century and one of the greatest astronomers who has ever lived, his reputation has not been paralleled by a series of translations of his works into languages other than German. Until now, not one of Kepler's major publications, in which his discoveries were disclosed, has appeared in a complete English version, nor are any of these writings available in French or Italian translations.[1] These works include — in addition to Kepler's *Mysterium cosmographicum* (1596; ed. 2, 1621) — the *Ad Vitellionem paralipomena* or *Astronomiae pars optica* (1604), the *Astronomia nova* (1609), the *Harmonice mundi* (1619), the *Epitome astronomiae Copernicanae* (1618-1620-1621), and the *Tabulae Rudolphinae* (1627). Until not very long ago, only a few short selections had been published in English,[2] of which the longest was Kepler's response to Galileo's *Sidereal messenger,* taken from the introduction to Kepler's *Dioptrice* (1611).[3]

The first translations of any usable length were published in 1952 as part of the series of "Great Books of the Western World": Book 4 and Book 5 of the *Epitome of Copernican astronomy* and Book 5 of the *Harmony of the world.*[4] Then, in 1965, Edward Rosen brought out his English version of Kepler's *Conversation with Galileo's sidereal messenger*[5] and John Lear presented *Kepler's dream,*[6] so that at last a work of Kepler's was completely translated and published in its entirety.[7] The next year, 1966, saw an English version of Kepler's little New Year's essay of 1611 on the snowflake[8] and in 1967 Rosen produced his English translation of Kepler's *Somnium* or *Dream,* the third time this work had been rendered in an English version.[9] At last there had been made available to English and American readers a complete work written by Kepler on an astronomical subject. In 1979 J.V. Field published an English version of Kepler's account of star polyhedra.[10]

Now, thanks to the intellectual labors of Alistair Duncan and Eric Aiton, we have an annotated translation of the *Mysterium cosmographicum,* Kepler's first full-dress essay in astronomy. May we hope that this same team will next produce a translation of the *Harmony of the world,* in which Kepler announced the third or harmonic law of planetary motion. How useful it would be if scholars and students could also have available an English version of Kepler's revolutionary *New astronomy,* in which Kepler disclosed the law of equal areas and the elliptical orbits of the planets!

Kepler's *Mysterium cosmographicum* of 1596 marked his public appearance as a major astronomer. It is a notable book for the student of Kepler's thought not only because it is his first significant astronomical work, but also because it was revised by Kepler for a new edition (1621), in which he introduced a series of annotations containing second thoughts

7

on the topics discussed and relating his ideas to the exciting new develop- ments made in astronomy between the two editions, of which the most remarkable had been the disclosures made by the use of the astronomical telescope.[11] A particularly valuable feature of the present edition is the in- clusion of these later notes.[12]

Kepler's *Mysterium cosmographicum* usually appears in the literature of the history of astronomy primarily because it contains the first astronomical "law" that Kepler discovered, a geometric relation among the average distances of the planets from the sun in the Copernican system. This "law" stated that the planetary orbits lie in a set of six imag- ined nested spheres, separated by the five regular solids, so placed that each solid circumscribed the immediate inner sphere and was circum- scribed by the next outer sphere.[13] The discovery of this "law" is often considered as a curiosity or as an aberration of Kepler's youth, but readers of *The secret of the universe* will see that Kepler displayed in this discovery the same traits of logical reasoning, geometric skill, and com- putational ability that characterize his later writings such as *The new astronomy*. Far from rejecting this "law" as an extravagance of his youth, Kepler continually reiterated his faith in its truth and importance. This is clear not only from the notes Kepler added to the second edition in 1621 (and which are given in English translation in the present edition), but also from the fact that he did not ever revise or retract the statement at the beginning of this book, in which he declared that its purpose was to demonstrate that when God created the universe and "determined the order of the heavenly bodies" or planets, He had in mind "the five regular bodies which have enjoyed such great distinction from the time of Pythagoras and Plato down to our own days." In the *Harmonice mundi* (1619), he reiterated his faith in what he had shown "in my *Mysterium cosmographicum,* published twenty-two years ago, that the number of planets, or spheres, surrounding the sun had been fixed by the all-wise creator as a function of the five regular solids on which Euclid wrote a book many centuries ago."[14] He even included a diagram indicating the relation of the planetary orbital spheres to the regular solids. This scheme also appears at great length in full display in Book 4 of Kepler's *Epitome of Copernican astronomy*.

This continuing belief in the planetary "law" of geometric solids ex- plains why Kepler was so disturbed when he heard that Galileo had discovered some new "planets." Since there are only five regular solids, Kepler's "law" provides for only six spheres, corresponding to six possible orbits and six planets. There is no place for any additional planets and, in fact, this was an argument that Kepler had used to justify belief in the Copernican system with its six planets (Mercury, Venus, earth, Mars, Jupiter, Saturn) as opposed to the Ptolemaic system with its seven planets (moon, Mercury, Venus, sun, Mars, Jupiter, Saturn).[15] Eventually, when Kepler got hold of a copy of Galileo's book, he discovered that the new

"planets" (or wandering astronomical bodies) were secondary planets or satellites and so did not disturb his system. He could thus continue to maintain his belief that God had had in mind the five regular solids when He had designed and created the solar system.

What may most commend this book to astronomers is that in it Kepler sets forth a goal, method, and program, which he was to follow successfully throughout the rest of his astronomical career. As he says in the beginning of the *Mysterium cosmographicum,* there were three things for which he chiefly sought the cause: the number and size of the planetary orbits and the motions of the planets in those orbits. In the *Harmonice mundi,* announcing the third law of planetary motion, Kepler could refer to "that which I prophesied two-and-twenty years ago, as soon as I discovered the five solids among the celestial orbits." As Eric Aiton points out below, almost all "the astronomical books written by Kepler (notably the *Astronomia nova* and the *Harmonice mundi*) are concerned with the further development and completion of themes that were introduced in the *Mysterium cosmographicum.*"

I. Bernard Cohen

NOTES ON PREFACE

1. On Kepler's reputation and the lack of availability of his writings in English versions, see my "Kepler's century: prelude to Newton's," *Vistas in Astronomy,* 1975, vol. 18, pp. 3-36, esp. pp. 34-35. On editions and translations, see Max Caspar: *Bibliographia Kepleriana* (Munich: C.H. Beck'sche Verlagsbuchhandlung, 1936; revised and updated by Martha List, Munich, 1968). See additionally, the bibliographical supplement to the account of Kepler by Owen Gingerich in the *Dictionary of Scientific Biography,* 1973, vol. 7, pp. 289-312, and Martha List's " 'Bibliographia Kepleriana' 1967-1975," *Vistas in Astronomy,* 1975, vol. 18, pp. 955-1010.

2. Notably John H. Walden's translation of a portion of the *Harmonice mundi* on pp. 30-40 of Harlow Shapley and Helen Z. Howarth (eds.): *A source book in astronomy* (New York: McGraw-Hill Book Company, 1929). This same volume also includes a two-page extract on the reconciliation of the texts of Scripture with the Copernican doctrine of the mobility of the earth, taken from Thomas Salusbury's translation of a portion of the *Astronomia nova,* originally published in 1661.

Additionally, in 1951, Carola Baumgardt made available in English a number of Kepler's letters; see her *Johannes Kepler: Life and letters,* with an introduction by Albert Einstein (New York: Philosophical Library, 1951).

3. *The Sidereal messenger of Galileo Galilei and a part of the preface to Kepler's Dioptrics containing the original account of Galileo's astronomical discoveries,* a translation with introduction and notes by E.S. Carlos (London, Oxford, Cambridge: Rivington's, 1880; facsimile reprint, London: Dawsons of Pall Mall, 1959).

4. Robert Maynard Hutchins (ed. in chief): *Great books of the western world,* vol. 15, Ptolemy, Copernicus, Kepler (Chicago, London, Toronto: Encyclopaedia Britannica, 1952). The translations from Kepler had been completed in 1939 by Charles Glenn Wallis.

5. *Kepler's conversation with Galileo's sidereal messenger,* first complete translation, with an introduction and notes, by Edward Rosen (New York, London: Johnson Reprint Corporation, 1965—The Sources of Science, No. 5).

6. John Lear: *Kepler's dream: with the full text and notes of Somnium, sive astronomia lunaris, Joannis Kepleri,* translated by Patricia Frueh Kirkwood (Berkeley, Los Angeles: University of California Press, 1965).

7. An earlier translation, made by Joseph Keith Lane, was submitted in June 1947 "in partial fulfillment of the requirements for the degree of Master of Arts in the Faculty of Philosophy, Columbia University."

8. Johannes Kepler: *The six-cornered snowflake,* translated by Colin Hardie, with the Latin text on facing pages, and essays by B.J. Mason and L.L. Whyte (Oxford: at the Clarendon Press, 1966).

9. *Kepler's Somnium, the dream, or posthumous work on lunar astronomy,* translated with a commentary by Edward Rosen (Madison, Milwaukee, London: The University of Wisconsin Press, 1967). For the earlier translations see notes 6 & 7 *supra.*

10. See J.V. Field's "Kepler's star polyhedra," *Vistas in Astronomy,* 1979, vol. 23, pp. 109-141.

It should be noted that Alexandre Koyré's *La révolution astronomique* (Paris: Hermann, 1961), translated as *The astronomical revolution* (London: Methuen; Ithaca: Cornell University Press; Paris: Hermann, 1973), contains such a wealth of extracts as to constitute a veritable Keplerian anthology; these were rendered into French by Koyré and given in English versions made by the translator of the volume, R.E.W. Maddison.

11. The first publication of the astronomical revelations of the telescope was made in Galileo's *Sidereus nuncius* (1610), for which see notes 3 & 5 *supra.*

12. Kepler's *Somnium* (see notes 6 & 9 *supra*) also contains a series of notes; these are very extensive and by far outweigh the short text.

13. The spheres are purely imaginary since Tycho's studies of the motion of a comet through the solar system had shown conclusively that real spheres (say of crystal) could not exist. Kepler says explicitly in the *Mysterium cosmographicum* that the spheres are not supposed to be real or physical, citing Tycho's work.

14. *Harmonice mundi,* Book 5, ch. 3; see Koyré's *Astronomical revolution* (cited in note 10 *supra*), pp. 330-331.

15. See I.B. Cohen: "Perfect numbers in the Copernican system: Rheticus and Huygens," *Studia Copernicana,* 1978, vol. 16 ("Science and History: Studies in honor of Edward Rosen"), pp. 419-425.

APPARATUS CRITICUS

The Latin text is a facsimile of the second edition (Frankfurt, 1621). The two editions (Tübingen, 1596 and Frankfurt, 1621) are designated by *T* and *F* respectively. Misprints and textual variations are printed in roman, editorial explanations in italic. In general, misprints in *T* have only been indicated when these also occur in *F*. However, the variant readings in *T* enable the reader to recognize those misprints which originate in *F*. Again only variants in words are indicated, so that differences in spelling or in abbreviated forms are not shown. The number references to Kepler's own notes were of course added in *F* but their absence in *T* has not been indicated in each individual case.

The following are a few examples of the notation used.

42) repererit

This means that, in line 42 of the relevant page (here p. 54), there is a misprint, the correct word being repererit.

1) *T:* Verum hoc pacto neque

This means that, on line 1 (p. 64), in place of Verum neque, *T* has Verum hoc pacto neque.

Notae Auctoris in F only.

This means that the collection of notes appended to the chapter is to be found only in *F*.

The following abbreviations are used by Kepler in the text.

Latin		*Greek*	
ℯ for ӕ		℮ for ρο	
& for et		ς for στ	
q; for que		ꝋ for ου	
ꝗ for que			
sc. for scilicet			
n. for enim			
á, é, ú for am, em, um			
ét for etiam			

Footnotes to Latin text, showing misprints, variants, etc. Numbers in brackets [] are pages in this edition.

[36] 1) *This epigram (Latin translation only) was printed on the title page in T.*
[38] *Epistola Dedicatoria in F only.*
  24) Astronomiae
[40] 32) aliis
[44] 17) lugentibus
[48] I.K. *added at the end of the poem in T.*
[50] *Notae in F only.*
[52] 24) atqui *T:* atqui
[54] 42) repererit
[56] *Notae Auctoris in F only.*
[60] 35) distinctis
[62] 1) *T:* Praefatio Ad Lectorum
  26) *T:* &c 95
  31) intermedii
[64] 1) *T:* Verum hoc pacto neque
[68] 7) Dodecaedron *T:* Dodecaedron
[70] *Notae Auctoris in F only.*
  18) fol. 437 & 438
  34) fol. 145

[72] 1) sit
  5) maximae
  15) fol. 26
[76] 19) *T:* apparet, quod
  21) *T:* demonstravit, et quod ex eo nulla sit causa, simile
[80] 35) angulo TGV
[82] 3) Planetarum. (14) Nam
[84] *Notae Auctoris in F only.*
[92] 5/6) ouo, qua *T:* ouo, qua
[98] 2/3) (ut ipse Rhetico dicere solitus est)
  T: (ut ipse Rhetico dicere solitus est)
  4) credidit ea *T:* credidit ea
[100] 4) *T:* Annotatio in Caput Secundum
  35) constituitur
  38) constituitur
  40) sane alijs
[102] *Notae Auctoris in F only.*
[110] 29) *T:* etiam basium non
[112] 34) multifacia *T:* multifacia
[114] *Notae Auctoris in F only.*
[118] *Notae Auctoris in F only.*

[120] *Notae Auctoris in F only.*
[122] 38) Dodecaedron *T:* Dodecaedron
       39) Dodecaedri
[124] 3/4) *T:* igitur per medium, decem
       lateribus, talem describit viam,
       transeunte
       Notae Auctoris in F only.
[126] 37) lateribus
[134] 30) repererit
[136] 20) patet, (32) quare
       Notae Auctoris in F only.
       39) in terna
[140] 42) Harmonicarum
[142] 7) uni
       13) ut in Harmon. lib. III
       22) rectanguli
       48/49) per sectionem
[144] 21) scripta
       25) Trisdiapason cum epidiapente
       44) repudiat
[148] 25) in O, HGL *T:* in O, HGL
       33) *T:* autem ex HI
       39) cubi habeat
[150] 43) OH
       44) OIH
[152] 5) *T:* Item Octaedron
       10) exsecto
       23/25) *Breviter...NM 36 in F only.*
       Notae Auctoris in F only.
[154] 43/44) repleatur
[156] 4) ob oculos *T:* ob oculos
       4/6) *Parentheses in F only.*
       25) *T:* terreni, et lunam comprehendens
[158] *Notae Auctoris in F only.*
       7) diversos
       9) 723
       40) eorum
[162] 6) 1 38 52
       29) +0 30
       Notae Auctoris in F only.
[168] *Notae Auctoris in F only.*
       8) solis
       19) spissitudinis
       19/20) estimae

[170] 37) impediunt
[174] 15) *T:* Porro, varietas
       26/27) ἐπιχείρημα *T:* ἐπιχείρημα
       Notae Auctoris in F only.
       37/38) aliorsum
[180] 6) post hanc *T:* post hanc
[182] 7) τῶν
[184] 8) anni 1551 *T:* anni 1551
       39) possemus
[186] *Notae Auctoris in F only.*
       15/16) Geometricas
       29) considerarent
[188] 13) specie *T:* specie
[190] 10) ab ipsis
       18) inferiorum theorijs *T:* inferiorum
       theorijs
[192] 1) sollicite
[194] *Notae Auctoris in F only.*
       10) ante 17. annos
[200] 11) Venerium
[202] 23) *T:* est ambo theoremata
       Notae Auctoris in F only.
       42) delapsam
[204] 23) 526⅛ dat distantiam Martis
[206] 30) lucens
       44) attestantur
[208] 5) orbes
       28/29) iugulare
       38) confessionem
[210] 16) 658 694 +36 *T:* 658 694 +26
       27) 36.43
[212] 16) 958
       41) falso
[214] 20) inter medias
[216] 12) tardus
[218] *Notae Auctoris in F only.*
[222] 8) boreo
       11) erit ♄ in ♎ *T:* erit ♄ in ♎
       21) extremam
       Notae Auctoris in F only.
       42) Astronomicis
[224] 14) si ineffabiles

TRANSLATOR'S NOTES

Kepler writes sound Renaissance Latin, which is on the whole correct Ciceronian Latin with the exception of a few constructions not found in the best classical prose. For instance, although he sometimes uses the accusative and infinitive for short indirect statements, he generally uses *quod* with the subjunctive, particularly for longer indirect statements. This construction is not found in formal classical Latin but later became normal.

Similarly, although he has a taste for slightly unusual and colorful words and phrases, Kepler's vocabulary is essentially that of classical Latin. There are occasional exceptions. For instance, in note 37 to chapter 12 in the second edition (page 146) he seems to use *causor* to mean "cause" rather than in its normal post-Augustan sense of "give as a pretext"; and like Copernicus he occasionally uses *ipse* simply as a definite article, especially with indeclinable nouns or with mathematical expressions used as nouns, where there is no other means but the declension of *ipse* to indicate the case, as in the phrase *ipsius AC* in note 4 to chapter 22 (page 218).

Like many other authors of his time, Kepler frequently uses Greek words, usually but not always in Greek script, where there is no obvious reason for not using the equivalent Latin word. For example, he uses the Greek for the title of Copernicus's *De revolutionibus* more than once (for instance on page 182), although there is no apparent reason for not using the Latin title, as he does elsewhere, and although the book was, after all, written in Latin. Caspar (KGW 1:430) explains that this particular Greek phrase is derived from Rheticus's *Ephemerides* (Leipzig, 1550). Originally no doubt authors used Greek words and phrases in this way because a knowledge of Greek was a recent acquisition of Renaissance scholars and its use seemed to add a touch of distinction to their styles; or possibly they may just have been showing off their knowledge. However, by Kepler's time the habit had become so common that Greek words might be used almost without thinking. The effect is rather like the use of occasional French words or phrases in English. For the sake of clarity, however, a French equivalent for a Greek word has been used in only one or two cases in this translation, where the meaning is obvious; and the English translation has usually been given.

Kepler is inclined to slight inaccuracies in the classical references with which he embellishes his prose. For instance, in chapter 18 (page 178) he misquotes Horace, *Epistles*, I, i, 32 as *"Est aliqua prodire tenus"* instead of *"Est quadam prodire tenus."* The misquoted version scans correctly, and is also found in *De cometa anni 1607* (KGW 8:139). However, such slips occur only in allusions which do not affect the sense of what Kepler is saying, and probably show merely that he did not always have the means of verifying such references at hand.

The following words need a particular comment.

Artifex. This word has been translated "practitioner." Kepler refers to Tycho Brahe as *summus Artifex* ("the leading practitioner") in the notes on the original dedication added to the second edition (page 58). It seems to mean a practicing astronomer who actually makes observations of the positions of stars and calculations based on a given cosmographical model. In the first paragraph of chapter 18 (page 176) he distinguishes them from the cosmographers or physicists, who are presumably those who construct cosmographical models of spheres and so forth without quantitative data. Clearly Kepler counts himself not in this class, but as a practitioner. In the second paragraph of chapter 18 he seems to distinguish practitioners also from astronomers, who are presumably those who use the results of

the practitioners in practice. However, in chapter 19, pages 188-90, astronomers and practitioners seem to be the same.

Corpus. Sometimes, as in chapter 2 (page 94), Kepler uses *corpus* to mean "matter," though not quite in the full Aristotelean sense. Elsewhere he uses it to mean a body such as a star. However, where he uses the word to refer to the five perfect solids it has been translated as "solid," as it would sound very odd in English to speak of "the five perfect bodies."

Demonstro and *demonstratio.* These words, which mean literally "show" or "demonstrate" and "demonstration," are sometimes used in that general sense, sometimes to mean "prove" or "proof," and once to mean the geometrical construction for particular figures, referring presumably to the demonstration that the construction does yield the figure concerned (in note 16 on chapter 12, page 142). Often, however, they mean the process of inferring a particular set of data, or formula for calculation, from the system of circles or other figures assumed as the hypothesis for the principles governing the phenomena concerned. This process is called in English "deriving" a formula or set of data, and the words "derive" and "derivation" have therefore been used to translate *demonstro* and *demonstratio* in such cases, where "prove" or "proof" would not represent them correctly. Copernicus habitually uses those words in this sense.

Exorno. When Kepler writes of a particular planet or group of planets being associated with a geometrical construction of spheres or circles which will account for its observed motions, he sometimes refers to it as being "fitted out," using the word *exorno* in its primary sense of equipping, fitting out, or furnishing. However, the word also carries with it something of the secondary but common sense of embellishing or decorating. It has therefore been paraphrased in this translation as "display," which is intended to convey the connotations of *exorno* rather than its literal meaning.

Mundus. In Kepler's time, as well as before him and long after, the universe was generally supposed to be bounded by the sphere of the fixed stars, and there was therefore little ambiguity in using *mundus* as if it were synonymous with *universum,* the universe. However, properly speaking, *mundus* in such writers as Copernicus and Kepler refers only to that part of the universe which includes the sun, moon, earth, and other planets, and is bounded by the visible stars. It is difficult to avoid translating the word as "universe," as has been done here; but no answer is implied to the question whether there is anything outside the fixed stars.

Orbis. This word can mean either a sphere, such as the hollow spheres which are supposed to make up the *mundus,* or the circle, passing through the thickness of the sphere, on which the planets move. This ambiguity has been pointed out by Edward Rosen in *Three Copernican Treatises,* (2nd edition [New York, 1959], pp. 13-21). Birkenmajer recommended therefore that *orbis* should be translated into Polish by the word *krag,* which had a similar ambiguity, though he did not follow his own recommendation (see Copernicus, *Opera Omnia,* Vol. II [Warsaw, 1975], p. 356). However, there is no suitable word in English which has such a convenient ambiguity. Where Kepler clearly means a two-dimensional circle, as in chapter 1 (Plate I) where he is referring to circles drawn in the diagram to represent the orbits, *orbis* has been translated as "orbit." Elsewhere *orbis* must refer to the solid sphere, as in the title of chapter 13 where Kepler discusses the spheres inscribed within the five perfect solids; and in those cases the word has been translated "sphere," though the same English term has been used where Kepler uses *sphaera.* Nevertheless, there are many instances where *orbis* is used with little indication, or no certain indication, of which sense is intended. In those instances "orbit" or "sphere" has been chosen according to which seemed the most probable.

Translator's Notes

Prosthaphaeresis. This is a Greek word, usually written by Kepler in Latin script, meaning a correction to be added to or subtracted from a mean position in calculating a true position. However, as the word *prosthaphaeresis* is used in the history of mathematics to mean a quite different process, the word has been translated here as "equation," representing the equivalent Latin term *aequatio.*

Proportio and **ratio.** In English the word "ratio" generally refers to the relationship between a pair of numbers, and "proportion" to the relationship between the various dimensions of a figure or between a series of more than two numbers. This distinction has generally been maintained in the present translation. However, the distinction between the two corresponding words in Kepler's Latin is not the same; and indeed it has often seemed appropriate to translate *proportio* as "ratio" and *ratio* as "proportion," though not invariably.

Quantitas. In some cases Kepler uses *quantitas* to mean merely "amount" or "quantity" in a general sense. However, in chapter 2 and occasionally elsewhere he uses the word in a special technical sense to mean whatever is capable of being represented quantitatively. Thus on page 94 he writes "We see three kinds of quantity in the universe: the shape, number, and extension of objects." The English word "quantity" has been used to translate *quantitas* in this sense, although its normal meaning is not quite as wide as Kepler's use of the Latin, since there is no exact English equivalent. Kepler also sometimes uses *quantum* as a noun equivalent to *quantitas* in either sense.

Species. Kepler sometimes uses *species* in its ordinary senses to mean "appearance" or "kind." However, in the particular phrase *species immateriata,* which he uses for instance in note 4 to chapter 16 and in note 3 to chapter 20, in the second edition, he clearly means something akin to the Neoplatonic sense of an emanation flowing from God. *Species* was occasionally used in classical Latin to mean "vision," as in Lucretius, *De rerum natura* IV:236, 242, and V:707, 724. However, it was also used to mean a very fine simulacrum of itself which was given off by a visible body, and by the effect of which on the eye it was seen. For the Epicureans, of course, such a simulacrum would be material; but a *species* might also be immaterial. In the later Middle Ages the word was used in this sense by Robert Grosseteste and Roger Bacon, under the influence of Neoplatonic philosophy, to mean a power propagated by a body, of which light was only one example. (See A.C. Crombie, *Robert Grosseteste and the Origins of Experimental Science,* 2nd ed. Oxford, 1962 (1953), pp. 104-16 and 144-47). The phrase *species immateriata* has therefore been translated here as "immaterial emanation."

Vale. The Romans put the names of both the writer of a letter and the person to whom it was addressed at the beginning of the letter, and at the end of it wrote *Vale* or *Valete* (literally "farewell"). This pattern was followed in Renaissance Latin. However, since the modern custom is to put the writer's name at the end of a letter, *Vale* has here been translated simply as "J. Kepler."

INTRODUCTION

Johannes Kepler was born in Weil der Stadt on 27 December 1571* and six years later moved with his parents to nearby Leonberg. In 1589, after attending the Cloister Schools in Adelberg and Maulbronn, he entered the theological college of the University of Tübingen, at that time one of the leading Lutheran centers of higher education. For the first two years Kepler studied in the faculty of arts, taking his M.A. in 1591, and then entered on his theological studies with the intention of becoming a pastor in the Lutheran Church. A few months before he was due to complete his theological studies in 1594, there occurred an event which proved to be a decisive turning point in Kepler's life. Georg Stadius, the mathematics teacher at the Protestant School in Graz, had died, and in response to a request from the school authorities, the theological faculty in Tübingen recommended Kepler for the post (KGW 19, 3).[1] Although Kepler had already begun to question some Lutheran doctrines — and indeed would have encountered great difficulties if he had become a Lutheran pastor — this lack of orthodoxy could not have been the reason why the Tübingen theologians sent him to Graz, for at this time, as Kepler himself relates,[2] he had, on account of his youth, kept his theological doubts to himself. There is no reason to suppose that the Tübingen theologians had any other motive than the desire to recommend the candidate with the best qualifications, who would do credit to the University. In accepting the post, Kepler expressed his wish to be allowed later to return to his theological studies (KGW 13, 9-11).

Kepler had been introduced to the Copernican system by his teacher Michael Maestlin at the University of Tübingen. On the basis of Maestlin's lectures and his own reflections, he gradually compiled a list of superiorities of Copernicus over Ptolemy from the mathematical point of view (KGW 1, 9). At the time of these early studies, Kepler had evidently not read the *Narratio prima,* for he remarked later that Rheticus, who had made the comparison briefly and clearly in his *Narratio prima,* could have saved him the trouble of compiling the list himself. In Graz, Kepler made use of both the *Narratio prima* of Rheticus and the *De revolutionibus* of Copernicus, of which he possessed his own copy (KGW 13, 45).[3]

Kepler was attracted to the Copernican system because, in his view, this system alone provided the reasons for things which in others provoked astonishment. In effect, each of the motions attributed to the earth by Copernicus explained some irregularity or apparent coincidence in the motions of the other planets (KGW 1, 17-18). Whereas Copernicus, however, had recognized the wonderful arrangement of the world *a posteriori* from the observations, Kepler claimed that this could have been proved *a priori* from the idea of creation. According to his own account, the decisive insight that led him to discover, as he thought, God's plan for the construction of the universe in the five regular polyhedra or Platonic

17

solids came to him on 19 July 1595 during the course of a lecture to his class on the great conjunctions of Jupiter and Saturn. (Cf. KGW 13, 28). The pattern of these conjunctions suggested to Kepler's receptive mind that the distances of Jupiter and Saturn might be approximated by the radii of the circumscribed and inscribed circles of an equilateral triangle. Reflecting that this figure was the first regular polygon, he tried to represent the distances of the other planets by means of a sequence of such polygons, inscribing a square in the circle of Jupiter, a circle in this square (to represent the orbit of Mars), a regular pentagon in this circle, and so on. But he found that this scheme failed to represent the distances of the planets in general, and he had to recognize that, in any case, it could not explain the restriction of the number of planets to six. He then reflected that two-dimensional figures were inappropriate to explain the arrangement of solid planets. Clearly a finite set of three-dimensional figures was needed, and this brought to mind the five Platonic solids.

On 2 August 1595 (o.s.) Kepler communicated to Maestlin the first results of his attempt to deduce the distances of the planets *a priori,* remarking that nothing is fashioned by God without a plan (KGW 13, 27), but making no mention of the polyhedral hypothesis. In a second letter, written on 14 September 1595, he gave a brief account of the polyhedral hypothesis and also of his attempt to explain the relation between the distances and periodic times of the planets. While the polyhedral hypothesis provided the reasons for the number, order and magnitudes of the planetary spheres, in order to explain the motions Kepler postulated an *anima movens* in the sun, whose efficacy (*vigor motus*) weakened with distance from the sun, in the same way that the intensity of light weakened with distance from the source. At this time, however, he believed that the weakening depended on the distance according to a relation involving the sine function. Kepler concluded his letter by asking Maestlin for his opinion concerning these ideas.

Without waiting for a reply to his last two letters, Kepler wrote again to Maestlin on 3 October 1595, giving the first full account of the polyhedral hypothesis and asking whether Maestlin could recommend Georg Gruppenbach in Tübingen as a suitable printer to be entrusted with the work (KGW 13, 39). Besides a detailed description of the polyhedral hypothesis and his view concerning the principles underlying the construction of the world, this letter also contains a clarification of his explanation of the effect of the *anima movens* in the sun. Having abandoned the formula involving the sine function, Kepler now supposed that this force, like the intensity of light (which he described as spreading out in a circle, not a sphere), weakened in proportion to the distance from the source (KGW 13, 38). Here we see the beginnings of the physical theory that led Kepler to the discovery of the area law and the elliptical orbits of the planets. Later he discovered the inverse-square law for the intensity of light

emanating from a source. This fundamental law of photometry is first stated in his *Ad Vitellionem paralipomena* (KGW 2, 22).

Describing his plans for the book, Kepler explained to Maestlin that, at the beginning, he intended to introduce some theses to show that the Copernican hypothesis was not opposed to Scripture (KGW 13, 34). Concerning the principles underlying the construction of the world, he maintained that these were to be sought not in the idea of pure numbers but in the concept of geometrical relations. For the properties of pure numbers (except those of the Trinity, which was God himself), Kepler regarded as accidental, whereas the properties of geometrical relations, such as those associated with the regular polyhedra and the musical ratios, he held to be grounded in nature (KGW 13, 35). Fundamental among geometrical relations, in Kepler's view, was the distinction between the curved and the straight, by means of which Nicholas of Cusa (who is not mentioned in the letter) had expressed the relation of God to the creation. Towards the end of the letter, Kepler declared, "I wished to be a theologian; for a long time I was troubled, but now see how God is also praised through my work in astronomy" (KGW 13, 40). Evidently, he had abandoned the idea of becoming a Lutheran pastor, having found his true vocation in astronomy. From this time, he regarded himself as a priest of the Book of Nature.[4]

Towards the end of January 1596 Kepler was given two months leave from his post in Graz in order to visit his ailing grandfathers. He took the opportunity afforded by this visit to consult Maestlin in person and arrange with Gruppenbach for the publication of his work. Kepler did not in fact return to Graz until August, though most of this time was spent in Stuttgart at the court of the Duke of Württemberg, negotiating the construction of a model of his new system in the form of a "Kredenzbecher," which the Duke had authorized (KGW 13, 50-54 and 74-75). On his return, the authorities in Graz accepted Kepler's explanations for his extended absence and granted his request for the payment of his salary in respect of this period (KGW 13, 94 and KGW 19, 11-12).

Among the problems that Kepler put to Maestlin was the following. Whereas Copernicus had taken the center of the earth's orbit as his reference point (in Kepler's view, so as not to confuse the reader by departing too much from Ptolemy), a valid test of the polyhedral hypothesis needed a comparison with the distances of the planets from the true sun. Maestlin calculated these distances for Kepler, after he had first computed the dimensions of the Copernican planetary representations anew from the *Prutenic tables*. The new dimensions are appended to a letter of 27 February 1596 (KGW 13, 56-65), which contains Maestlin's first written comments on Kepler's discovery, and also printed as an appendix to the *Mysterium cosmographicum* (KGW 1, 132-145).[5] The distances from the sun and the table of apogees and aphelions, together with illustrative diagrams prepared by Maestlin, appear only in the *Mysterium cosmo-*

graphicum itself (KGW 1, 52-53). They were probably handed to Kepler personally when he visited Maestlin in March, since they are referred to by Kepler in a letter of 1 April 1596 (KGW 13, 75). Martin Crusius, the professor of Greek, records in his diary that Kepler was a guest for dinner at the university on 12 and 28 March. Under the first date, Crusius noted: "He has discovered something new in astronomy."[6]

In his letter of 27 February 1596, Maestlin expressed approval of the polyhedral hypothesis, which permitted the calculation of the planetary distances *a priori,* but claimed that Kepler had not allowed room for the epicycle (KGW 13, 54-55), so that, to accommodate the epicycle-on-eccentric representations of Copernicus, the spheres should have twice the thickness he had given them. This difficulty could not be avoided. Maestlin chose as the basis of his diagrams and calculation of planetary distances the representation of planetary motion called by Copernicus eccentric-on-eccentric, to which the criticism would also apply (*De revolutionibus,* Book 5, chapter 4). In the *Mysterium cosmographicum,* Kepler pointed out that, even in the case of the epicycle-on-eccentric representation, Maestlin's objection would only apply if the spheres were supposed to be real, whereas real spheres, he added, had already been rejected by Tycho Brahe (KGW 1, 75-76). Clearly, Kepler's spheres were designed to accommodate the real paths of the planets rather than the various Copernican representations.

Maestlin's method of calculating the distances of the planets from the sun is set out in his letter of 11 April 1596 (KGW 13, 77-79), written in reply to a query of Kepler concerning the calculation of the distances of Mercury (KGW 13, 75-77). Throughout the *Mysterium cosmographicum* Kepler used the distances of Mercury given by Maestlin in this letter; they are based on a position of the apogee of Mercury (for the time of Ptolemy) computed from the *Prutenic tables.* Although the letter is printed in the *Mysterium cosmographicum* (KGW 1, 67-68), it appears there in a new version edited by Maestlin himself. The distances of Mercury given in Maestlin's tables, as printed in the *Mysterium cosmographicum* (and presumably, like the letter, edited by him), are based, however, on Ptolemy's own position for the apogee. Indeed, all the distances in Maestlin's table are based on the data for the time of Ptolemy (and in particular on Ptolemy's values for the positions of the apogees), so that it is for the dimensions of the planetary spheres in the time of Ptolemy that the polyhedral hypothesis is tested in the *Mysterium cosmographicum* (KGW 1, 52-53). Maestlin considered that the data did not exist on which a test could be made for the contemporary postions of the planets, since the eccentricities in the time of Copernicus were in doubt (KGW 1, 68), and he accordingly advised Kepler (alluding to a remark of Rheticus that Copernicus had congratulated himself when he came within 10′ of the true positions of the planets) that he should be content with an approximate confirmation of his hypothesis.

On 1 May 1596 Kepler petitioned the Rector and Senate of the University of Tübingen for permission to publish his book with their recommendation (KGW 13, 81). This was a step he had to take in order to satisfy Gruppenbach. Maestlin, whose opinion was sought by the University, recommended publication of Kepler's book, in which, he remarked, the number, order and magnitudes of the spheres were deduced *a priori* (KGW 13, 84), but suggested that, in addition to removing obscurities, Kepler should write a preface comprising a clear explanation of the Copernican system, with the aid of a diagram, and a description of the principal properties of the regular polyhedra, including the method of calculating the radii of the circumscribed and inscribed spheres (KGW 13, 85). On 6 June 1596, Matthias Hafenreffer, the Pro-rector of the University, communicated a summary of Maestlin's report to Kepler, together with the unanimous approval of the Senate (KGW 13, 86).

Kepler included the requested description of the Copernican system in chapter one and appended a piece on the properties of the Platonic solids to chapter two. The radii of the inscribed and circumscribed spheres were discussed in chapter thirteen, as Kepler pointed out to Maestlin in his letter of 11 June 1596, and he hoped that this would suffice (KGW 13, 90). Maestlin however thought that more clarification of the Copernican system was needed, and having taken charge of the printing when Kepler returned to Graz, himself added the *Narratio prima* of Rheticus as an appendix.

Although the title page is dated 1596, the printing was completed and the work was published at the beginning of March 1597. At the time of his visits to Tübingen, Kepler had explained to Maestlin that this book was not a work of mathematics but of cosmography, like Aristotle's *De caelo* (KGW 13, 70). About a year after publication, he clarified his intentions further, remarking to Herwart von Hohenburg that this Prodromus (forerunner) would serve as an introduction to a series of cosmographical treatises on the subjects of Aristotle's *De caelo* and *De generatione* (KGW 13, 190-191).

In his report to the University, Maestlin did not mention Kepler's plan to include at the beginning of his book some theses on the harmonization of the Copernican hypothesis with the Bible. These theses did not in fact appear in the printed work; at the beginning of chapter one, Kepler simply stated that he promised to say nothing contrary to Scripture and that, if Copernicus were convicted of such offense, he would consider him finished (KGW 1, 14). Correspondence following the publication of the *Mysterium cosmographicum* reveals that discussions had taken place on this subject during Kepler's visits to Tübingen in 1596, when Hafenreffer, in order to prevent the possibility of theological objections, had recommended the omission of the chapter (which he recalled was chapter five) in which Kepler had attempted to reconcile the Copernican system with the

Bible (KGW 13, 203). The substance of the omitted chapter was later published in the introduction to the *Astronomia nova.*

Writing to Maestlin on 9 April 1597, soon after the publication of the *Mysterium cosmographicum,* Kepler expressed relief that the defenders of Scripture had raised no objections against his book (KGW 13, 113). Six months later, Maestlin informed him that some theologians were not pleased with the book and that Hafenreffer, in the course of a sermon, had declared that God did not hang up the sun in the middle of the universe like a lantern in the middle of a room (KGW 13, 151). Maestlin added, however, that the critics were inhibited from open hostility by Kepler's dedication of the key diagram to the Duke (see frontispiece). The Duke's attitude was no doubt influenced decisively by Maestlin's statement (in his letter of 12 March 1596) that, while the ancient hypotheses were easier to understand, and for that reason were taught to beginners, nevertheless all practitioners (*artifices*) agreed with the demonstrations of Copernicus (KGW 13, 68). Thus the Duke had Maestlin's authoritative confirmation for the statement of Kepler himself (in his letter of 29 February 1596), that all the famous astronomers (*berhümbte astronomi*) of their time followed Copernicus rather than Ptolemy and Alfonso (KGW 13, 66). Maestlin's more carefully worded statement, which emphasizes a concern with the technical aspects of Copernican astronomy, would truthfully include Tycho Brahe, who regarded the system of Copernicus as mathematically admirable, although not in accord with physical principles. In particular, he followed Copernicus in rejecting Ptolemy's equant. Concerning Hafenreffer, Kepler expressed the view to Maestlin that he was really a secret Copernican, whose advice to treat this system as a mathematical hypothesis was prompted simply by his desire to avoid dissension in the Lutheran Church (KGW 13, 231).

Although Maestlin himself was committed to the Copernican system and in sympathy with Kepler's *a priori* reasons, such as the polyhedral hypothesis, he was critical of Kepler's speculations concerning the *anima movens* in the sun, suggesting that this idea would be the ruin of astronomy (KGW 13, 111). For Maestlin, the distinction was between mathematical and physical hypotheses. It is curious that, even after the clear success of physical reasoning in Kepler's *Astronomia nova,* Maestlin explicitly rejected phsyical astronomy in a letter of 21 September 1616 (KGW 17, 187).

On 13 December 1597 Kepler sent a presentation copy of his *Mysterium cosmographicum* to Tycho Brahe in Denmark (KGW 13, 154-155). When this reached Tycho in March of the following year in Wandsbek, he replied to Kepler in friendly terms, inviting him to make a visit that would allow personal discussion. Acknowledging that, without doubt, God had a harmonius plan for the creation, Tycho suggested that a better test of the polyhedral hypothesis would be possible, if the true values of the eccentricities, which he had sought to determine over a number of years,

were substituted for those used by Kepler (KGW 13, 197-200). It was for the purpose of obtaining these values of the eccentricities, in order to make such a test, that Kepler visited Tycho in Prague early in 1600 (KGW 14, 128). Tycho's interest was attracted by Kepler's ability rather than by the polyhedral hypothesis, for in a letter to Maestlin, written soon after his invitation to Kepler, he made clear his view that progress in astronomy could not be expected from *a priori* deductions but only from more accurate observations (KGW 13, 204-205).

In his preface to the reader, at the beginning of the *Mysterium cosmographicum,* Kepler explains that there were above all three things whose causes he sought; namely, the number, magnitudes and motions of the planetary spheres. From the beginning, as is evident from his correspondence with Maestlin, Kepler envisaged two types of causes, exemplified by the polyhedral hypothesis and the *anima movens* that he postulated in the sun. The first may be described as a final cause, for it reflects God's purpose to create the most beautiful and perfect world, while the second has the character of an efficient cause. In thus combining final and efficient causes Kepler was in fact following Plato. For in the *Timaeus* (46D-E), Plato emphasizes that, in explaining the origins of individual things, both mechanical causes and divine purposes must be considered, and moreover, if we wish to attain a true scientific explanation satisfying to the human reason, we must be primarily concerned with the causes that lie outside the material in the realm of the spiritual. Aesthetic principles, such as those of beauty and perfection, will serve as guides in the search for *a priori* causes; for Kepler claims, quoting Cicero's translation of the *Timaeus,* that it was not possible for the perfect architect to create anything other than the most beautiful (KGW 1, 23-24. Cf. *Timaeus* 30A).

The general idea of the world as the visible image of God, which we find at the end of the *Timaeus* (92C), is in keeping with many passages of the Bible (e.g. Romans 1, v. 20) and came to be transformed by Christian writers into the concept of the Book of Nature. In his *Compendium theologiae,* the Tübingen theologian Jakob Heerbrand described this concept as embracing "the whole universe, the world and everything that is in it," and he also took the beauty of the universe as the basis of his first argument for the existence of God. These ideas are so closely paralleled in Kepler's thought that a direct influence seems likely.[7]

At the beginning of chapter two, where he outlined his principal thesis, Kepler raised the question why God had first created material bodies. The key to the solution of this problem he found in the comparison of God with the "curved" and of created nature with the "straight," which had been made by Nicholas of Cusa and others (KGW 1, 23). Kepler saw the harmony between the things at rest, in the order sun, sphere of fixed stars and intervening space, as a symbol of that between the three Persons of the Trinity (KGW 1, 9 and 23). In seeking a similar harmony for the things

in motion, namely the planetary spheres, he was led to speculate on the divine cosmological intention. In Kepler's view, God intended that we should discover the plan of creation by sharing in his thoughts (KGW 13, 309). First, it seemed to Kepler that such a useful idea as the distinction between the curved and the straight could not have arisen by accident, but must have been contrived in the beginning by God, according to his decrees. Then, in order that the world should be the best and most beautiful and reveal his image, Kepler supposed, God had created magnitudes and designed quantities whose nature was locked in the distinction between the curved and the straight, and to bring these quantities into being, he created bodies before all other things (KGW 1, 24). As the eye is for colors and the ear for sounds, Kepler wrote to Maestlin on 9 April 1597, so is the mind or intellect for the knowledge of quantity (KGW 13, 113. Cf. *Timaeus* 46D-47E).[8] Seeking to comprehend God's thoughts through human thoughts, however, was like trying to reach the curved through the straight, so that, in Kepler's view, certainty was impossible. Consequently, his *a priori* reasons were only probable and needed to be tested against observations (KGW 1, 24 and 71).

Kepler first ordered the solids by comparing the differences between the radii of their circumscribed and inscribed spheres with the intervals between the planets (KGW 1, 27). Then, in chapters three to eight, he gave the *a priori* reasons for the order thus indicated by the data. This provides an example of Kepler's methodological principle that hypotheses must be "built upon and confirmed by observations" (KGW 14, 412). Of special interest in his ordering of the solids is the position of the earth. In Kepler's view, the regular solids fall naturally into two classes. The first class comprised the cube, tetrahedron and dodecahedron, possessing faces of different shapes and vertices common to three faces, while the second consisted of the octahedron and icosahedron, possessing faces of the same shape and vertices common to four or five faces (KGW 1, 29). As the abode of man, the earth occupied a privileged place between the two classes (a kind of geocentrism of importance), and had also been provided, unlike the other planets, with a satellite of similar nature (KGW 1, 29-30).

Before introducing the test of the polyhedral hypothesis against the empirical data, Kepler interpolated four chapters on astrological and harmonic questions. Although he knew of the existence of Ptolemy's ·*Harmonica* and the commentary by Porphyry, he had not yet read them. Moreover, his ideas on astrology had probably not yet completely crystallized, so that it is perhaps not surprising that he later became dissatisfied with these chapters. In the second edition, chapter nine, on the astrological properties of the planets, is described as a digression (KGW 8, 59), while chapter eleven, on the origin of the zodiac, is dismissed as meaningless (KGW 8, 62), and in chapter ten, the primary source of the *numeri numerati* is transferred from the regular polyhedra

to the division of the circle into plane polygons (KGW 8, 60). Chapter twelve, in which Kepler discussed the division of the zodiac and the astrological aspects, is especially interesting for the introduction of the idea of a correlation between the properties of the regular solids and the aspects on the one hand, and the musical harmonies on the other (KGW 1, 39-43). At the beginning of this chapter, which is heavily annotated in the second edition, Kepler remarks that many were of the opinion that the division of the zodiac into twelve signs was arbitrary, a view he himself vigorously defends in *De stella nova* (KGW 1, 168-172). Indeed, as he points out in a letter to Herwart (KGW 15, 453), the only part of traditional astrology he had retained in this work was that relating to the aspects. These derived their efficacy from their grounding in the geometrical structure underlying the natural world. Kepler's definitive account of the efficacy of the aspects and their relation to the musical harmonies is given in the *Harmonice mundi* (Book 4, chapters 6 and 7).

Kepler's first test of the polyhedral hypothesis, described in chapter fourteen, compares the ratio of the least distance of each planet and the greatest distance of the one immediately below, as predicted by the hypothesis, with the corresponding ratio of the distances given by Copernicus (KGW 1, 98). In this test, the distances are taken from the center of the earth's orbit, except for the pairs Mars-earth and earth-Venus, where they are taken from the sun, so as to give the earth's sphere a thickness equal to the eccentricity of the earth's orbit.

In chapter fifteen we reach the core of the *Mysterium cosmographicum,* for this contains Maestlin's diagrams and tables together with Kepler's test of the polyhedral hypothesis using the distances from the sun calculated by Maestlin. To some extent, the test is marred by major errors in Maestlin's calculation of the distances of Venus and Saturn, and as an added confusion, the distances of Mercury on which Kepler had based his calculation (given to him by Maestlin in his letter of 11 April 1596) had been revised by Maestlin (presumably without Kepler's knowledge) during the printing of the *Mysterium cosmographicum.*

To calculate the distances of the planets acccording to the polyhedral hypothesis, Kepler started with the earth's sphere, taking for the radii of the inner and outer surfaces the least and greatest distances of the earth from the sun. The outer surface of the earth's sphere was then taken as the inscribed sphere of the dodecahedron, whose circumscribed sphere became the inner surface of the sphere of Mars. The radius of the outer sphere of Mars (that is, the theoretical greatest distance of Mars) was then calculated from the known radius of the inner sphere, using the Copernican value of the eccentricity. This process was continued upwards to Saturn and downwards to Mercury. In the case of Mercury, however, Kepler found that a better fit was obtained by taking the circle in the octahedron-square instead of the inscribed sphere as the outer bound of the orbit. He adduced *a priori* reasons to justify this exception from the

general rule. For example, Mercury was special among the planets in having an eccentric whose radius varied according to the position of the earth in relation to the apsides (*De revolutionibus,* Book 5, chapter 27), while the octahedron was similarly exceptional among the regular polyhedra in having the possibility of an unobstructed circular path outside the inscribed sphere (KGW 1, 58). In the second edition, he explained that the reason for the peculiarity of Mercury did not, after all, lie in the octahedron (KGW 8, 97. Cf. KGW 7, 435).

Kepler constructed two versions of the polyhedral hypothesis. In the first, the earth's sphere was given a thickness equal to the eccentricity of the earth's orbit, while in the second, the thickness of the earth's sphere was increased to include the moon's orbit. As he had no *a priori* reasons for preferring either version, Kepler expressed his willingness to choose whichever gave the better fit. If the first version were chosen, no difficulty would arise in relation to the moon's orbit cutting the solids, because these, as Kepler emphasized, were not material (KGW 1, 55). It was the apparent connection of the earth and the moon that led Kepler to his earliest speculations on gravity. Inclining to the Platonic view, described in the *Timaeus* (63C-E), according to which bodies of the same nature are drawn together, Kepler explained that the moon, being of the same nature as the earth (an idea he attributed to Maestlin), follows or is drawn wherever the earth goes (KGW 1, 55).

Besides computing tables of distances of the planets according to the polyhedral hypothesis, Kepler also computed tables of angles which bring out clearly the correspondence with the Copernican data. For Venus and Mercury, the sine of the angle is taken to be the greatest distance of the planet from the sun when the mean distance of the earth from the sun is taken as a unit, so that the angle represents approximately the maximum elongation of the planet from the sun. For Mars, Jupiter and Saturn, the sine of the angle is taken to be the mean distance of the earth from the sun when the greatest distance of the planet is taken as a unit, so that the angle represents approximately the prosthaphaeresis in the apogee (KGW 1, 54). Tables I and II show, respectively, the distances and angles as given by Kepler and (in parentheses) the corrected values, calculated from the same Copernican data.

Although the errors in Kepler's values prevented him from making a choice between the two versions, the corrected angles show clearly that the hypothesis in which the moon's orbit is included in the earth's sphere gives the better fit and is indeed remarkably close to the Copernican data. Kepler himself pointed out that the differences in the angles did not exceed the margin of error in the longitudes of the planets calculated from the *Prutenic tables* (KGW 1, 65). Again, the small discrepancies could arise from errors in the eccentricities. For Kepler, lacking a knowledge of the *a priori* reasons of the eccentricities and their differences, had to use the Copernican values, which were known to be unreliable (KGW 1, 60).

TABLE I

Greatest and least distances of the planets from the sun, taking the mean distance of the earth = 1000.

According to	Copernicus		Polyhedral Hypothesis		Polyhedral Hypothesis (moon inc.)	
Saturn	9987	(9727)	10599	(10011)	11304	(10588)
	8342	(8602)	8852	(8854)	9441	(9364)
Jupiter	5492	(5492)	5111	(5109)	5451	(5403)
	4999	(4999)	4652	(4650)	4951	(4918)
Mars	1649	(1648)	1551	(1550)	1658	(1639)
	1393	(1393)	1311	(1310)	1398	(1386)
Earth	1042	(1042)	1042	(1042)	1102	(1102)
	958	(958)	958	(958)	898	(898)
Venus	741	(721)	761	(762)	714	(714)
	696	(717)	715	(757)	671	(710)
Mercury	489	(481)	506	(535)	474	(502)
	233	(233)	233	(260)	219	(242)

TABLE II

	Polyhedral Hypothesis	Diff.	Copernicus	Diff.	Polyhedral Hypothesis (moon inc.)
Saturn	5° 25'	− 20'	5° 45'	− 41'	5° 4'
	(5° 44')	(− 10')	(5° 54')	(− 29')	(5° 25')
Jupiter	10° 17'	− 12'	10° 29'	− 6'	10° 23'
	(11° 17')	(+ 48')	(10° 29')	(+ 11')	(10° 40')
Mars	40° 9'	+ 2° 47'	37° 22'	+ 30'	37° 52'
	(40° 10')	(+ 2° 48')	(37° 22')	(+ 14')	(37° 36')
Venus	49° 36'	+ 1° 45'	47° 51'	− 2° 18'	45° 33'
	(49° 38')	(+ 3° 27')	(46° 11')	(− 37')	(45° 34')
Mercury[9]	30° 23'	+ 1° 4'	29° 19'	− 1° 1'	28° 18'
	(32° 22')	(+ 3° 36')	(28° 46')	(+ 1° 21')	(30° 7')

The tables show Kepler's values and (in parentheses) the correct values calculated from the Copernican data used by Kepler.

These *a priori* reasons he eventually located in the ideas of musical harmony described in the *Harmonice mundi,* a work he started to plan in 1599, while the search for more accurate empirical values of the eccentricities, as we have already remarked, led him to Tycho Brahe.

Having completed his proof that the *a priori* reasons of the distances of the planets in the Copernican system were to be found in the five regular solids, Kepler turned his attention from final causes to efficient causes, seeking a confirmation of the distances in the effect of the moving soul (*anima motrix*) in the sun on the motions of the planets (KGW 1, 68).

Thus, in chapter 20, he introduced the concept of the solar force (weakening in proportion to the distance from the sun) and the theory of the motions of the planets that he had described to Maestlin in the letter of 3 October 1595.

TABLE III

Interpolation of the polyhedra so as to obtain the best fit with the distances predicted by the theory of the motions and with the Copernican data.

	Copernicus max mean distance min	Motions mean distance	Polyhedra circum-radius in-radius
Saturn	9987 9164 8341	9163	
Cube			9163 5261
Jupiter	5492 5246 5000a	5261	
Tetrahedron			5000a 1648b
Mars	1648b 1520 1393c	1440	
Dodecahedron			1393c 1102d
Earth	1042 with 1102d 1000 moon 1000 958e 898	1000	
Icosahedron			958e 762f
Venus	741h 719 696	762f	
Octahedron			741h 429g
Mercury	489 360 231	429g	

In chapter twenty-one, Kepler attempted to bring the two causes, final and efficient, together in a comparison with the Copernican data. The results are shown in Table III. Kepler's values of the Copernican distances, in the first column, have been retained without correction, but the arrangement of the third column has been changed to clarify Kepler's intention to interpolate the polyhedra so as to obtain the best fit (KGW 8,

109 and 117). The mean distances calculated from the motions, given in the second column, are free of arithmetical errors. Kepler regarded these distances as more reliable than the Copernican data. The cube is found to fit between the mean distances (based on the motions) of Saturn and Jupiter, while the method of fitting the other solids is indicated by the letters marking the starting and finishing points in each case.

Kepler's preference for the distances based on the motions, and his application, in chapter twenty-two, of the theory of the moving soul in the sun to explain the Ptolemaic equant (and other representations used by Copernicus) already point the way to the achievements of the *Astronomia nova,* where physical reasoning (in the form of a search for efficient causes) was to play a decisive role in the discovery of the first and second laws of planetary motion.[10]

Almost all the astronomical books written by Kepler (notably the *Astronomia nova* and the *Harmonice mundi*) are concerned with the further development and completion of themes that were introduced in the *Mysterium cosmographicum.* The ideas of this work did not constitute just a passing fancy of youth but rather the seeds from which Kepler's mature astronomy grew. When a new edition was called for, he decided against changing the text itself, for a complete revision would have required the inclusion of all the main ideas of his other books (KGW 8, 10). Instead, he simply added explanatory notes and references to his definitive accounts of various topics given elsewhere, especially in the *Harmonice mundi* and the *Epitome astronomiae copernicanae.* Kepler's correspondence gives no clues concerning the composition of these notes; the only reference to them is contained in a letter to Bernegger of 11 August 1621, where Kepler remarks that Gottfried Tampach (the Frankfurt publisher) was preparing a new printing of the *Mysterium cosmographicum* with his notes (KGW 18, 75). These notes were probably written hurriedly — no attempt was made to correct the arithmetical errors of the first edition — shortly before the book was published in Frankfurt in 1621.

Before Kepler was born, the French humanist Pierre de la Ramée (Ramus) had called for a reform of astronomy by the rejection of hypotheses — that is, mathematical fictions such as epicycles having no basis in nature — and he expressed the hope that one of the celebrated schools of Germany would provide the philosopher and mathematician capable of constructing this astronomy without hypotheses.[11] Writing to Maestlin at the beginning of October 1597, Kepler claimed that he (and Copernicus also) had answered the challenge of Ramus, for he supposed that Ramus had proposed only the rejection of fictitious hypotheses and not those that were natural and true (KGW 13, 141. Cf. 165).[12] Kepler returned to this theme in the *Astronomia nova,* where he presented his claim on the verso of the title page, and again, in the preface to the *Tabulae Rudolphinae,* he mentioned among the causes for the long delay in

publication, "the transfer of the whole of astronomy from fictitious circles to natural causes." Traditional astronomy had sought to "save the appearances presented by the planets,"[13] using mathematical hypotheses of the kind condemned by Ramus. In place of this, Kepler substituted a concept of astronomy as a science which sought to describe and explain physical reality in terms of both final (aesthetic) and efficient (mechanical) causes, by the invention of hypotheses based upon and confirmed by observations.[14] From our vantage point we can see that, when Kepler made the discovery forming the basis of the *Mysterium cosmographicum,* he had not just "discovered something new in astronomy," as Martin Crusius noted in his diary, but a new way of doing astronomy, which may be seen (at least in part) as a return to the authentic teaching of Plato in the *Timaeus* (in the sense of explanation in terms of both final and efficient causes), thereby effecting a revolution in method which has earned him the title of founder of modern astronomy.

NOTES ON INTRODUCTION

*This introduction was published in an earlier version as an essay dedicated to Bernhard Sticker (on his seventieth birthday), leader of the International Symposium held in Weil der Stadt in 1971 to commemorate the quatercentenary of Kepler's birth. E. Aiton, Johannes Kepler and the 'Mysterium cosmographicum,' *Sudhoffs Archiv,* 61 (1977), 173-194.

1. KGW = Johannes Kepler, *Gesammelte Werke,* edited by Walther von Dyck, Max Caspar, Franz Hammer and Martha List, Munich, 1937 – .

2. Johannes Kepler, *Selbstzeugnisse,* edited by Franz Hammer and translated by Esther Hammer, Stuttgart-Bad Cannstatt, 1971, p. 63.

3. A facsimile reprint of Kepler's copy of *De revolutionibus,* with introduction by Johannes Müller, has been published by Johnson Reprint Corporation, New York and London, 1965. There are two new English translations: Copernicus, *On the revolutions of the heavenly spheres,* translated by A. M. Duncan, London, Vancouver and New York, 1976; Copernicus, *On the revolutions,* translated by Edward Rosen, Warsaw and London, 1978.

4. For Kepler's own account of his theological development, see Johannes Kepler, *Selbstzeugnisse* (see note 2 above), pp. 61-65. See also Jürgen Hübner, Naturwissenschaft als Lobpreis des Schöpfers, in *Internationales Kepler-Symposium Weil der Stadt 1971,* edited by Fritz Krafft, Karl Meyer and Bernhard Sticker, Hildesheim, 1973, pp. 335-356, and Martha List, Kepler und die Gegenreformation, in *Kepler Festschrift 1971,* edited by E. Preuss, Regensburg, 1971, pp. 45-63. On Kepler's theology, see Jürgen Hübner, *Die Theologie Johannes Keplers zwischen Orthodoxie und Naturwissenschaft,* Tübingen, 1975.

5. There is an English translation by A. Grafton in *Symposium on Copernicus,* Philadelphia, 1973 (= *Proceedings of the American Philosophical Society,* 117 (1973), 413-552).

6. *Kepler und Tübingen* (Tübingen Kataloge Nummer 13), published by the Kulturamt der Stadt Tübingen, 1971, p. 29.

7. *Internationales Kepler-Symposium* (see note 4), pp. 338-340.

8. According to Rheticus, the function of the human mind was to understand harmony and number. E. Rosen, *Three Copernican treatises*, New York, 1971, p. 196.

9. The value 29° 19′ given in the middle column is inconsistent with the value taken by Kepler for the distance of Mercury from the sun, namely 29′ 19″ (with the mean distance of the earth as 1°). The correct value is 29° 15′.

10. See C. Wilson, Kepler's derivation of the elliptical path, *Isis*, 59 (1968), 5-25 and E. J. Aiton, Kepler's second law of planetary motion, *Isis*, 60 (1969), 75-90.

11. See R. Hooykaas, *Humanisme, science et réforme: Pierre de la Ramée*, Leiden, 1968, p. 67.

12. See E. Aiton, Johannes Kepler and the astronomy without hypotheses, *Japanese studies in the history of science*, 14 (1975), 49-71.

13. Following a misinterpretation of Simplicius in his commentary on Aristotle's *De caelo*, this concept of astronomical method has been mistakenly attributed to Plato. According to two recent analyses of this problem, it would seem that the idea of saving the appearances originated either with the Stoics of the time of Posidonius or with Eudoxus. For the arguments relating the idea to Posidonius, see Fritz Krafft, Physikalische Realität oder mathematische Hypothese? *Philosophia naturalis*, 14 (1973), 243-275. Cf. *Internationales Kepler-Symposium* (see note 4), pp. 64-66. On the attribution to Eudoxus, see Jürgen Mittelstrass, *Die Rettung der Phänomene. Ursprung und Geschichte eines antiken Forschungsprinzips*, Berlin, 1962. Cf. Jürgen Mittelstrass, *Neuzeit und Aufklärung*, Berlin, 1970, pp. 250-263. See also E. J. Aiton, Celestial spheres and circles, *History of Science* (on press).

14. On Kepler's methodology see J. Mittelstrass, Wissenschaftsliche Elemente der keplerschen Astronomie; R.S. Westmen, Kepler's theory of hypothesis and the realist dilemma; G. Buchdahl, Methodological aspects of Kepler's theory of refraction. In *Internationales Kepler-Symposium* (see note 4). These papers have been reprinted (that of Mittelstrass in English translation) in *Studies in history and philosophy of science*, 3 (1972), 203-298. See also J. L. Russell, Kepler and scientific method, *Vistas in astronomy*, 18 (1975), 733-745.

Prodromus

DISSERTATIONVM COSMOGRA PHICARVM, CONTINENS MYSTE- RIVM COSMOGRAPHI- CVM,

DE ADMIRABILI

PrOPORTIONE ORBIVM COELESTIVM, DEQVE CAVSIS

cœlorum numeri, magnitudinis, motuumque periodicorum genuinis & pro-
prijs,

DEMONSTRATVM, PER QVINQVE
regularia corpora Geometrica,

A

*M. IOANNE KEPLERO, VVIRTEM-
bergico, Illuſtrium Styriæ prouincialium Mathematico.*

Quotidiè morior, fateorque: ſed inter Olympi
Dum tenet aſſiduas me mea cura vias:
Non pedibus terram contingo: ſed ante Tonantem
Nectare, diuina paſcor & ambroſiâ.

Addita eſt erudita NARRATIO M. GEORGII IOACHIMI
RHETICI, *de Libris Reuolutionum, atq; admirandis de numero, ordine, & diſtantijs Sphararum Mundi hypotheſibus, excellentiſſimi Mathematici, totiusq; Aſtronomiæ Reſtauratoris* D. NICOLAI
COPERNICI.

TVBINGÆ
Excudebat Georgius Gruppenbachius,
ANNO M. D. XCVI.

Prodromus
DISSERTATIONVM COSMOGRAPHICARVM,
continens
MYSTERIVM
COSMOGRAPHICVM
DE ADMIRABILI PROPORTIONE OR-
bium cœlestium: deque caufis cœlorum numeri, magnitudinis, motuumque periodicorum ge-
nuinis & propriis,

Demonftratum per quinque regularia corpora Geometrica.

Libellus primum Tûbingæ in lucem datus Anno Chrifti
M. D X C V I.
à

M. IOANNE KEPLERO VVIRTEMBERGICO, TVNC TEMPO-
ris Illuftrium Styriæ Prouincialium Mathematico.

Nunc vero poft annos 25. ab eodem authore recognitus, & Notis notabiliffimis
partim emendatus, partim explicatus, partim confirmatus: deniq; omnibus fuis
membris collatus ad alia cognati argumenti opera, quæ Author ex illo tempore fub duorum Impp. Rudolphi & Matthiæ aufpiciis; etiamq; in
Illuftr. Ord. Auftriæ Supr-Anifanæ clientela
diuerfis locis edidit.

Potiffimum ad illuftrandas occafiones Operis, Harmonice Mundi, dicti, eiuf-
que progreffuum in materia & methodo.

Addita eft erudita NARRATIO M. GEORGII IOACHIMI RHETICI, de
Libris Reuolutionum, atque admirandis de numero, ordine, & diftantiis Sphærarum Mundi hypothefibus, excellentiffimi Mathematici, totiufque Aftronomiæ Re-
ftauratoris D. NICOLAI COPERNICI.
ITEM,
Eiufdem IOANNIS KEPLERI *pro fuo Opere Harmonices Mundi* APOLOGIA *aduer-*
fus Demonftrationem Analyticam Cl. V. D. Roberti de Fluctibus, Me-
dici Oxonienfis.

Cum Priuilegio Cæfareo ad annos XV.

FRANCOFVRTI,
Recufus Typis ERASMI KEMPFERI, fumptibus
GODEFRIDI TAMPACHII.
Anno M. DC. XLI.

Forerunner of the Cosmological Essays, which contains

THE SECRET OF THE UNIVERSE

On the Marvelous Proportion of the Celestial Spheres, and on the true and particular causes of the number, size, and periodic motions of the heavens,

Established by means of the five regular Geometric solids.

A little book first brought into the light of day at Tübingen in the Year of Christ

1596

by

Master Johannes Kepler of Württemberg, at that time Mathematician of the Illustrious Districts of Styria;

Now after 25 years revised by the same author, partly emended, partly explained, and partly confirmed by most remarkable notes, and lastly compared in all its parts with other works having a similar argument, which the author since that time has published in various places under the auspices of the two Emperors Rudolph and Matthias, and also under the patronage of the Illustrious Orders of Austria over the Enns.

Especially to illustrate the relevance of the work entitled *Harmonice Mundi*, and of its advances in matter and method.

In addition, the learned *Narratio* of Master George Joachim Rheticus, on the Books of the Revolutions, and the wonderful hypotheses on the number, order and distances of the Spheres of the Universe of the most excellent Mathematician and Restorer of the whole of Astronomy, Dr. Nicolaus Copernicus.

Also, the same Johannes Kepler's *Defense* for his Work *Harmonice Mundi*, against the *Analytical Description* of the famous Dr. Robert Fludde, Physician of Oxford.

With Imperial Privilege for Fifteen Years.

Frankfurt,
Printed at the Press of Erasmus Kempfer, at the expense of Godefried Tampach.
In the Year 1621.

Epigramma Ptolemæo adscriptum.

Οἶδ', ὅτι θνατὸς ἐγὼ καὶ ἐφάμερος· ἀλλ' ὅταν ἄστρων
Μαστεύω πυκινὰς ἀμφιδρόμους ἕλικας,
Οὐκέτ' ἐπιψαύω ποσὶ γαίης, ἀλλὰ παρ' αὐτῷ
Ζηνὶ διοτρεφέος πίμπλαμαι ἀμβροσίης.

LATINE.

Quotidie morior, fateorque: sed inter Olympi
  Dum tenet aßiduas me mea cura vias:
Non pedibus terram contingo: sed ante Tonantem
  Nectare, diuinâ pascor & ambrosiâ.

I. K.

Epigram ascribed to Ptolemy

*I know that I am mortal and ephemeral.
But when I search for the close-knit en-
compassing convolutions of the stars,
my feet no longer touch the earth, but in
the presence of Zeus himself I take my
fill of ambrosia which the gods
produce.*[1]

REVERENDISSIMO PRIN-
CIPI, ADMODVM REVERENDIS PRÆ-
SVLIBVS; ILLVSTRIBVS, GEVEROSIS, LL. BA-
ronibus; Nobilibus, Strenuis, Equeſtris Ordinis, DD. Pro-
uincialibus vniuerſis Splendidiſſimi Ducatus Styriæ; Dominis meis gratioſiſ-
ſimis.

EVERENDISSIME *Princeps* ; *Admodum Reue-*
rendi, *Illuſtres*, *Generoſi*; *Nobiles*, *Strenui*; *Domini*
gratioſiſimi. *Annus hic eſt viceſimusquintus*, *ex quo*
libellum ego præſentem, *Myſterium Coſmographicum*
indigetatum; Magiſtratibus illius temporis, *de veſtræ*
communitatis honoratiſimo corpore lectis, *inſcriptum*
inter homines vulgaui. *Etſi vero tunc oppidò iuuenis*
eram, *primumque hoc Aſtronomicæ profeßionis tyrocinium edebam: ſucceſſus*
tamen ipſi conſecutorum temporum elata voce teſtantur; nullum admirabilius,
nullum felicius, *nullum ſcilicet in materia digniori poſitum eſſe vnquam à quo-*
quam tyrocinium. *Non enim haberi debet illud nudum ingenij mei commen-*
tum (abſit huius rei iactantia à meis, *admiratio à lectoris ſenſibus*, *dum ſapien-*
tiæ creatricis tangimus Pſalterium heptaechordum) quandoquidem, *non ſecus*,
ac ſi dictatum mihi fuiſſet ad calamum, *oraculum cælitus delapſum*, *ita omnia*
vulgati libelli capita præcipua, *& veriſima ſtatim (quod ſolent opera Dei ma-*
nifeſta) fuerunt agnita ab intelligentibus: & per hos viginti quinque annos
mihi telam pertexenti reſtaurationis Aſtronomicæ (cœptam à Tychone Brahe è
Nobilitate Danica celebratiſimo Aſtronomo) facem non vnam prætulerunt:
denique quicquid ſere librorum Aſtronomicorum ex illo tempore edidi, *id ad*
vnum aliquod præcipuorum capitum, *hoc libello propoſitorum*, *reſerri potuit*,
cuius aut illuſtrationem aut integrationem contineret ; non equidem amore
mearum inuentionum, *abſit iterum hæc inſania; ſed quia rebus ipſis*, *& obſer-*

):(2 *uatio-*

TO THE MOST REVERED PRINCE,

TO MY GREATLY REVERED PATRONS:
THE ILLUSTRIOUS, EMINENT BARONS;
THE NOBLE AND ENERGETIC MEMBERS OF THE
ORDER OF KNIGHTS;
TO MY LORDS THE INHABITANTS OF THE MOST
SPLENDID DUCHY OF STYRIA, ONE AND ALL:
MY MOST DEAR LORDS.

Most revered Prince; greatly revered, illustrious, eminent, noble, energetic, most dear lords. This is the twenty-fifth year since I made public among men the present little book, entitled *The Secret of the Universe*, and dedicated to the elected magistrates of that time of the most honorable corporation of your community. Although indeed I was very much a young man, and was producing it as the first apprentice piece of my vocation to astronomy, yet its successes in the times which have followed bear witness at the tops of their voices that no apprentice piece has ever been more remarkable, more successful, or of course carried out on worthier material by anyone. For it should not be considered as a mere contrivance of my own intellect (may there be no boasting of this affair in my feelings, nor wonder in the reader's, while we touch the seven-stringed Psaltery of the Creative Wisdom) since, just as if it had been dictated to my pen, an oracle fallen from heaven, every chapter of the little book was recognized at once, by those who understood it, as important and quite true (as the manifest works of God usually are). Throughout these twenty-five years, while I have been weaving the fabric of the reform of astronomy (started by Tycho Brahe of the Danish nobility, the very celebrated astronomer), they have carried a torch before me more than once. And, finally, almost every book on astronomy which I have published since that time could be referred to one or another of the important chapters set out in this little book, and would contain either an illustration or a completion of it. I say this not out of love of my own discoveries — again may there be no such madness in me — but because from the subject itself, and from the observations of Tycho Brahe, which deserve complete trust, I have thoroughly learnt that

uationibus Tychonis Brahei fide omni dignissimis edoctus fui, nullam aliam inueniri posse viam ad perfectionem Astronomiæ, certitudinemque calculi; nullam ad constituendam scientiam huius seu partis Metaphysicæ de cœlo, seu Physicæ cœlestis; quam quæ hoc libello vel expresse præscripta, vel timidis saltem opinionibus, & rudi Minerua adumbrata esset. Testes sisto illic commentaria Martis anno 1609. edita, quæque adhuc domi premo commentaria de motibus cæterorum Planetarum; hic vero Harmonices Mundi libros V. anno 1619. vulgatos, & Epitomes Astronomiæ librum IV. anno 1620. absolutum: testes tot numero lectores, qui, ex quo nacti sunt opera dicta, iam ab annis bene multis exemplaria flagitant, dudum distracta, huius primi mei libelli; vt ex quo tam multa vident deriuata theoremata.

Cum igitur instarent amici, non Librarij tantum, sed etiam Philosophiæ periti, vt secundam editionem adornarem: officij quidem mei putaui, non diutius repugnare; de modo tamen editionis aliquantulum contradixi. Erant enim, qui consulerent, libellum emendarem, augerem, perficerem: morem scilicet cæterorum Authorum, quem tenent in excolendis libris proprijs, & ipse obseruarem. Mihi contra sic visum, nec perfici libellum posse, nisi transcriptis in illum plerisque meorum operum, quæ per hos vigintiquinque annos edidi, pene integris; nec hoc iam tempus amplius esse, librum aliquem hoc titulo, post editos alios, veluti de nouo publicandi: denique libellum ipsum propter successum admirabilem, pro meo non reputandum, quem arbitratu meo mutem, augeamve; quin potius interesse lectoris, vt intelligat, à quibus initijs, quousque perductæ à me fuerint contemplationes Mundanæ. Vincentibus ergo rationibus istis, formam editionis talem elegi, quæ solet obseruari in libris alienis recudendis; vbi nihil mutamus; quæ vero loca emendatione egent, aut explicatione, aut integratione, ea commentarijs adiuuamus, differenti typo exaratis. Seruiuit hæc forma & religioni & breuitati, vt errores quidem de mentis meæ tenebris ortos, interspersosque materiæ de operibus Dei perfectissimis, ipse coarguerem ingenue, expungeremque: quæ vero capita libelli, acie mentis irretorta, in lumen illud operum diuinorum ineffabile directa, clare percepissem; aut vbi viam quidem rectam ingressus, nimium tamen propere substitissem, ea secernerem, & quibus alij operum meorum locis ad scopum tandem peruenerim, lectori significarem.

Vt igitur libellum in hac altera editione, etiam quoad ipsam dedicationem, relinquerem intactum, vt ipsum etiam vestibulum responderet opusculo reliquo: videtis, opinor, Proceres Reuerendissimi, Generosissimi, aliter mihi non faciendum fuisse, quin etiam hanc editionem ad primos patronos, quos in sequenti dedicatione sum alloquutus, aut, si qui ex hoc tempore rebus huma-

nis

no other way can be found to the perfection of astronomy and accuracy in its calculations, no other way to establish knowledge of this metaphysical aspect of the heaven, or heavenly physics, than what had been written already in this little book either expressly, or at least in timid conjectures, and in a rough and ready way. I cite as witnesses on the former point the *Commentaries on Mars* published in the year 1609, and the *Commentaries* on the motions of the other planets which up till now I have kept to myself, and in the latter case the five books of the *Harmony of the Universe* made public in the year 1619 and Book IV of the *Epitome of Astronomy*, completed in the year 1620, and I count as witnesses so many readers who have, for very many years since the time when they obtained the works mentioned, been demanding copies, long since scattered, of this my first little book, as they see so many theorems derived from it.

Then since my friends, not only booksellers, but also those versed in philosophy, were pressing me to prepare a second edition, I did indeed think it my duty not to object any longer; yet I disagreed with them a little over the character of the edition. For there were some who advised me to emend, enlarge and complete the book; that is to say, that I should myself adopt the custom of other authors, which they observe in refining their own books. It seemed to me on the contrary that I could not complete the book, except by transcribing into it several of my works, which I have published during these twenty-five years, almost in their entirety; that this was no longer the time for putting out a book with this title, after I had published others, as if it were new; and lastly that the little book itself, on account of its remarkable success, should not be thought of as my own, to alter or enlarge at will, but that it was rather of interest for the reader to understand from what beginnings, and to what point my studies of the universe have been brought. This reasoning won, then, and I chose the form of edition which is usually adopted in reprinting other people's books, in which we change nothing. Those places which need emendation, or explanation, or completion, we reinforce with commentaries, set in different type. This form assisted both religion and brevity, so that I could frankly refute and expunge the errors which had sprung indeed from the darkness of my mind, and were scattered among material on the most perfect works of God; I could distinguish those chapters in the book which had not been deflected by the action of my mind but which I had perceived clearly because they were turned towards the unutterable light of the divine works, and those where, although I had set off on the right path, I had yet stopped too soon; and I could indicate to the reader the other places in my works in which I have at last reached the goal.

Then in order to leave the little book intact in this second edition, even including the actual dedication, so that it should serve as an anteroom for the remainder of my little work, you see, I believe, most revered, most eminent nobles, that I could do no other than submit this edition also with a new dedication to my first patrons, whom I addressed in the

nis exempti sunt: ad eorum filios, aut successores, (quorum nonnullos interea Terrarum Orbis Monarchæ, virtutem remunerati, ad summum dignitatis culmen euexerunt) denique ad hoc idem corpus communitatis honoratißimum, cuius stipendiis suffultus, olim libellum conscripsi, noua dedicatione remitterem.

Nec leuia mihi hoc agitanti præbuit incitamenta, inde Styriæ modernæ, hinc prouinciarum circumiacentium respectus. Illinc namque multos è nobilitate videbam, qui me vel audiuere docentem, vel communi mensa aut contubernio meo vsi, me propius cognouerunt, exque eo tempore beneuolentiam à patribus in se deriuatam erga me conseruant, quibusque pollent copiis, demonstrant, dignitatis & gratiæ Cæsareæ fructum per beneficentiam exigentes: nec desunt ex Ecclesiasticorum numero, qui non minus, quam antecessores sui, & artes Mathematicas & me cultorem amant, meque ad se inuisendos, si turbæ conquieuissent, de propinquo se euocaturos nuntiarunt. Dignum igitur erat mea in vtrosque gratitudine, vt quibus possem mutuis officiis tantos fautores percolerem, ampliusque demereri studerem.

Hinc vero ex parte Austriæ, pauidam imbellemque Astronomiam circumstantia pericula, terrores, calamitates, ærumnæ subinde admonent, de circumspiciendis auxiliis. Transiuit illa anno 1600. è Styria in Bohemiam, vt quæ sub Austriacæ domus vmbra primas radices egerat, eadem sub illa & maturesceret. Ibi varie iactata à tempestatibus bellorum, tam intestinorum, quam externorum, tandem post excessum Rudolphi Imperatoris anno 1612. constanti domus Austriacæ studio, recurrit in Austriam: vbi vtinam quam benigne excepta & fota, tam impensa generosarum mentium occupatione (non minus atque à me eius instauratore) percoli potuisset. Verum, eheu, quantis sese mutuo bonis exuunt mortales miseri, per scabiem contentionum turpißimam? Quam profunda, sic meritos, obruit ignorantia fati? Quam lamentabili consilio Ignem dum fugimus, medios incurrimus ignes?

Vtinam vero etiam nunc, post consequutam rerum Austriacarum conuersionem, locus supersit illi Platonis oraculo; qui, cum Græcia longo & ciuili bello arderet vndique, malisque vexaretur omnibus, quæ ciuile bellum comitari solent, consultus super Problemate Deliaco; quæsito prætextu, ad suggerenda populis consilia salutaria; ita demum tranquillam ex Apollinis sententia Græciam futuram respondit: si se ad Geometriam cæteraq; philosophica studia Græci conuertissent: quia hæc studia animos ab ambitione & reliquis cupiditatibus, ex quibus bella & cætera mala existunt, ad amorem pacis & moderationem in omnibus rebus adducerent.

dedication which follows, or, if any since that time have been removed from human affairs, to their sons, or their successors (some of whom the monarchs of this earthly sphere have meanwhile raised up, as the reward of their excellence, to the loftiest peak of honor), and finally to that same most honorable corporation of your community, by whose stipend I was supported when I wrote this little book long ago.

Also, when I was engaged on it, considerable incentives were supplied to me by the regard of modern Styria on the one hand, and the surrounding provinces on the other. For in the former I saw many from the nobility who either listened to my teaching, or made closer acquaintance with me through sharing the same table or dwelling, and since that time have maintained the generosity passed on to them by their fathers, and have shown it with all the resources at their command, claiming imperial honor and gratitude as the result of their kindness. There has also been no lack of churchmen, who are no less fond than their predecessors of the mathematical arts and of myself as fostering them, and have stated that they would invite me to visit them, if the disorders had abated, at close quarters. It was therefore fitting in view of my debt of gratitude to both of these that I should honor such generous patrons as far as I was able with reciprocal courtesies, and strive to deserve more.

In the latter, however, on the Austrian side, timorous and unwarlike astronomy is warned by the conditions, dangers, terrors, disasters, and troubles to look round for assistance. She crossed in the year 1600 from Styria into Bohemia, so that just as she had put her first roots under the shelter of the Austrian house she might also grow to maturity under it. After being tossed to and fro there by the tempests of both civil and foreign wars, in the end after the death of the Emperor Rudolph in the year 1612, with unceasing zeal for the Austrian house, she returned to Austria. Would that she could have been honored there with the devoted attention of eminent minds (no less than by myself, who restored her) as much as she was accepted and favored with goodwill. Yet, alas, of what great goods do miserable mortals despoil one another, by their shameful itching for quarrels. How profound an ignorance of their fate overwhelms them, as they have deserved. With what deplorable perverseness do we rush into the midst of the flames, in fleeing from the fire.

Would that even now indeed there may still, after the reversal of Austrian affairs which followed, be a place for Plato's oracular saying. For when Greece was on fire on all sides with a long civil war, and was troubled with all the evils which usually accompany civil war, he was consulted about a Delian Riddle, and was seeking a pretext for suggesting salutary advice to the peoples. At length he replied that, according to Apollo's opinion Greece would be peaceful if the Greeks turned to geometry and other philosophical studies, as these studies would lead their spirits from ambition and other forms of greed, out of which wars and other evils arise, to the love of peace and to moderation in all things.

Vtinam denique iam suppressis armis tantum detur induciarum à mise-riis, vt viris bonis vacet, simile quippiam Ciceroniani illius consilij commini-sci: qui, euersa Republica Romana, cum esset vix consolabilis dolor, in tanta omnium rerum amissione & desperatione recuperandi, post-quam illi arti, cui studuerat, nihil esse loci, neque in Curia, neque in foro, vidit: omnem suam curam atque operam ad Philosophiam con-tulit: monens Sulpitium suum, in iisdem versari rebus, quæ, etiamsi minus prodessent, animum tamen à sollicitudine abducerent; áque molestiis leuarent.

Quibus votis si Deus annuat, non equidem indignas homine Christiano voluptates, ærumnarum solatia, Mathematice mea vel ex astronomicis exer-citiis, vel ex contemplatione diuinorum operum, exque Harmonice Mundi (fatali illa occupatione, in durißimis exacti biennij dissonantiis) proponere pa-rata semper erit. At quia in id est incepta hæc occupatio Astronomica, vt per-ficiatur: quid igitur hoc Austriæ statu calamitosißimo potius agat, quàm vt præsidia, quilus ipsa indiget, ad opera inter homines vulganda, adque nomen Rudolphi, Tabulis perpetuis asserendum; pudore cohibita ne ab afflictis vel iu-bentibus omnia petat; potius inde corroget, quorsum clades istæ, quorsum pro-digiorum cælestium expiationes horribilißimæ non pertigerunt: denique ad pristinos patronos, ad quos dimidio viæ iam anno 1612. *appropinquauerat, reliquo etiam dimidio excurrat? E Styria quondam, vti dixi, ad Braheum, id est, ad Opus Tabularum Rudolphinarum maturandum, profectus est libel-lus iste, me latore: quid insolens, quid adeo alienum à pristino instituto vestro, Proceres, quid denique non gratum Ferdinando Imperatori Augusto, Rudol-phi post Matthiam successori, feceritis; si reuertentem nunc libellum, veterem clientem vestrum, de rebus interea gestis, audiatis, si Tabularum Opus laborio-sum & sollicitum, si delicias humani generis, si Rudolphi Imperatoris Nomen honoresque, modica liberalitate promouendos suscipiatis; si hanc vetustißimam Mathematicarum disciplinarum clientelam domus Austriaca, ne hoc quidem grauißimo motu concussa, intercedente vestra succenturiata prouidentia, di-mittat, exterisve cedat?*

Hic igitur dedicationis huius repetitæ scopus esto, quem si à vestra, Proce-res, magnificentia fuero consequutus, id omen mihi maximum erit, fore, vt, priusquam ego Rudolphinas in lucem proferam, colophone hoc restaurationi Astronomicæ imposito: restauratus sub Ferdinando II. post annos ab excessu Ferdinandi primi minus sexaginta, prouinciarum Austriacarum, antiquus ille quinarius, repressis bellis ciuilibus, & pace rerum optima reducta, denuo pristi-num in nitorem efflorescat; quod omen, angoribus ob mala præsentia non le-

niter

Lastly, may arms now be abandoned and enough respite from miseries be granted for good men to have leisure to compose such advice as Cicero gave. When the Roman republic had been overthrown, "as his sorrow was scarcely consolable at such complete loss of everything and despair of recovery, when he saw that there was no place either in the Senate house or in the lawcourt, for the art which had been his study, he devoted his whole attention and effort to philosophy, advising his friend Sulpicius to occupy himself with the same subject, as although it would be less profitable, yet it would divert the spirit from anxiety, and relieve it of troubles."

If God were to consent to these wishes, my mathematics would always be ready to propose, either from astronomical exercises, or from the contemplation of the works of God, or from the harmony of the universe (that destined occupation during the harsh discords of the past two years), pleasures certainly not unworthy of a Christian man, as consolations for his troubles. But because this astronomical occupation was undertaken with the intention of completing it, what in the present calamitous state of Austria should she rather do, than, restrained by decency, seek all the assistance which she needs, to make public her works among men, and to claim the name of Rudolph for her perpetual tables, not from those who are afflicted or in need, but rather entreat them from quarters to which those misfortunes, those horrible expiations of heavenly portents, have not penetrated, and lastly hasten over the remaining half of her journey to her original patrons, to whom she had already approached halfway in the year 1612? As I have said, this little book borne by me had already set out long ago from Styria to Brahe, that is, to expedite the work of the Rudolphine Tables. What would be unprecedented, what so foreign to your original undertaking, noble sirs, and lastly what unwelcome to the Emperor Ferdinand Augustus, the next successor of Rudolph after Matthias, if you were to listen to the things which the little book now returning, of which you used to be the patrons, has to say about what has been achieved in the meantime; if you were to undertake the promotion with modest liberality of the anxious and laborious work of the Tables, of the delight of mankind, and of the honor and repute of the Emperor Rudolph; and if the Austrian house, unshaken even by this most grievous disturbance, at the intercession of your own provision as replacement, were to part with its ancient patronage of the mathematical disciplines, or yield it to strangers?

Then let this be the goal of this renewed dedication; and if by your magnanimity, noble sirs, I achieve it, that will be a most important omen to me, that before I bring the Rudolphine Tables into the light of day, with the addition of this finishing touch to the restoration of astronomy, the ancient fivefold confederation of the Austrian provinces, restored under Ferdinand the Second less than sixty years after the decease of Ferdinand the First, with the suppression of civil wars, and the return of the best of all blessings, peace, will finally blossom forth into her original splendor. May that omen, though considerably impaired by anxieties on

uiter quaſſatum, DEVS OPT. MAX. *miſeratione Eccleſiæ, Filij ſu;*
ne redemptæ, propitius firmet, iram ſuam, tandem à nobis auerſam in gentes
Eccleſiam vaſtantes conuertat, Imperium Ferdinandi II. Imperatoris Auguſti, extinctis irarum incentiuis, ſalutari Clementiæ aura mitigatum proſferet, qua ratione & Styria, fortunæ meæ prima incunabula, cumque illa & vos Reuerendiſſimi Generoſiſſimique Proceres, ſub alis Aquilæ tuti à vulture limitaneo, rerumque omnium copia locupletes, in annos innumeros, perduretis: quibus debita cum veneratione me commendo. Valete. Dabam Francofurti ⅔. *Junij, Anno* M. DC. XXI.

Reu^{ſz} & Gen^{mæ} Mag^{æ} V^{æ}

Deditiſſimus Cliens

Iohannes Keplerus, olim Styriæ Procerum, poſt Impp. Cæſſ. Rudolphi & Matthiæ, l. m. Ordd. q; Auſtriæ Supr-Aniſanæ Mathematicus.

account of the present evils, be favorably confirmed by God the most excellent and greatest, in compassion for the Church, redeemed by the blood of his Son; may his anger at length be averted from us, and turned against the nations which are laying waste the Church; may the empire of the Emperor Ferdinand II Augustus, all incitements to anger being quenched, prosper in the mildness of the health-giving breeze of clemency; and by the same token may both Styria, the first cradle of my fortune, and with her you also, most revered and eminent nobles, continue for countless years, safe under the wings of the Eagle from the vulture on her borders, and rich in abundance of all things. To you with due respect I commend myself.

Frankfurt, 20/30 June, in the year 1621.

'Your Most Revered and Eminent Magnanimity's

Most Devoted Adherent,

Johannes Kepler,
formerly Mathematician to the Nobles of Styria,
and later to the Emperors Rudolph and Matthias
and to the Orders of Austria
over the Enns.

QVID mundus, quæ caufa Deo, ratioque creandi,
Vnde Deo numeri, quæ tantæ regula moli,
Quid faciat fex circuitus, quo quælibet orbe
Interualla cadant, cur tanto Iupiter & Mars,
Orbibus haud primis, interftinguantur hiatu:
Hic te Pythagoras docet omnia quinque figuris.
Scilicet exemplo docuit, nos poffe renafci,
Bis mille erratis, dum fit Copernicus annis,
Hoc, melior Mundi fpeculator, nominis. At tu
Glandibus inuentas noli poftponere fruges.

NOTÆ

GREETINGS, FRIENDLY READER

The nature of the universe, God's motive and plan for creating it, God's source for the numbers, the law for such a great mass, the reason why there are six orbits, the spaces which fall between all the spheres, the cause of the great gap separating Jupiter and Mars, though they are not in the first spheres — here Pythagoras reveals all this to you by five figures. Clearly he has revealed by this example that we can be born again after two thousand years of error, until the appearance of Copernicus, in virtue of this name, a better explorer of the universe. But hold back no longer from the fruits found within these rinds.

N O T Æ
IN LIBELLVM, CVI TITVLVS
DE ADMIRABILI PROPORTIONE
ORBIVM COELESTIVM, &c.

In Titulum libri Notæ Auctoris.

*P*RODROMVS.] *Poſtquam ad Philoſophiæ ſtudium acceſſi, anno ætatis 18:*
Anno Chriſti 1589. verſabantur in manibus iuuentutis exercitationes exotericæ
Iulij C. Scaligeri: cuius ego libri occaſione cœpi ſucceſſiue varia comminiſci de va-
riis quæſtionibus, vt de Cœlo, de Animis, de Geniis, de Elementis, de Ignis natura,
de fontium origine, de fluxu & refluxu maris , de figura continentium terrarum,
interfuſorumque marium, & ſimilia. Verum cum inuentio iſta proportionis Or-
bium cœleſtium mihi videretur eximia; non expectandum mihi ſum ratus, donec omnes naturæ par-
tes peruagarer , nec hoc inuentum obiter euulgandum , coniectum in cumulum quæſtionum cætera-
rum, leui quadam probabilitate vtentium. Quin potius ab huius inuenti editione initium diſſerta-
tionum mearum facere placuit: auſusque ſum in omnibus reliquis quæſtionibus ſimilem ſperare ſuc-
ceſſum: ſed fruſtra, Cœlum enim, principium operum Dei, longe præſtantiorem ornatum habet,
quam reliqua minuta & vilia. Itaque Prodromus quidem egregius fuit : Epidromus vero, qualem
ego tunc propoſueram, nullus eſt ſecutus: quia in reliquis quæſtionibus nequaquam mihi æque ſatisfa-
ciebam. Lector tamen opera mea Aſtronomica, & in primis libros Harmonicorum, pro genuino &
proprio epidromo habere poterit huius libelli; quia eadem vtrinque via curritur; quæque tunc impe-
dita ſatis erat , facta nunc eſt tritiſſima, & quæ tunc breuis nec ad ſcopum pertingens; illa & conti-
nuatur in Harmonicis, & currus circa metam agitur. Talis fuit Prodromus, nauigatio prima Ame-
rici Veſpucij; tales Epidromi nauigationes hodiernæ annuæ in Americam.

Myſterium Coſmographicum.] Extant apud Germanos Coſmographiæ, Munſteri a-
liorumque, vbi de toto quidem mundo partibuſque cœleſtibus fit initium, ſed breuibus illa paginis ab-
ſoluuntur; præcipua vero libri moles complectitur deſcriptiones regionum & vrbium. Itaque vulgus
Coſmographia pro Geographiæ dictione vtitur: impoſuitque vox iſta, à mundo licet deducta, offici-
nis librariis, iisque qui Catalogos librorum conſcribunt, vt libellum meum inter Geographi-
ca referrent. Myſterium autem pro Arcano vſurpaui, & pro tali venditaui
inuentum hoc: quippe in nullius Philoſophi libro talia
vnquam legeram.

A ILLV-

NOTES ON THE LITTLE BOOK ENTITLED
'ON THE MARVELOUS PROPORTION OF THE CELESTIAL SPHERES, etc.'

Notes of the author on the title of the book.

Forerunner.] After I came to the study of Philosophy, in my eighteenth year, the year of Christ 1589, the *Exercitationes Exotericae* of Julius C. Scaliger were passing through the hands of the younger generation; and taking the opportunity offered by that book I began to devise various views on various enquiries, such as on the heaven, on souls, on characters, on the elements, on the nature of fire, on the origin of springs, on the ebb and flow of the sea, on the shape of the continents of the Earth, and the seas that flow between them, and the like. Yet since the discovery of the proportion of the heavenly spheres seemed to me outstanding, I thought I should not wait until I could traverse all the parts of Nature, and that this discovery should not be published incidentally, thrown onto a pile of other inquiries, achieving but a slight probability. I decided rather to make the publication of this discovery the starting point of my dissertations, and dared to hope for a similar success in all the remaining inquiries; but in vain. For the heaven, the chief of the works of God, is much more notably embellished than the rest, which are paltry and mean. So the forerunner was indeed excellent; but no successor, of the kind which I had then intended, followed it, because in the rest of the inquiries I did not achieve anything which gave me equal satisfaction. However the reader will be able to have my astronomical works, and especially the books of the *Harmonice,* as the authentic and appropriate successor of this little book; because the same course is run in both cases; what was then rather obscure has now been made easily accessible; and not only is what was then brief and short of the goal now continued in the *Harmonice* but the chariot is rounding the turning point.[2] The forerunner was like the first voyage of Amerigo Vespucci; the successors are like today's annual voyages to America.

The Secret of the Universe.] There exist in Germany cosmographies by Munster and others, in which indeed the beginning is about the whole universe and the heavenly regions, but they are finished off in a few pages. The main bulk of the book, however, comprises descriptions of territories and cities. Thus the word cosmography is commonly used to mean geography; and that title, though it is drawn from the universe, has induced bookshops and those who compose catalogues of books, to include my little book under geography. Nevertheless I have taken the mystery as a secret, and marketed this discovery as such: and indeed I had never read anything of the sort in any philosopher's book.

ILLVSTRIBVS, GE-NEROSIS, NOBILISSIMIS ET STRENVIS, DOMINO SIGISMVN-

do Friderico, Libero Baroni ab Herberſtein, in Neuperg & Guetten-
haag, Domino in Lancovviz, Camerario & Dapifero Carinthiæ
hæreditario, Cæfareæ Maieſtati & fereniſſimo Ar-
chiduci Auſtriæ Ferdinando à conſiliis,
Capitaneo Prouinciæ
Styriæ:

&

DOMINIS N.N. ILLVSTRIVM STYRIÆ ORDINVM
Quinque-viris Ordinariis, Viris ampliſſimis, Dominis meis clementibus
& beneficis, falutem & mea feruitia.

VOD ante (1) *feptem menfes promifi, opus doctorum teſtimonio
pulchrum, & iucundum, longéque præferendum annuis prognoſti-
cis: tandem aliquando Coronæ veſtræ fiſto, Ampliſſimi Viri; Opus,
inquam, exigua mole, labore modico, materia vndiquaque mira-
bili. Nam fiue quis antiquitatem fpectet;* (2) *tentata fuit ante bis
mille annos à Pythagora: fiue nouitatem, primum nunc à me inter
homines vulgatur. Placet moles? Nihil eſt hoc vniuerfo mundo maius neque amplius. Defideratur dignitas? Nihil preciofius, nihil pulchrius hoc lucidiſſimo Dei tem-
plo. Lubet fecreti quid cognofcere? Nihil eſt aut fuit in rerum natura occultius; So-
lùm hac in re non omnibus fatisfacit, quod vtilitas eius incogitantibus obfcura eſt.
Atque hic eſt ille liber Naturæ, tantopere facris celebratus fermonibus; quem Pau-
lus gentibus proponit, in quo Deum, ceu Solem in aqua vel fpeculo, contemplentur.
Nam cur Chriſtiani minus hac contemplatione nos oblectaremus; quorum proprium
eſt, Deum vero cultu celebrare, venerari, admirari: id quod tanto deuotiori animo
fit, quanto rectius, quæ & quanta condiderit noſter Deus, intelligimus. Sane quam
plurimos hymnos in Conditorem, verum Deum, cecinit verus Dei cultor Dauides;
quibus argumenta ex admiratione cælorum deducit. Cæli enarrant, inquit, glo-
riam* DEI. *Videbo cœlos tuos, opera digitorum tuorum, Lunam &
ſtellas, quæ tu fundaſti: Magnus Dominus noſter, & magna virtus eius;
qui numerat multitudinem ſtellarum, & omnibus nomina vocat. Alicubi plenus fpiritu, plenus facra lætitia exclamat, ipfumque mundum acclamat; Laudate cœli Dominum, laudate eum Sol & Luna, &c. Quæ vox cœlo? quæ
ſtellis? qua Deum laudent inſtar hominis? Nifi quod, dum argumenta fuppeditant
hominibus laudandi Dei, Deum ipfæ laudare dicuntur? Quam vocem, cœlis & Naturæ rerum dum aperimus his pagellis, clarioremque efficimus: nemo nos vanitatis,
aut inutiliter fumpti laboris arguat.*

*Taceo, quod hæc materia Creationis, quam negarunt Philofophi, magnum
argumentum eſt: dum cernimus, vti Deus more alicuius ex noſtratibus Archite-
ctis, or-*

TO THE ILLUSTRIOUS, EMINENT,

MOST NOBLE AND ENERGETIC LORD
SIGISMUND FREDERICK,

FREE BARON OF HERBERSTEIN, NEUBERG AND GUETTEN-
HAAG, LORD OF LANKOWITZ,
HEREDITARY CHAMBERLAIN AND STEWARD OF CARINTHIA,
COUNSELOR TO HIS IMPERIAL MAJESTY AND TO THE
MOST SERENE ARCHDUKE OF AUSTRIA,
FERDINAND, CAPTAIN OF THE PROVINCE OF STYRIA;
AND
TO THEIR LORDSHIPS THE MOST NOBLE FIVE COMMISSIONERS OF
THE ILLUSTRIOUS ORDERS OF STYRIA, MOST GENEROUS OF MEN,
MY KINDLY AND LIBERAL LORDS, GREETINGS AND MY HOMAGE.

(1) Seven months ago I promised you a work which would be acknowledged by the learned as handsome, and pleasing, and far preferable to annual predictions. Now at last I add it to your crown, most generous of men — a work, I say, of tiny bulk, of modest effort, of contents in every way remarkable. For if we look to ancient times, it had been attempted (2) two thousand years before by Pythagoras; if we look to modern times, it is now published among men by me for the first time. Do you want something bulky? Nothing in the whole universe is greater or more ample than this. Do you require something important? Nothing is more precious, nothing more splendid than this in the brilliant temple of God. Do you wish to know something secret? Nothing in the nature of things is or has been more closely concealed. The only thing in which it does not satisfy everybody is that its usefulness is not clear to the unreflecting. Yet here we are concerned with the book of Nature, so greatly celebrated in sacred writings. It is in this that Paul proposes to the Gentiles that they should contemplate God like the Sun in water or in a mirror. Why then as Christians should we take any less delight in its contemplation, since it is for us with true worship to honor God, to venerate him, to wonder at him? The more rightly we understand the nature and scope of what our God has founded, the more devoted the spirit in which that is done. How many indeed are the hymns which were sung to the Creator, the true God, by the true worshiper of God, David, in which he draws arguments from the marvels of the heavens.[3] "The Heavens are telling," says he, "the glory of God. I shall see thy heavens, the work of thy fingers, the Moon and stars, which thou hast created. Great is our Lord, and great is his excellence, who numbers the multitude of the stars, and calls them all by name." Elsewhere, full of the spirit, full of holy joy, he exclaims, and acclaims the very universe, "Praise the Lord, ye heavens, praise him ye Sun and Moon, etc." What voice has the heaven, what voice have the stars, to praise God as a man does? Unless, when they supply men with cause to praise God, they themselves are said to praise God. And if we reveal this voice for the heavens and for the Nature of things in these pages, and make it clearer, no one should charge us with a vain deed or with undertaking useless toil.

I pass over in silence the fact that this very matter, of Creation, which the philosophers denied, is a strong argument, when we perceive how God, like one of our own architects, approached the task of constructing the universe with

Etis: ordine & norma ad mundi molitionem accesserit, singulag, sit ita dimensus: quasi non ars naturam imitaretur, sed Deus ipse ad hominis futuri morem ædificandi,
respexisset.

Quanquam quid necesse est, diuinarum rerum vsus instar obsonij nummo æ-
stimare? Nam quid quæso prodest ventri famelico cognitio rerum naturalium, quid
tota reliqua Astronomia? Neque tamen audiunt cordati homines illam barbariem,
quæ deserenda propterea ista studia clamitat. Pictores ferimus, qui oculos, Sympho-
niacos, qui aures oblectant: quamuis nullum rebus nostris emolumentum afferant.
Et non tantum humana, sed etiam honesta censetur voluptas, quæ ex vtrorumque
operibus capitur. Quæ igitur inhumanitas, quæ stultitia, menti suum inuidere hone-
stum gaudium, oculis & auribus non inuidere? Rerum naturæ repugnat, qui cum his
pugnat recreationibus. Nam qui nihil in naturam introduxit, Creator Optimus, cui
non cùm ad necessitatem, tum ad pulchritudinem & voluptatem abunde prospexe-
rit: is mentem hominis, totius naturæ dominam, suam ipsius imaginem, solam nulla
voluptate beauerit? Imo vti non quærimus, qua spe commodi cantillet auicula, cum
sciamus inesse voluptatem in cantu, propterea, quia ad cantum istum facta est: ita
nec hoc quærendum, cur mens humana tantum sumat laboris in perquirendis hisce
cælorum arcanis. Est enim ideò mens adiuncta sensibus ab Opifice nostro; non tantù
vt seipsum homo sustentaret, quod longè solertiùs possunt vel brutæ mentis ministe-
rio multa animantium genera: sed etiam, vt ab iis, quæ, quod sint, oculis cernimus,
ad causas quare sint & fiant, contenderemus: quamuis nihil aliud vtilitatis inde
caperemus. Atq adeò vt animalia cætera, corpusq humanum cibo potuq sustentan-
tur: sic animus ipse hominis, (3) diuersum quiddam ab homine, vegetatur, auge-
tur, & adolescit quodammodo, cognitionis isthoc pabulo: mortuoq, quàm viuo simi-
lior est, si harum rerum desiderio nullo tangitur. Quare vti Naturæ prouidentiâ pabulum animantibus nunquam deficit: ita non immeritò dicere possumus, propterea
tantam in rebus inesse varietatem, tamq reconditos in cælorum fabrica thesauros;
vt nunquam deesset humanæ meti recens pabulum, ne fastidiret obsoletum, neu quiesceret, (4) sed haberet in hoc mundo perpetuam exercendi sui officinam.

Neq verò harum epularum, quas ex ditissimo Conditoris penu in hoc libello,
velut in mensa depromo, propterea minor est nobilitas: quod à maxima vulgi parte
vel non gustabuntur, vel respuetur. Anserem laudant plures, quàm phasianum, quia
ille communis est, iste rarior. Neque tamen vllius Apitij palatus hunc illi postponet.
Sic huius materiæ dignitas tantò maior erit; quo pauciores laudatores, intelligentes
modo sint, reperiet. Non eadem vulgo conueniunt & principibus: neque hæc cælestia
promiscuè omnium, sed generosi saltem animi pabulum sunt: non meo voto, vel opera, non sua natura, non Dei inuidiâ: sed plurimorum hominum, vel stupiditate, vel
ignauia. Solent principes aliqua magni precij inter secundas habere mensas, quibus vtantur non nisi saturi, leuandi fastidij causa. Sic hæc & huiusmodi studia
generosissimo & sapientissimo cuique tum demum sapient, vbi è casa per pagos, oppi-
da, prouincias, regna ad orbis imperium ascenderit, omnia probè perspexerit; neq,
vt sunt humana, quicquam vllibi reperierit beatum, diuturnum, & tale, quo finiri
& saturari queat eius appetitus. Tunc enim incipiet meliora quærere, tunc à terra
huc in cælum ascendet, tunc animum fessum curis inanibus ad hanc quietem trans-
feret, tunc dicet:

Felices animas, quibus hæc cognoscere primum
Inq; domos superas scandere cura fuit,
quare contemnere incipiet, quæ olim præstantissima censuit, sola hæc Dei opera ma-

A 2   gnifaciet,

order and pattern, and laid out the individual parts accordingly, as if it were not art which imitated Nature, but God himself had looked to the mode of building of Man who was to be.

Though why is it necessary to reckon the value of divine things in cash like victuals? Or what use, I ask, is knowledge of the things of Nature to a hungry belly, what use is the whole of the rest of astronomy? Yet men of sense do not listen to the barbarism which clamors for these studies to be abandoned on that account. We accept painters, who delight our eyes, musicians, who delight our ears, though they bring no profit to our business. And the pleasure which is drawn from the work of each of these is considered not only civilized, but even honorable. Then how uncivilized, how foolish, to grudge the mind its own honorable pleasure, and not the eyes and ears. It is a denial of the nature of things to deny these recreations. For would that excellent Creator, who has introduced nothing into Nature without thoroughly foreseeing not only its necessity but its beauty and power to delight, have left only the mind of Man, the lord of all Nature, made in his own image, without any delight? Rather, as we do not ask what hope of gain makes a little bird warble, since we know that it takes delight in singing because it is for that very singing that the bird was made, so there is no need to ask why the human mind undertakes such toil in seeking out these secrets of the heavens. For the reason why the mind was joined to the senses by our Maker is not only so that Man should maintain himself, which many species of living things can do far more cleverly with the aid of even an irrational mind, but also so that from those things which we perceive with our eyes to exist we should strive towards the causes of their being and becoming, although we should get nothing else useful from them. And just as other animals, and the human body, are sustained by food and drink, so the very spirit of Man, (3) which is something distinct from Man, is nourished, is increased, and in a sense grows up on this diet of knowledge, and is more like the dead than the living if it is touched by no desire for these things. Therefore as by the providence of Nature nourishment is never lacking for living things, so we can say with justice that the reason why there is such great variety in things, and treasuries so well concealed in the fabric of the heavens, is so that fresh nourishment should never be lacking for the human mind, and it should never disdain it as stale, nor be inactive, but (4) should have in this universe an inexhaustible workshop in which to busy itself.

Yet the nobility of this banquet which from the Creator's sumptuous store I set forth in this book, as on a table, is no less because by the majority of the people is will not be savored, or will be spat out. More men praise the goose than the pheasant, because the former is common, the latter rarer; and yet no Apicius's palate will rank pheasant lower than duck. Similarly the fewer there are found to praise this subject, provided they are intelligent, the greater will be its merit. The same things do not suit the people and the princes, and these heavenly matters are not nourishment for everyone indiscriminately, but just for a noble mind — not by my wish, or efforts, not by its own nature, not from God's jealousy, but by the stupidity or ignorance of the majority of men. Princes usually have something very expensive kept for the dessert course, which they use only if they are satiated, to relieve the monotony. So these subjects, and those like them, will appeal to the wisest and most eminent of men only when he has ascended from the cottage through country, towns, provinces, kingdoms to dominion over the world, and has fully explored all possibilities, yet, as these things are human, he has found nothing anywhere which is blessed with happiness, everlasting, and able to satisfy and satiate his appetites. For then he will begin to seek for better things, then he will ascend from the Earth below to heaven, then he will lift up his spirit, tired with empty cares, to that tranquility, then he will say:[4]

Happy the souls whose first concern it was
To gain this knowledge and soar to heavenly homes;

and therefore he will begin to despise what once he thought most important, he will value only these works of God, and he will derive pure and sincere delight at

gnificiet, atque meram & finceram tandem voluptatem ex his contemplationibus capiet. Contemnant igitur hæc & huiusmodi meletemata, quicunque quantumcunq; volent, quærantque fibi vndiquaque commoda, diuitias, thefauros: Aftronomis ifthæc gloria fufficiat, quod Philofophis fua fcribunt, non rabulis, Regibus non paftoribus. Prædico intrepide, futuros tamen aliquos, qui fuæ fibi fenectutis hinc comparent folatium; tales nempe, qui quoad Magiftratus gefferunt, ita fe gefferunt, vt liberi morfibus confcientiæ, habiles effe poffint fruendis hifce deliciis.

(5) Exiftet iterum Carolus aliquis, qui, cum Europa, quoad imperauerit, non caperetur; feffus imperiis, exigua S. Iufti cellula capiatur: cuique inter tot fpectacula, titulos, triumphos, tot diuitias, vrbes, regna; vnica Turrianica, vel iam (6) Copernicopythagorea Sphæra Planetaria tantopere placeat, vt orbem terrarum cum ea commutet, digitoque circulos, quam populos imperiis regere malit.

Non hæc eo dico, viri ampliffimi, vt nouum paradoxon, fenes difcipulos, in fcenam, feu in fcholas producam; fed vt appareat quodnam genuinum tempus fit meffem de his ftudiis colligendi. Cur enim de femente facienda aliter ego fentiam, atque viri prudentiffimi de veftra Corona; qui hæc ftudia inter præcipua cenfuerunt, quæ iuuenilibus Nobilitatis animis in veftra fchola proponerentur. Sic enim exiftimant, neque aptius effe genus hominum ad colenda Mathemata, Nobilitate: vt quibus artes aliæ ad victum comparandum non ita neceffariæ; nec aptiora Nobilitati ftudia, Mathematicis: propterea, quod occulta & mirifica quadam facultate polleant præceteris; feroces animos ad humanitatem, adque fobrium rerum terrenarum contemptum inftituendi. Qui fructus etfi difficultate & infolentia materiei iuuenibus obfcuratur: fenibus tamen, vti modo dictum, fuo tempore fefe patefacit. Atque hæc ego hactenus, cum de præfentibus pagellis, tum de omni Aftronomia, ad vos Aftronomiæ & Literaturæ totius amatores, Viri ampliffimi: vt eius vos admoneam, quod pridem tenetis: neque nulli vfui fore hoc, quod humilis offero & dedico, opufculum, vobis, qui vere generofi, vere nobiles eftis: & fi quam laudem meretur inuentio, illam magna ex parte ad vos pertinere; qui veftra liberalitate, veftroque ftipendio mihi occafiones & otium hæc ita commentandi feciftis: Accipite igitur, Viri Ampliffimi, hoc grati animi fymbolum, meque humilem clientem in veftram gratiam fufcipite, & denique (7) affuefcite inter Atlantes, Perfeas, Oriones, Cæfares, Alphonfos, Rodolphos, cæterofque Aftronomiæ promotores accenferi. Valete. Idibus Maii: qui dies ante annum initium fuit huius laboris.

Ampl. V.

Humilis in Schola veftra Græciana
Mathematicus

M. IOANNES KEPLERVS
VVirtemberg.

IN DEDICATIONEM ANTIQVAM
Notæ Auctoris.

(1) *Ante feptem menfes.*] *Anno* 1595. *die* 9/19 *Iulij poftridie natalis decimioctaui Sereniffimi Ferdinandi Archiducis, Roman. nunc Imperatoris Augufti, Hungariæque & Bohemiæ Regis: cuius in ditione hæreditaria Styria tunc merebam ftipendia, inueni hoc fecretum: ftatimque ad illud*

last from these studies. Accordingly let these and like occupations be despised by whoever wishes, and as much as they wish, and let them seek for themselves everywhere profit, wealth, and treasures: for astronomers let it be glory enough that they write for philosophers, not for pettifoggers, for kings, not shepherds. I predict without dismay that there will nevertheless be men who will draw from here solace for their old age, such men indeed who have conducted not only great offices but also themselves in such a way that, free from the remorse of conscience, they can be fit to enjoy these delights.

(5) There will arise another Charles, who, as he was not captivated by Europe, as long as he held dominion over it, will, tired of dominion, be captivated by the narrow cell of the monastery of Yuste; who, among so many spectacles, titles, triumphs, so many riches, cities, kingdoms, will be so pleased by a Torrianan, or (as it would be now) (6) a Copernico-Pythagorean planetarium alone that he will exchange the whole round world for it, and prefer to rule circles with his finger rather than nations with his dominion.[5]

I do not say this, most generous sirs, so that by a new paradox I may bring old men as students to schools or college; but so that it may be seen what is the natural time for reaping the harvest from these studies. For on sowing the seed why should my opinion differ from that of the sagacious members of your assembly, who have decreed that these studies should be among those most prominently offered to the young spirits of the nobility in your school? For it is their view, both that no kind of men is more fit for the pursuit of mathematics than the nobility, as for them other skills are not so necessary for earning a living; and that no studies are more fit for the nobility than mathematics, because more strongly than any others it possesses some hidden and wonderful power of civilizing fierce spirits and instilling into them a sober contempt for earthly things. Although this harvest is concealed from the young by the difficulty and unfamiliarity of the subject, yet to the old, as has just been said, it reveals itself in its own time.

This, then, is what I have to say, both about the pages before you and about the whole of astronomy, to you who are lovers of astronomy and the whole of learning, most generous sirs: to inform you of what you have long understood, that this little work which I humbly offer and dedicate to you will not be valueless to you, who are truly eminent, truly noble; and that if the discovery deserves any praise, it belongs to you to a great extent, as by your generosity and by your salary you have granted me the time and opportunity for this account. Accept, then, most generous sirs, this symbol of a grateful spirit, take me as a humble dependent into your favor, and finally (7) accustom yourselves to being included among the Atlases, Perseuses, Orions, Caesars, Alfonsos and Rudolphs, and the other benefactors of astronomy.

15th May (the anniversary of the beginning of this task).

I am, most generous Sirs,

the Humble Mathematician in your School of Graz,

Master Johannes Kepler
of Württemberg

AUTHOR'S NOTES ON THE ORIGINAL DEDICATION

(1) *Seven months ago.*] In the year 1595, on the 9/19th of July, the day after the eighteenth birthday of his Serene Highness the Archduke Ferdinand, now Roman Emperor and King of Hungary and Bohemia, in whose hereditary dominion of Styria I was then earning my living, I discovered this secret; and turning

57

illud excolendum conuersus, Octobri sequente, in dedicatione prognostici anniuersarij, quod erat ex officio scribendum, editionem libelli promisi, vt significarem publice, quam grauis mihi, philosophiam amanti, esset ista coniectandi necessitas. Ex eo profectus in VVirtembergiam, inter domestica negotia, nihil æque pensi habui, ac editionem libelli, quæ mihi iuuenculo, nulla eruditionis fama publica, typographiis sibi de damno metuentibus, plurimum exhibuit molestiarum: & erant qui absurditate moti dogmatis Copernicani, conatibus meis intercederent. Itaq, scripta dedicatione ista Idibus Maij Stuccardiæ. post duos menses reuersus sum in Styriam, relicta Mæstlino Præceptori meo editionis cura penè desperata. Ille vero ad exornandum, commendandum, & inter homines vulgandum opusculum, quod ingenti cum gratulatione primum aspexerat, nihil fecit reliqui: perfecitque prudentia & industria sua, vt libellus tandem ederetur, fine anni 1596. & sequentibus Nundinis Vernalibus anni 1597. catalogo Francofurtensi insereretur: duro nominis mei fato; Nam pro Keplero, expresserunt Repleum. Quo ipso tempore, flagrante bello Hungarico cum Turcis, de prouinciis limitaneis Ferdinando hæredi tradendis arduis deliberationibus actum, quippe exactis annis hæredis tutelaribus.

Cum igitur casus quidam, oppidò quam pulcher, initia speculationum istarum, cum gubernationis Ferdinandinæ primordiis connexuerit: quis vetet etiam successus reliquos commemorando exsequi: vnde fides firmetur, spei optima plena, non casu cæcum, sed genium perspicacissimum & vigilantissimum fuisse, qui hanc vitem imbellem, humique serpentem, vlmis illis sublimibus coaptauerit.

Etenim factum est illo ipso anno 1597. vt Tycho Brahæus, vir illustri stemmate Danico prognatus, consiliisque restaurandæ Astronomiæ susceptis celebratissimus, successu, quoad vixit, fælicissimus, vt hic inquam Dania patria relicta, cum omni apparatu Astronomico transiret in Germaniam. Cum autem huius viri instituta mihi ex relatu, & prælectionibus Mæstlini dudum essent nota, cum mentionem illius, vt summi Artificis, passim in ipso libello secissem: pulchrum æquumque mihi visum est, primum atque libellum meum Catalogo Francofurtensi insertum sciui, inter cæteros Matheseos Professores, etiam Tychonem, vt antesignanum, consulere super materia libelli, quam cum proprio, tum Mæstlini iudicio, maximi momenti rem esse rebar. Et cæteri quidem promptè responderunt, Galilæus Patauio, Vrsus Praga, Limnæus Icna: ad Tychonem vero Epistola mea tardius delata, quod is locum inscriptum interea mutasset, voluptatem, ex responso tanti viri secituram, per integrũ annum detinuit: hausi tandem illam affatim, adiunxique lætitiæ publicæ, quæ tunc Styriam tenebat ob exordia gubernationis Ferdinandi, florentissimi Principis. Quanquam Eclipsis magna Solis in Dodecatemorio Piscium, qui locus Ferdinando culminat, multoque magis intemperies hominum certorum, iam signa meo iudicio prætulissent ærumnis, paulo post per prouincias illas consecutis.

Argumentum literarum Brahei hoc erat, vti suspensis speculationibus à priori descendentib. animũ potius ad obseruationes quas simul offerebat, considerandas adijcerem: inq, ijs primo gradu facto, postea demũ ad causas ascenderem, & tale quid in sua potius Hypothesi, quam ipse Copernicanâ veriorem censebat, comminiscerer: denique, vt ad ipsum me conferrem, quippe qui iam mare transisset. Cumque non statim ego responderem, Brahæus eodem argumento plures ad me per annum sequentem scripsit epistolas, quarum vna post aliam, sua qualibet mora interposita, mihi sunt redditæ. Interim Græcij dissipato nostro cœtu discretim, ipse salarium, quod capiebam à Proceribus Prouinciæ sine opera, bene collocaturus, consilium tandem cepi, Tychonem Brahe visitandi, toties inuitantem. Venerat ille anno 1598. VVitebergam, iturus ad Imperatorem: vbi cum aliquandiu substitisset; anno sequenti 1599. in Bohemiam se contulit; cui Benatica arx Regia, quinque milliaribus Praga distans, habitanda concessa fuit: cum Rudolphus Imp. Pilsnæ commoraretur, ob pestem Pragæ grassantem. Hæc omnia mihi Fridericus Hofmannus L. B. Styrus, Imp. Rudolphi Consiliarius aulicus, qui tunc Pragæ venerat, retulit: me ad capessendum iter adhortatus est, loco mihi oblato in comitatu suo. Ita factum vt ad Brahæum venirem initio anni 1600. quando Ferdinandus Archidux nuptias Græcij celebrauit cum consobrina sua Bauarica: breuique captis, Braheanorum laborum, vicissimque exhibitis ingenij mei experimentis, pactus conditiones, cum ipso commorandi, quas Styriæ Proceres ratas haberent, post aliquot mensium conuersationem, reuersus sum Grætium. Receptis autem breui aliquot Brahei epistolis (quibus ille me vacillantem in proposito propter difficultates ortas, confirmauit, addita commemoratione, quid ipse cum Imperatore de me aduocando, iam egisset) denique mense Octobri familiam Pragam transtuli. Nec diutius anno vno potitus Magistro superstite, post eius obitum ab Imperat. Rudolpho curator Operis Tabularum, quibus Brahæus à Rudolpho nomen esse voluit, surrogatus sum: in quo perficiendo per hos 20. annos desudaui. Ita omnis mihi vita, studiorum, operumque meorum

A 3   meorum

at once to the refinement of it, in the following October, in the dedication of the almanac for the year, which it was my duty to write, I promised to publish a little book, so as to signify publicly how oppressive that obligation to make forecasts was to me as a lover of philosophy. When after that I set out for Württemberg, in the midst of domestic troubles, I considered nothing to be as important as the publication of this book, though for me, a stripling, with no reputation for erudition, and with the printers afraid of making a loss, it produced a great deal of vexation. There were also those who were impelled by the absurdity of the Copernican doctrine to interfere with my attempts. Therefore after writing the dedication on May 15 at Stuttgart, two months later I returned to Styria, leaving to my teacher Maestlin the almost hopeless responsibility for publication. He indeed left nothing undone by way of embellishing, recommending, and spreading abroad among men my little work, which he had greeted at first sight with great enthusiasm; and by his own judgment and diligence he brought it about that the little book was at length published at the end of the year 1596, and at the following spring quarter day of the year 1597 was registered in the Frankfurt catalogue — though with an unfortunate mishap to my name, which they entered as Repleus instead of Kepler. At that very time, when the war between Hungary and the Turks was raging, a decision was made by arduous negotiations on handing over the frontier provinces to Ferdinand who was heir to them, as the years of his minority as heir had been completed.

Since, then, an exceedingly happy chance had linked the start of these speculations of mine with the beginning of Ferdinand's reign, who would object to my following that up by recounting my further results, so that confidence may be strengthened, full of splendid hope, that it was not blind chance, but a most perspicacious and watchful guardian spirit which has grafted this feeble vine, that crept on the ground, to those lofty elms.

For it came about in that very year 1597 that Tycho Brahe, a man who had kinship with the illustrious nobility of Denmark, and was greatly celebrated for the designs which he had undertaken for the reformation of astronomy, and was most fortunate with his results, while he lived, it came about, I say, that he left his fatherland of Denmark, and crossed over with all his astronomical equipment to Germany. However, since the great man's intentions had long been known to me from the reports and lectures of Maestlin, and since I had frequently mentioned him as the leading practitioner in the actual book, it seemed to me a right and excellent thing, as soon as I knew that my little book had been registered in the Frankfurt catalogue, to consult about the contents of the book, among the rest of the professors of mathematics, Tycho also as their standard bearer, as both by my own judgment and Maestlin's I thought it a matter of the greatest importance. The others did indeed reply promptly, Galileo from Padua, Ursus from Prague, Limnaeus from Jena. However, the rather late delivery of my letter to Tycho, because he had changed his address in the meantime, delayed for a whole year the pleasure which was to follow from the reply of so great a man.[6] In the end, I regaled myself on it abundantly, and added it to the public joy which then prevailed in Styria because of the opening of the reign of Ferdinand, that most flourishing prince. Nevertheless the great eclipse of the Sun in the House of Pisces, a position of the utmost significance for Ferdinand, and much more the immoderate behavior of certain men, had already in my opinion given the warning signs of the troubles which followed soon after in those provinces.

The burden of Brahe's letters was that I should hold in abeyance my speculations which were derived *a priori*, and apply my mind instead to consideration of the observations which he was simultaneously offering; that I should take the first step on that foundation, proceed to induction of causes thereafter, and work out something of that sort on the basis rather of his own hypothesis, which he believed to be truer than that of Copernicus; and finally, that I should travel to him, as after all he had already crossed the sea. And since I did not reply at once, Brahe sent me more letters with the same burden during the following year, which were delivered to me one after another, each of them with its own period of delay. Meanwhile our band of students at Graz was scattered, and to make good use of the salary which I was drawing from the nobles of the province without duties, in the end I adopted the plan of visiting Tycho Brahe who so often invited me. He had come in the year 1598 to Wittenberg, on his way to the emperor; and after remaining there for some time, in the following year, 1599, he moved on to Bohemia. The royal castle of Benatek, five miles distant from Prague, was granted to him to live in, as the Emperor Rudolph was staying at Pilsen, on account of the plague which was then raging in Prague. I was told all this by Friedrich Hofmann, a Styrian nobleman and court counselor to the Emperor Rudolph, who had come to Prague at that time: he urged me to undertake the journey, offering me a place in his own company. Thus it turned out that I came to Brahe at the beginning of the year 1600, when the Archduke Ferdinand celebrated his wedding at Graz with his Bavarian cousin; and having soon received proofs of Brahe's labors, and in return shown proofs of my own talent, and having agreed on the conditions for my staying with him, which the nobles of Styria ratified, after an association of several months I returned to Graz. However I soon received several letters from Brahe (in which he stiffened my determination in my undertaking, over which I had been wavering on account of difficulties which had arisen, and added a mention of what he had done by way of speaking for me to the emperor), and finally I moved my household to Prague in the month of October. The master whom I thus acquired survived for no longer than a single year, and after

meorum ratio ab hoc vno libello consurrexit. Et cur non magnificè me iactem, dum recolo memoria, quod demonstratis iam planetarum omnium motibus, tandem ad absoluendam telam hoc libello captam, ad opus sc. Harmonicum, illo ipso anno, quo Ferdinandus Archidux in regem Bohemiæ susceptus est, animum adiecerim, quòd anno sequenti 1618. quo anno Ferdinandus Diadema regni Vngariæ suscepit; ego librum V. Harmonicorum absoluerim : quod denique anno 1619. quo Ferdinando summa dignitas Imperialis accessit, Harmonicen ipse meam eodem & loco & mense coronationis eius publicauerim. Faxit Deus, vt extinctis dissidiorum ciuilium dissonantiis, in toto Monarchæ huius imperio, inque Austria superiore, moderno meo domicilio, suauissima pacis Harmonia, quæ in æquitate imperiorum & promptitudine obsequiorum consistit, ab hoc ipso tempore restauretur, quo ego primum hunc meum libellum Notis emendatum integratumque denuò in publicum edo. Sic enim fieri poterit, vt vastitati prouinciarum cicatricibus obductis, vt siccatis aquis horrendi diluuij; solibusque reuersis reflorescens copiæ cornu, etiam mihi destinatos à Rudolpho Imp. sumptus (impeditos per superiorum temporum turbulentiam) denique ad tabularum Astronomicarum opus edendum affundat.

(2) Ante bis mille annos.] Quia dogma de quinque figuris Geometricis, inter Mundana corpora distributis, refertur ad Pythagoram, à quo Plato hanc Philosophiam est mutuatus. Vide Harmonices lib. I. fol. 3. 4. Item lib. II fol. 58. 59. Nam eadem quidem, & illis & mihi figuræ quinque erant propositæ, idem & illis & mihi Mundus, at non eadem vtrinque Mundi partes, si literam solam spectes; nec eadem applicandi ratio.

(3) Diuersum quiddam ab homine.] Condona lector tyroni locutionem minus emendatam. Corpus equidem agnoscit philosophia quodammodò diuersum quiddam ab homine, quia illud continuæ mutationi est obnoxium; cùm homo semper idem sit: Animum vero perhibet id, quò homo sit homo: adeo nò est animus diuersum quid ab homine. Verùm illatio manet eadem, esse suum etiam animo pabulum, seorsim à pabulo corporis, suas etiam seorsim delicias.

(4) Sed haberet in hoc mundo.] Non legeram Senecam, qui penè eandem sententiam Eloquentiæ Romanæ flosculis sic exornauit, Pusilla res mundus est, nisi in eo, quod quærat, omnis mundus inueniat.

(5) Existet iterum Carolus aliquis.] Non equidem cogitaueram tunc fore, vt in Imp. Rudolphi aulam vocarer. Namque hunc Monarcham vere alterum Carolum hic deprehendi, non abdicatione quidem: at profectò fastidio actionum iniquissimarum, domi forisque occurrentium, reductione mentis ab iis, & beato, (quantum ad naturales contemplationes,) recreationum exercitio; vt æquius fuerit, subditos suis potius importunitatibus, quam Regis sui fastidio irasci.

(6) Copernico-pythagorêa.] Ad sphæram allusi Systematis Planetarij, constructam ex Orbibus planetariis, & Corporibus quinque regularibus Pythagoricis, suis quoque coloribus à cæteris distincto, orbibus cæruleis, limbis vero, in quibus planetas decurrere significabatur, albis : perlucidis omnibus, sic vt Sol in centro pendulus videri posset. Saturni orbis, sex circulis, repræsentabatur, qui mutuo concursu, terni quidem, angulo Cubi locum signabant. Bini verò, centro plani cubici superstabant; Iouis orbium extimus tribus, intimus sex circulis, Martis extimus iterum sex; intimus vero, non minus quam Telluris vterque, Venerisque extimus; singuli denis circulis adumbrabantur, quorum quini duodecies, terni vicies, bini tricies concurrebant. Veneris Orbis intimus, æqualis erat Iouis extimo, Mercurij orbis, Iouis intimo: spectaculum non inamœnum, cuius rudimentum quidem, at non plane genuinum, videre est in figura tertia sequenti ex ære.

(7) Assuescite inter Astronomiæ promotores.] Locum inuenit adhortatio mea, commodo meo non exiguo ; quam commemorationem honori Procerum ex gratitudinis lege tribuo. Illustris D. Capitaneus de proprio, statim; cæteri, vt erant loco corporis Prouincialium, expectato eorum conuentu anni 1600. magnificam mihi tunc in Bohemia absenti renunciationem, quanquam exhausto continuis bellis limitaneis ærario impetrarunt. Ita Cælorum conditor mihi operum suorum præconi, tunc de viatico prospexit, familiam in Bohemiam translaturo.

P R AE-

his death I was appointed in his place by the Emperor Rudolph to supervise the work of the Tables, which Brahe wished to be named after Rudolph. In completing it I have sweated for these twenty years. Thus the whole scheme of my life, studies, and works arose from this one little book. And why should I not make a splendid boast, when I recall to memory that having already derived the motions of all the planets, I eventually turned my mind to completing the fabric which had been begun in this little book, that is, my work on the Harmony, in the very year in which the Archduke Ferdinand was accepted as king of Bohemia; that in the following year, 1618, the year in which Ferdinand accepted the crown of the kingdom of Hungary, I completed Book V of the *Harmonice*; and lastly, in the year 1619, in which Ferdinand achieved the highest office in the empire, I made public my *Harmonice* in the same place and month as his coronation. May God bring it about, that the discords of civil conflicts may be extinguished throughout the empire of this monarch, and in upper Austria, my present abode, and that the delightful harmony of peace, which consists in justice of rule and readiness of obedience, may be restored from the very moment at which I finally issue to the public this my first little book corrected and completed with notes. For thus it may come to pass that scars grow over the devastation of the provinces, that the waters of this horrible flood dry up, and that with the return of sunshine the horn of plenty flourishes again and even pours forth the funds intended for me by the Emperor Rudolph (held up by the turbulence of times past) for the eventual publication of the work of the astronomical tables.

(2) *Two thousand years before.*] Because the doctrine of the five geometrical figures' being distributed among the bodies of the universe is traced back to Pythagoras, from whom Plato borrowed this part of his philosophy. See *Harmonice*, Book I, pages 3-4, also Book II, pages 58-59. For they and I had the same five figures in mind, and the same universe, but not the same parts of the universe in each case, if you look only at the letter; nor the same method of applying them.

(3) *Which is something distinct from Man.*] Pardon a novice, reader, for this ill-considered expression. Philosophy indeed recognizes that the body is in a way something distinct from the man, because the former is subject to continual change, while the man is always the same; but it asserts that the spirit is that in virtue of which a man is a man, so that the soul is not something distinct from the man. However the conclusion remains the same, that the soul has its own food too, separate from the food of the body, and also its own separate delights.

(4) *But should have in this universe.*] I had not read Seneca, who elegantly expressed almost the same sentiment as follows in his anthology of Roman eloquence: "The universe is a tiny thing, if the whole universe does not find in it whatever it seeks."[7]

(5) *There will arise another Charles.*] I certainly did not suppose then that I should be called to the court of the Emperor Rudolph. For I discovered this monarch to be truly a second Charles, not indeed in abdicating, but certainly in his disgust for the evil activities which he found at home and abroad, in his withdrawal of his mind from them, and in his happy enjoyment (as far as contemplation of Nature went) of his recreations; so that his subjects would have done better to be angry at their own insolence than at the disgust of their king.

(6) *Copernico-pythagorean.*] I alluded to the sphere of the planetary system, constructed of the planetary spheres, and the five regular Pythagorean solids, each distinguished from the others by their own colors, the orbits sky-blue, and the bands, in which it was implied that the planets ran round, white; all transparent, so that the Sun could be seen suspended in the center. The sphere of Saturn was represented by six circles, which by their common intersection, three at a time, signified the position for the vertex of the cube, but intersected two at a time over the position of the center of a face of the cube. The outermost of the spheres of Jupiter was shown by three circles, its innermost by six circles, and the outermost of Mars again by six; but the innermost of Mars, just as were both those of the Earth, and the outermost of Venus, were each sketched out by ten circles, of which every five met twelve times, every three twenty times, and each pair thirty times. The innermost sphere of Venus coincided with the outermost of Jupiter, that of Mercury with the innermost of Jupiter. It was a not unpleasing spectacle, of which the elements, though not an exact likeness, may be seen in the third engraved figure which follows.

(7) *Accustom yourselves to being included.*] My exhortation found its mark to my not inconsiderable advantage; and under the obligation of gratitude I assign the credit for this attention to the nobles. The illustrious captain immediately from his own funds, and the rest, as they represented the corporate body of the inhabitants of the province, at their expected assembly of the year 1600, although their treasury was drained by the continuous frontier wars, obtained for me a magnificence remittance, at a time when I was absent in Bohemia. Thus the Founder of the Heavens provided for me as herald of his works the expenses of my journey, when I was about to move my household to Bohemia.[8]

PRAEFATIO ANTIQVA AD LECTOREM.

ROPOSITVM eſt mihi, Lector, hoc libello demonſtrare, quod Creator OptimusMaximus, in creatione mundi huius mobilis,& diſpoſitione cœlorum, (1) ad illa quinque regularia corpora, inde à Pythagora
& Platone, ad nos vſque,celebratiſſima reſpexerit,atque ad illorum naturam cœlorum numerum,proportiones,& motuum rationem accom-
modauerit.Sed antequam te ad rem ipſam venire patiar: cum de occaſione huius libelli, tum de ratione mei inſtituti,aliqua tecum agam : quæ & ad tuum intellectum,
& ad meam famam pertinere arbitratus fuero.

Quo tempore Tubingæ,ab hinc ſexennio clariſſimo viro M.Michaeli Mæſtlino operam dabam : motus multiplici incommoditate vſitatæ de mundo opinionis,
adeo delectatus ſum Copernico,cuius ille in prælectionibus ſuis plurimam mentionem faciebat: vt non tantum crebro eius placita in phyſicis diſputationibus candi-
datorum defenderem: ſed etiam accuratam (2) diſputationem de motu primo,
quod Terræ volutione accidat , conſcriberem. Iamque in eo eram,vt eidem etiam
(3) Telluri motum Solarem , vt Copernicus Mathematicis , ſic ego Phyſicis,
ſeu mauis,Metaphyſicis rationibus aſcriberem. Atque in hunc vſum partim ex ore
Mæſtlini,partim meo Marte, quas Copernicus in Matheſi præ Ptolemæo habet cõ-
moditates,paulatim collegi:quo labore me facile liberare potuiſſet IoachimusRhe-
ticus,qui ſingula breuiter,& perſpicue prima ſua Narratione perſecutus eſt. Interea
dum illud ſaxum voluo, ſed παρέργως,ſecus Theologiam:commode accidit,vt Grætium venirem,atque ibi Georgio Stadio, p.m.ſuccederem: vbi officii ratio me arctius
his ſtudiis obſtrinxit.Ibi in explicatione principiorum Aſtronomiæ,magno mihi vſui
fuerunt omnia illa,quæ antea vel à Mæſtlino audiueram,vel ipſe affectaueram.Atq;
vt in Virgilio,fama Mobilitate viget,vireſque acquirit eundo : ſic mihi harum rerum
diligens cogitatio, cogitationis vlterioris cauſa fuit. Donec tandem anno 1595.cum
ocium à lectionibus cuperem bene, & ex officii ratione tranſigere,toto animi impetu in hanc materiam incubui.

Et tria potiſſimum erant, quorum ego cauſas, cur ita, non aliter eſſent, perti-
naciter quærebam, Numerus,Quantitas,& Motus Orbium. Vt hoc auderem,effe-
cit illa pulchra quieſcentium harmonia, Solis,fixarum & inter medii, cum Deo Pa-
tre, & Filio, & ſancto Spiritu: (4) quam ſimilitudinem ego in Coſmographia per-
ſequar amplius.Cum igitur ita haberent quieſcentia,non dubitabam de mobilibus,
quin ſe præbitura ſint.Initio rem numeris aggreſſus ſum,& conſideraui,vtrum vnus
orbis alius duplum,triplum,quadruplum , aut quid tandem haberet; quantumque
quilibet à quolibet in Copernico diſſideret. Plurimum temporis iſto labore,quaſi
luſu,perdidi;cùm nulla, neque ipſarum proportionum, neque incrementorum ap-
pareret æqualitas:nihilq; vtilitatis inde percepi,quam quod diſtantias ipſas,vt à Copernico proditæ ſunt , altiſſime memoriæ inſculpſi: quodque hæc variorum cona-
tuum commemoratio tuum aſſenſum,lector, quaſi marinis fluctibus , anxie hinc inde iactare poteſt,quibus fatigatus,denique tanto libentius ad cauſas hoc libello ex-
poſitas,tanquam ad tutum portum te recipias. Conſolabantur me tamen ſubinde,
& in ſpem meliorem erigebant , cum aliæ rationes,quæ infra ſequentur, tum quod
ſemper motus diſtantiam pone ſequi videbatur, atque vbi magnus hiatus erat inter
orbes,erat & inter motus. Quod ſi (cogitabam) Deus motus ad diſtantiarum præſcriptum aptauit orbibus: vtique & ipſas diſtantias ad alicuius rei præſcriptum ac-
commodauit.

Cum igitur hac non ſuccederet, alia via, mirum quam audaci,tentaui aditum.
(5) Inter Iouem & Martem interpoſui nouum Planetam , itemque alium inter Venerem & Mercurium,quos duos forte ob exilitatem non videamus,iiſque ſua tempora περιοδικὰ aſcripſi.Sic enim exiſtimabam me aliquam æqualitatem proportionũ
effecturum, quæ proportiones inter binos verſus Solem ordine minuerentur,verſus
fixas augeſcerent: vt propior eſt Terra Veneri in quantitate orbis terreſtris , quam
Mars

It is my intention, reader, to show in this little book that the most great and good Creator, in the creation of this moving universe, and the arrangement of the heavens, looked (1) to those five regular solids,[1] which have been so celebrated from the time of Pythagoras and Plato down to our own, and that he fitted to the nature of those solids, the number of the heavens, their proportions, and the law of their motions. But before permitting you to come to the actual subject, I shall discuss briefly both what occasioned this book and my reason for undertaking it, which I think will affect not only your understanding but my reputation.

At the time, six years ago, when I was studying under the distinguished Master Michael Maestlin at Tübingen, I was disturbed by the many difficulties of the usual conception of the universe, and I was so delighted by Copernicus, whom Mr. Maestlin often mentioned in his lectures, that I not only frequently defended his opinions at the disputations of candidates in physics but even wrote out a thorough (2) disputation on the first motion, arguing that it comes about by the Earth's revolution. I had then reached the point of ascribing to this same Earth (3) the motion of the Sun, but where Copernicus did so through mathematical arguments, mine were physical, or rather metaphysical. And for this purpose I collected together little by little, partly from the words of Maestlin, partly by my own efforts, the advantages which Copernicus has mathematically over Ptolemy.[2] I could easily have been relieved of this toil by Joachim Rheticus, who has briefly and penetratingly treated the particular points in his *Narratio Prima*.[3] In the meantime, while I was rolling that rock, but as a sideline apart from theology, by a lucky chance I came to Graz, where I succeeded the late George Stadius; and there the duties of my post obliged me to attend more closely to those studies. For expounding the principles of astronomy there, everything I had previously either heard from Maestlin or worked out for myself was of great value. And as in Virgil "the report Grows by travelling and gains strength as it goes,"[4] so for me the careful contemplation of these topics was the cause of further contemplation. Finally in the year, etc., '95, when I had a strong desire to rest from my lectures, and to have done with the duties of my post, I threw myself with the whole force of my mind into this subject.

There were three things in particular about which I persistently sought the reasons why they were such and not otherwise: the number, the size, and the motion of the circles. That I dared so much was due to the splendid harmony of those things which are at rest, the Sun, the fixed stars and the intermediate space, with God the Father, and the Son, and the Holy Spirit.[5] (4) This resemblance I shall pursue at greater length in my *Cosmographia*. Accordingly, since this was the case with those things which are at rest, I had no doubt that for things which move, similar resemblances would reveal themselves. In the beginning I attacked the business by numbers,[6] and considered whether one circle was twice another, or three times, or four times, or whatever, and how far any one was separated from another according to Copernicus. I wasted a great deal of time on that toil, as if at a game, since no agreement appeared either in the proportions themselves or in the differences; and I derived nothing of value from that except that I engraved deeply on my memory the distances which were published by Copernicus. But as this recital of my various attempts may toss your approval, reader, anxiously to and fro as if on the sea's waves, which will tire it, you will at last come all the more gladly to the causes explained in this little book, as though to a safe harbor. Yet I was comforted repeatedly, and my hopes were raised, not only by the other arguments which will follow below, but also by the fact that the motion always seemed to be in step with the distance, and where there was a great gap between the spheres, there was also one between the motions. But if (thought I) God allotted motions to the spheres to correspond with their distances, similarly he made the distances themselves correspond with something.

Since, then, this method was not a success, I tried an approach by another way, of remarkable boldness. (5) Between Jupiter and Mars I placed a new planet,[7] and also another between Venus and Mercury, which were to be invisible perhaps on account of their tiny size, and I assigned periodic times to them. For I thought that in this way I should produce some agreement between the ratios, as the ratios

Mars Terræ, in quantitate orbis Martii. Verum neque vnius planetæ interpofitio fufficiebat ingenti hiatui ♃ & ♂. Manebat enim maior Iouis ad illum nouum proportio, quam eft Saturni ad Iouem: Et hoc pacto quamuis obtinerem qualemcunque proportionem, nullus tamen cum ratione finis, nullus certus numerus mobilium futurus erat, neque verfus fixas, vfque dum illæ ipfæ occurrerent: neque verfus Solem vnquam, quia diuifio fpatii poft Mercurium refidui per hanc proportionem in infinitum procederet. (6) Neque enim ab vllius numeri nobilitate coniectari poteram, cur pro infinitis tã pauca mobilia extitifsét: Neque verifimilia dicit Rheticus in fua Narratione, cum à fanctitate fenarii argumentatur ad numerũSexCœlorum mobilium. Nam qui de ipfius mundi conditu difputat, non debet rationes ab iis numeris ducere, (7) qui ex rebus mundo pofterioribus dignitatem aliquam adepti funt.

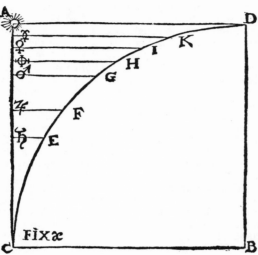

Rurfum alio modo exploraui, vtrum non in eodé quadrante diftantia cuiuslibet Planetæ fit refiduum ex finu, & motus eius fit refiduum ex eius complementi finu. Fingatur quadratum A B, à Semidiametro totius Vniuerfi A C defcriptum. Ex angulo igitur B Soli fiue Centro Mundi A oppofito, fcribatur cum radio B C Quadrans CED. Deinde in vero mundi radio AC notentur Sol, Fixæ & Mobilia pro ratione diftantiarum: à quibus punctis excitentur rectæ, vfque ad obuerfum Soli Quadrantem porrectæ. Quæ igitur eft proportio parallelorum, eandem virtutis mouentis finxi penes fingulos planetas. In Solis lineâ infinitas permanet, quia A D tangitur non fecatur à Quadrãte. Infinita igitur vis motus in Sole, nempe nil nifi motus ipfiffimo actu. In Mercurio infinita linea in Kabfciffa eft. Quare eius motus iam eft ad cæteros comparabilis. In fixis amiffa eft omninò linea, & compreffa in merum punctum C. Nulla igitur ibi virtus ad motum. Hoc theorema fuit, quod calculo erat examinandum. Quod fi quis probè ponderat, duo mihi defuiffe; primũ, quod ignoraui finum totum, fiue magnitudinem illius propofiti quadrantis: alterum, quòd motuum vigores non fuerunt aliter expreffi quàm in proportione vnius ad alium: qui, inquam, hæc probè ponderat, non immeritò dubitabit, vtrum aliquatenus hac difficili via peruenire potuerim necne. Et tamen continuo labore, atque infinita finuum & arcuum reciprocatione tantum effeci, vt intelligerem, locum habere non poffe hanc fententiam.

Æftas penè tota hac cruce perdita. Denique leui quadam occafione propius in rem ipfam incidi. Diuinitus id mihi obtigiffe arbitrabar, vt fortuitò nancifcerer, quod nullo vnquam labore affequi poteram: idq́; eò magis credebam: quòd Deum femper oraueram, fiquidem Copernicus vera dixiffet, vti ifta fuccederent. Igitur die 9. vel 19. Iulii anni 1595. monftraturus Auditoribus meis coniunctionum magnarum faltus per octona figna, & quomodo illæ pedetentim ex vno trigono tranfeant in alium, infcripfi multa triangula, vel quafi triangula, eidem circulo, fic vt finis vnius effet initium alterius. Igitur quibus punctis latera triangulorum fe mutuò fecabant, iis minor circellus adumbrabatur. Nam circuli triangulo infcripti radius, eft cir-

Hæc vides fequenti fchemate.

between the pairs would be respectively reduced in the direction of the Sun and increased in the direction of the fixed stars, as the Earth is nearer to Venus, relative to the size of the Earth's circle, than Mars is to the Earth, relative to the size of the circle of Mars. Yet the interposition of a single planet was not sufficient for the huge gap between Jupiter and Mars; for the ratio of Jupiter to the new planet remained greater than that of Saturn to Jupiter; and on this basis whatever ratio I obtained, in whatever way, yet there would be no end to the calculation, no definite tally of the moving circles, either in the direction of the fixed stars, until they themselves were encountered, or at all in the direction of the Sun, because the division of the space remaining after Mercury in this ratio would continue to infinity. (6) And I could not conjecture from the nobility of any number why so few moving stars existed in proportion to the infinite number (of the fixed stars). What Rheticus says in his *Narratio*, when he argues, from the sanctity of the number six, for six as the number of the moving heavens, is unlikely. For in discussing the foundation of the universe itself, one ought not to draw explanations from those numbers which have acquired some special (7) significance from things which follow after the creation of the universe.[8]

Again I investigated by another method whether the distance of any planet in the same quadrant may not be the remainder of the sine, and its motion the remainder of the complement of the sine.[9] Construct a square AB, described on the semidiameter of the whole universe AC. Then from the vertex B opposite to the Sun or the center of the universe A, a quadrant CED with radius BC. Next, on the true radius of the universe AC mark the Sun, the fixed stars and the moving stars in accordance with the ratio of their distances, and from these points erect perpendiculars reaching to the quadrant turned towards the Sun. Then the proportion of the parallel lines has been constructed as the proportion of the power of the motion belonging to the individual planets.[10] In the case of the Sun the line remains infinite, as AD is touched but not cut by the quadrant. Therefore the motive power in the Sun is infinite, though of course only in the actual bringing about the motion. In the case of Mercury the infinite line is cut off at K, so that its motion may now be compared with the others. In the case of the fixed stars the line completely disappears, and is compressed into the mere point C. Thus there is no power of causing motion here. This was the theorem which was to be tested by calculation. But on due consideration, I lacked two things. First, I did not know the total sine, or the size of the quadrant in question; secondly, the strengths of the motions were expressed only in terms of a ratio of one to another. On due consideration, as I say, it will properly be doubted whether I could reach any conclusion by this difficult route or not. However, by continuous toil and endless reiteration of the sines and arcs, I managed to determine that this proportion could not be maintained.

Almost a whole summer was wasted on this ordeal. Eventually by a certain mere accident I chanced to come closer to the actual state of affairs. I thought it was by divine intervention that I gained fortuitously what I was never able to obtain by any amount of toil; and I believed that all the more because I had always prayed to God that if Copernicus had told the truth things should proceed in this way. Therefore on the 9/19th of July in the year 1595 when I was going to show my audience the leaps of the great conjunctions through eight signs at a time, and how they cross step by step from one triangle to another, I inscribed many triangles, or quasi-triangles, in the same circle, so that the end of one was the beginning of another.[11] Hence the points at which the sides of the triangles intersected each other sketched out a smaller circle. For the radius of a circle inscribed in a triangle is half the radius of the circumscribed circle. The ratio of the *You see them in the following diagram*

Diagram opposite:
FIXÆ: *Fixed Stars*

est circumscripti radij dimidium. Proportio inter vtrumque circulum videbatur ad oculum penè similis illi, quæ est inter Saturnum & Iouem : & triangulum prima erat figurarum, sicut Saturnus & Iupiter primi Planetæ. Tentaui statim quadrangulo distantiam secundam Martis & Iouis, quinquangulo tertiam, sexangulo quartam. Cumque etiam oculi reclamarent in secunda distantia, quæ est inter Iouem & Martem quadratum triangulo & quinquangulo adiunxi. Infinitum est singula persequi.

Et finis huius irriti conatus fuit idem, qui postremi & felicis initium. Nempe cogitaui, hac via, siquidem ordinem inter figuras velim seruare, nunquam me peruenturum vsque ad Solem, neq; causam habiturum, cur potius sint sex, quàm viginti vel centum orbes mobiles. Et tamen placebant figuræ, vtpote quantitates, & res cœlis prior. (8) Quantitas enim initio cum corpore creata; cœli altero die. Quòd si(cogitabam.)pro Quantitate & proportione sex Cœlorum, quos statuit Copernicus, Quinque tantùm figuræ inter infinitas reliquas reperiri possent, quæ præ cæteris peculiares quasdam proprietates haberent: ex voto res esset. Atqui rursum instabam. Quid figuræ planæ inter solidos orbes? Solida potiùs corpora adeantur. Ecce, Lector, inuentum hoc & materiam totius huius opusculi. Nam si quis leuiter Geometriæ peritus totidem verbis moneatur, illi statim in promptu sunt Quinque regularia corpora cum proportione orbium circumscriptorum ad inscriptos: illi statim ob oculos versatur, scholion illud Euclideum ad propositionê 18.lib.13. Quo

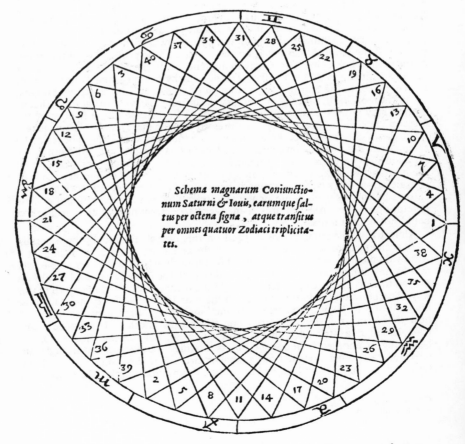

Schema magnarum Coniunctionum Saturni & Iouis, earumque saltus per octena signa, atque transitus per omnes quatuor Zodiaci triplicitates.

B demon-

circles to each other appeared to the eye almost the same as that between Saturn and Jupiter; and the triangle was the first of figures, just as Saturn and Jupiter are the first planets. At once I tried the second interval, between Mars and Jupiter, in a four-sided figure, the third in a five-sided figure, the fourth in a six-sided figure. Since that was obviously wrong at sight, in the second interval, which is between Jupiter and Mars, I added a four-sided figure to the three-sided and the five-sided figure. It is an infinite task to follow up individual cases.

The end of this useless attempt was also the beginning of the last, and successful, one. I naturally concluded that by this method, if I wished to keep an order among the figures, I should never reach the Sun, nor have an explanation why there should be six moving circles rather than twenty or a hundred. However, the figures were satisfactory, as they represent quantities, and so something prior to the heavens. For (8) quantity was created in the beginning along with matter, but the heavens on the second day. But if (thought I) corresponding with the size and proportion of the six heavens, as Copernicus established them, there could be found only five figures, among the infinite number of others, which had certain special properties distinct from the rest, it would be the answer to my prayer. Again I set to. Why should there be plane figures between solid spheres? It would be more appropriate to try solid bodies. Behold, reader, this is my discovery and the subject matter of the whole of this little work. For if anyone having a slight acquaintance with geometry were informed of this in so many words, there would immediately come to his mind the five regular solids with the proportion of their circumscribed spheres to those inscribed; there would immediately appear before his eyes the scholium to Euclid's Proposition 18 of Book XIII, in which it is

Diagram opposite:
Diagram of the great conjunctions of Saturn and Jupiter, and their leaps through eight signs, and crossings through all four quartiles of the Zodiac.

demonſtratur impoſſibile eſſe , vt plura ſint aut excogitentur regularia corpora, quam quinque. Res admiratione digna, cum nondum conſtaret mihi de ſingulorum corporum prærogatiuis in ordine, vſum me minime arguta coniectura ex notis Planetarum diſtantiis deducta, adeo feliciter ſcopum tetigiſſe in ordine corporum, vt nihil in illis poſtea, cum exquiſitis agerem rationibus, immutare potuerim. Ad rei memoriam aſcribo tibi ſententiam, ita vti incidit , & eo momento verbis conceptam. (9) *Terra eſt Circulus menſor omnium: Illi circumſcribe Dodecaetron: Circulus hoc comprehendens erit Mars. Marti circumſcribe Tetraedron: Circulus hoc comprehendens erit Iupiter. Ioui circumſcribe Cubum: Circulus hunc comprehendens erit Saturnus. Iam terræ inſcribe Icoſaedron: illi inſcriptus Circulus erit Venus. Veneri inſcribe Octaedron: Illi inſcriptus Circulus erit Mercurius.* Habes rationem numeri planetarum.

Hæc occaſio & ſucceſſus huius laboris: Vide nunc etiam meum in hoc libro propoſitum. Et quidem quantam ex inuentione voluptatem perceperim, núquam verbis expreſſero. Non me perditi temporis pœnitebat amplius, non pertæſum eſt laboris, moleſtias calculi nullas ſubterfugi, dies nocteſque computando conſumpſi: donec cernerem, vtrum concepta verbis ſententia cum Copernici orbibus conſentiret, an vero ferrent mea gaudia venti. Quod ſi rem, vti eſſe putabam, deprehenderem, votum Deo Opt. Max. feci, me prima occaſione hoc admirabile ſuæ ſapientiæ ſpecimen publicis typis inter homines enunciaturum: vt quamuis neque hæc vndiquaque abſoluta ſint, & forte reſtent nonnulla, quæ ex his fluant principiis, quorum inuentionem mihi reſeruare poſſem: tamen alii, qui valent ingenio, quam plurima, ad illuſtrationem Nominis diuini, primo quoque tempore iuxta me proferrent, & laudem ſapientiſſimo Creatori vno ore accinerent. Cum igitur paucis poſt diebus res ſuccederet, atque ego deprehenderem, quam apte vnum corpus, poſt aliud inter ſuos Planetas ſederet, totumque negocium in formam præſentis opuſculi redigerem: atque id Mæſtlino celebri Mathematico probaretur: intelligis, amice Lector, me voti reum, neque poſſe morem Satyrico gerere , qui nonum in annum iubet libros detinere.

Hæc vna cauſa eſt meæ maturationis: cui vt omnem tibi ſcrupulum, (10) ſiniſtre ſuſpicandi eximam, addo lubens & alteram, & recito tibi, illud Architæ ex Cicerone : *Si cœlum ipſum aſcendiſſem , Naturamque mundi, & pulchritudinem ſiderum penitus perſpexiſſem, inſuauis illa mihi foret admiratio, niſi te Lectorem æquum, attentum & cupidum, cui narrarem, haberem.* Hæc vbi cognoueris, ſi æquus es , abſtinebis à reprehenſionibus, quas non ſine cauſa præſagio: Sin autem ſuo quidem loco relinquis iſta: metus tamen, vt certa ſint, atque vt ego triumphum cecinerim ante victoriam : ergo vel tandem pagellas ipſas accede; & rem, qua de pridem agimus, cognoſce. Non reperies nouos & incognitos Planetas, vt paulo antea, interpoſitos, non ea mihi probatur audacia; ſed illos veteres parum admodum luxatos, interiectu vero rectilineorum corporum, quantumuis abſurdo, ita munitos, vt porro, quibus vncis cœlum quo minus ruat, ſuſpendatur, quærenti ruſtico reſpondere poſſis. Vale.

      * *
       *

IN

shown that it is impossible for there to be or to be conceived more than five regular solids. It is a wonderful thing that when I had not yet settled the properties of the individual bodies within their arrangement, yet using such a clumsy conjecture drawn from the known distances of the planets, I should so successfully have hit the target over the arrangement of the bodies that there was nothing which I could later change in them when I was working with the ratios calculated in detail. As an aid to memory I give you the proposition, conceived in words just as it came to me and at that very moment: (9) "The Earth is the circle which is the measure of all. Construct a dodecahedron round it. The circle surrounding that will be Mars. Round Mars construct a tetrahedron. The circle surrounding that will be Jupiter. Round Jupiter construct a cube. The circle surrounding that will be Saturn. Now construct an icosahedron inside the Earth. The circle inscribed within that will be Venus. Inside Venus inscribe an octahedron. The circle inscribed within that will be Mercury." There you have the explanation of the number of the planets.

This accident was also the happy ending of my toil. You can now also see my scheme for this book. What delight I have found in this discovery I shall never be able to express in words. No longer did I regret the wasted time; I was no longer sick of the toil; I did not avoid any of the tedium of the calculation; I devoted my days and nights to computation, until such time as I could see whether the proposition which I had conceived in words would agree with the circles of Copernicus, or whether my joy would be scattered to the winds.[12] But if I found out that I was right, I made a vow to Almighty God that at the first opportunity I would proclaim among men in public print this wonderful example of his wisdom, so that although the work is not in every way complete, and there may perhaps remain some points to emerge from these beginnings, the discovery of which I could reserve for myself, yet others, with powerful talents, would bring out as many of them as possible, to the glory of God's name, and at the earliest possible moment after me, and would sing the praise of our most wise Creator with a single voice. Therefore, since the success came after a few days, and I found out how neatly one body fitted after another among their planets, and I reduced the whole business to the form of the present work, and had it approved by Maestlin the celebrated mathematician, you will understand, dear reader, that I am bound by my vow, and cannot oblige the satirist who tells us to delay books for nine years.[13]

That is the sole cause of my haste; but to relieve you of any lingering (10) adverse suspicion, I gladly add another as well, the saying of Architas from Cicero: "If I had ascended the very heaven, and beheld completely the nature of the universe, and the beauty of the stars, the wonder of it would give me no pleasure, if I did not have you as a friendly, attentive, and eager reader to whom to tell it."[14] When you know that, if you are fair, you will refrain from the criticisms which I have good reason to expect. However, if you leave all that for its proper place, yet you are afraid that these things are not certain, and that I have sung my song of triumph before I have won the victory, then go on at last to the actual pages, and find out about what I have long been discussing. You will not find any new and undiscovered planets interpolated, as I did a little while ago; I do not favor that piece of audacity. You will find the old ones, very little disturbed, though so secured by the insertion between them of rectilinear bodies, however absurdly, that you will have an answer for the peasant who asks what hooks the sky is hung on to prevent it from falling.[15]

J. Kepler

IN PRÆFATIONEM AD LECTOREM
Notæ Auctoris.

(1) ADilla(quinque corpora) cœlorum numerum, proportiones & motuum rationem,&c.] *Etsi omnia omnibus coharent. Numerus tamen sex primariorum Orbium propriè desumptus est ex quinque Corporibus Solis: proportio, potiore quidem parte à Corporibus quinque Geometricis, sed quæ tamen circa minima concessit motibus, vt causa finali in Ideam operis recept est atim initio. Et hoc quidem intelligendum est de Motibus cuiusque planeta tardissimo vno, altero velocissimo, de Motibus, scilicet causa sua proprietatis consideratis. Motus vero periodici: Hoc est, numerus dierum in vnius cuiusque Planeta circuitu derivatus, tam à proportione orbium, quam ab Eccentricitatibus (quæ ex Harmoniis sunt constituta) longius à 5.corporibus recesserunt.*

(2) Disputationem de motu primo,quod is terræ volutione accidat.] *Habes illam disputationem cumulatam in Epitomes Astronomia lib.I.*

(3) Tellurimotum Solarem.] *Inserta est hæc disputatio Commentariis meis de motu Martis, præsertim in Introductione: reperitur vero accurate libro IV.Epitomes,fol.542. Argumenta plane demonstratiua ex penitissima Astronomiæ restauratione concinnata sunt.*

(4) Quam similitudinem ego in Cosmographia prosequar.] *Cosmographia quidem titulo nullum ex illo tempore librum edidi: at similitudo ista relata est à me in Epitomes libr. I.fol.42.vbi de Mundi figura extima, inque librum IV.eiusdem fol.437.& 448. vbi de tribus primariis Mundi membris disputo. Nec pro similitudine inani est habenda; sed inter causas accensenda, vt Mundi forma & Archetypus.*

(5) Inter Iouem & Martem interposui nouum Planetam.] *Non qui circa Iouem curreret, vt sidera Galilæi Medicea; ne fallaris, nunquam de iis cogitaui; sed qui vt ipsi primarii planeta, Solem in centro Systematis positum curriculo suo cingeret.*

(6) Neque ab vllius Numeri.] *En iam tunc reiectos à me numeros numerantes, vt appellant. Eosdem etiam abiicere à fundamentis Harmonicis, inter præcipua habui in illo Opere.*

(7) Ex rebus mundo posterioribus dignitatem.] *Senarius tamen habet aliquid abstractum à creaturis, quod scilicet primus est inter perfectos: perfectum autem id habetur, si tot sunt in partibus aliquotis vnitates, quot in toto. An hæc igitur proprietas conciliet Numero numeranti dignitatem aliquam? Consideretur & qualis hæc sit dignitas, & quomodo competat Numero. Primum hæc dignitas videtur esse nulla. Nam si dignitas esset aliqua,videtur Harmonica disciplina testimonium præbitura fuiss: omnibus numeris perfectis. At illa nullum recipit præter senarium. Reliqui enim perfecti, sunt primorum multiplices, vt patet ex Euclidis libro IX.prop. vltima. Quare (per Ax.III.libri mei III.Harmonices,fol.11.poster.& per Prop.VIII.libri IV. fol.145. quæ nituntur propp.XLV.XLVI. XLVII.libri I.)omnes perfecti, sic dicti, numeri, præter senarium, exulant à terminis, concordantias constituentibus; attestante etiam sensu auditus: idque propter Primos, vt septenarium,&c. à quibus deriuantur.Etsi enim sectiones Harmonica libro III.prop.XIX.fol. 26.poster.numerantur septem,qui numerus primus est : at nulli earum dat hoc, septenarius iste, vt sit Harmonica; sed prius quælibet per se est Harmonica, postea demum accidit illis iam constitutis vniuersis,vt sint numero septem: Sed neque hæc ipsa conditio,qua definiuntur numeri perfecti, in seipsa considerata, dignitatis quicquam habet: vt scilicet numeri omnes, qui vnum aliquem emetiuntur, in vnum conflati, æquent mensuratum. Est quidem æqualitas pulchrum quippiam, sed hæc æqualitas numeris ipsis, ratione sui ipsorum singulorum est accidentaria ; nec quicquam affert ad eorum constitutionem, sed resultat necessitate Geometrica ex iam constitutis; nec dat ipsis hoc, vt sint magis articulati; cum tamen circa ipsam hanc articulationem occupetur, & ea quodammodo definiatur: quin potius, qui iubetur hanc sic dictam perfectionem affectare, is hoc ipso circumscribitur, ne possit sumere articulatissimos.Et vt prius sumus ratiocinati de sectionibus, sic nunc etiam de numeris emetientibus vnum aliquem dicere possumus : quod scilicet prius quilibet illorum pro seipso emetiatur propositum numerum, non accipiens hanc naturam ab æqualitate prætensa, sed postea demum accidat illis singulis, vt vniuersi æquent mensuratum.Vide lib.meo III.Harmon.sub finem capitis III.fol. 31.poster.locum similem; de occursu ternarij,pro quo hic est occursus æqualitatis. Non plus igitur virtutis & dexteritatis confert numeris hæc æqualitas, quam agricola, inuentio thesauri; vt credib.le*

B 2 *nequa-*

Original Preface

AUTHOR'S NOTES ON THE PREFACE TO THE READER

(1) *To those (five solids)...the number of the heavens, their proportions, and the law of their motions, etc.*] Although all things are consistent with all things, yet the number of the six primary spheres has properly been taken from the five solids alone, their proportion principally from the five geometrical solids; but it has conceded very small amounts all round to the motions, as it was the final cause which was accepted for the Idea of the operation right from the start. And this is to be understood of the motions of each planet, its slowest on the one hand and its fastest on the other, that is of the motions considered as the cause of its particular properties. Indeed the periodic motions, that is to say, the number of days assigned to the revolutions of each individual planet, have both on account of the proportion of the orbits and on account of the eccentricities (which have been established from the harmonies) regressed further from the five solids.

(2) *Disputation on the first motion, arguing that it comes about by the Earth's revolution.*] You will find that disputation augmented in Book I of the *Epitome of Astronomy.*[16]

(3) *To...Earth the motion of the Sun.*] This disputation was inserted into my *Commentaries on the motion of Mars*, particularly in the introduction; however it is found in detail in Book IV of the *Epitome*, page 542.[17] Arguments which plainly indicate this were constructed from the total reform of astronomy.

(4) *This resemblance I shall pursue at greater length in my Cosmographia.*] I have not indeed published any book with the title of Cosmographia since that time; but this resemblance was dealt with by me in Book I of the *Epitome*, page 42, where I discuss the outside shape of the universe, and in Book IV of the same, pages 437 and 438, where I discuss the three primary parts of the universe.[18] Nor should it be taken as a meaningless resemblance, but it should be reckoned as one of the causes, as a form and archetype of the universe.

(5) *Between Jupiter and Mars I placed a new planet.*] Not to travel round Jupiter, like Galileo's Medicean stars — make no mistake, I had no thought of them — but, like the primary planets themselves, to circle in its path about the Sun, placed at the center of the system.

(6) *Not...from (the nobility) of any number.*] Look how I had already rejected the "counting numbers," as they are called.[19] To expel them also from the basic principles of harmony was also among my chief aims in that work.

(7) *Significance from things which follow after the creation of the universe.*] However, the number six has an attribute which is unconnected with created things, namely that it is first among the perfect numbers. A number is considered perfect if the sum of its proper divisors is the same as its whole. Does, then, this property impart some significance to a "counting number"? Let us consider both the nature of this significance and the way in which it belongs to the number. In the first place this significance appears to be nothing. For if the significance were anything, it is apparent that the discipline of harmony would have afforded some evidence of it for all the perfect numbers. But in fact it accepts none except the number six. For the remaining perfect numbers are multiples of prime numbers, as is evident from the last proposition of Euclid, Book IX. Consequently (by Axiom 3 of Book III of my *Harmonice*, the second page 11, and by Proposition 8 of Book IV, page 143, which depend on Propositions 45, 46 and 47 of Book I) all the perfect numbers, so-called, except for six, are banished from the territory which constitutes the concords (as the sense of hearing attests); and that is on account of the prime numbers (e.g., seven etc.) from which they are derived. For although in Book III, Proposition 19, the second page 26, the harmonic divisions amount to seven, which is a prime number, yet to none of them does the number seven impart the property of being a harmony; but each of them is on prior grounds a harmony in its own right, and only after that does it happen when they have all already been constituted that they are seven in number. However, this particular circumstance, by which they are designated perfect numbers, considered on its own, does not have any significance either, that is, the circumstance that they are all numbers which are proper divisors of some particular number, and if added together are equal to the one of which they are divisors. Equality is indeed a fine thing; but this equality is an accidental property of those numbers if each one is considered separately, and does not relate in any way to their constitution, but is the result of geometrical necessity which follows after they have been constituted. Nor does it impart to them the property of being more readily divisible, since it is a feature of this very divisibility, and is in a sense defined by it. On the contrary, a number which is ordained to take on this so-called perfection is restricted by that very fact from adopting the most readily divisible numbers as its divisors. And as we previously argued in dealing with the divisions, so in the present case also we can say of the proper divisors of some particular number, that each of them is a factor of the number in question on prior grounds, not receiving this characteristic from a previously existing quality; but only after that does it happen to them individually that all together they are equal to the number of which they are proper divisors. See in Book III of my *Harmonice* at the end of Chapter 3, the second page 31, similar remarks on three notes forming a chord, for which this is a meeting forming an equality. Therefore this equality

71

nequaquam insit senarium DEO Creatori placuisse propter hanc indolem. Dico secundo,hanc affectionem non competere Numeris,vt numerantibus. Id facile probatur ex Euclidis lib. VII. VIII. IX. Vt enim auctor ille demonstret, inesse quibusdam hanc perfectionem, cogitur vti numeris figuratis,id est, vt scholæ loquuntur, Numeris numeratis,seu parallelogrammis, æquali mensura diuisis in longum & latum. Quare si qua maxime nobilitatis nota esset, hæc sic dicta perfectio, illa primo competeret Geometricis figuris. Etsi vero senarius veram suam & realem nobilitatem habet ex sexangulo, quæ figura ipsum prouehit in disciplina harmonica: non ideo tamen etiam ad constituendum numerum primariorum Mundi corporum fit aptus. Figura enim illa circulum, vt continuam quantitatem in sex partes diuidit: corpora Mundana non sunt partes vnius continuæ quantitatis. Illa figura inter planas est: corporibus vero mundi solida, seu trium dimensionum spacia data sunt peragranda. Recte igitur repudiaui senarij ipsius per se considerati dotes, ne adsciscerem illas inter caussas senarij Cælorum: recte censui, oportuisse præcedere caussas aliquas euidentes, ex quibus deinde senarius iste Cælorum vltro resultaret; sicut in Harmonica disciplina, causis prægressis idoneis, resultat & ternarius consonantium in idem sonorum fol.31. poster.& septenarius diuisionum Harmonicarum fol.27.poster.

(8) Quantitas enim initio cum corpore.] *Imo Ideæ quantitatum sunt erantque Deo coæterna, Deus ipse; suntque adhuc exemplariter in animis ad imaginem Dei (etiam essentia sua)factis;qua in re consentiunt gentiles Philosophi, & Doctores Ecclesiæ.*

(9) Terra est Circulus.] *Scripseram ista mihi soli; intelligebam igitur pro Terra, Orbem, quo illa vehitur, Magnum à Copernico dictum: sic pro quolibet Planeta, orbem ipsius. Et pertinet vltimum comma;* Habes rationem, &c. *etiam ad hanc ex schedis exscriptam sententiam.*

(10) Siniftre suspicandi,&c.] *Laboraui pueriliter, ne quis mihi imputaret, me nouatorem esse,ostentandi solum ingenij causa librum scripsisse: his opposui & votum & penitissimam persuasionem de veritate eorum, quæ liber contineret, & denique ardorem conferendi cum alijs de his inuentis. Et erant, opinor, idonea causa,profligandi pudoris inepti.*

confers no more power and flexibility on numbers than the finding of treasure does on a farmer. Thus it is quite unbelievable that the number six should have pleased God the Creator on account of this quality. Secondly I say that this characteristic does not belong to the numbers insofar as they are "counting numbers." That is easily proved from Euclid, Book VII, VIII, IX. For in order to demonstrate that certain numbers possess this perfection, that author is obliged to use numbers represented by diagrams, that is, as the schools say, "counted numbers," or parallelograms, divided in equal proportions in length and breadth. Consequently if this so-called perfection were any mark of the highest nobility, it would belong primarily to the geometrical figures. Although indeed the number six has its own true and real nobility from the hexagon, the figure which makes it important in the discipline of harmony, yet it is not on that account also fitted to constitute the number of primary solids of the universe. For that figure divides the circle as a continuous quantity into six parts, but the bodies which make up the universe are not parts of one continuous quantity. That figure is one of the plane figures: but the bodies of the universe are allotted spaces which are solid, or of three dimensions, to traverse. I was therefore right to repudiate the endowments of the number six considered on its own, and not to count them among the causes of the heavens' being six in number. I rightly judged that there ought to be some preceding causes in evidence, of which it would afterwards be the spontaneous result that the heavens would be six in number; just as in the discipline of harmony appropriate causes come first, and the result is that a set of three notes form a concord together (the second page 31), and there are seven harmonic divisions (the second page 26).

(8) *For quantity. . .in the beginning along with matter.*] Rather the Ideas of quantities are and were coeternal with God, and God himself; and they are still like a pattern in souls made in the image of God (also from his essence). On this matter the pagan philosophers and the Doctors of the Church agree.

(9) *The Earth is the circle.*] I had written this for myself alone, and so by the Earth I understood the orbit on which it travels, called the Great by Copernicus, and similarly by each planet, its own orbit. The last sentence, "There you have the explanation, etc.," also belongs to this statement copied from my notes.

(10) *Adverse suspicion, etc.*] I struggled in my youthful way to avoid the imputation of being a radical, of having written a book just for the sake of showing off my own cleverness. I resisted that both by this avowal and by a most thorough defense of the truth of what the book contained, and lastly by zeal in discussing these discoveries with other people. Indeed there were, I believe, sufficient reasons for casting aside any absurd modesty.

CAPVT I.

Quibus rationibus Copernici hypotheses fiant consentaneæ. Et explicatio hypothesium Copernici.

Ts 1 pium eſt, ſtatim ab initio huius de Natura diſpu-tationis videre, an nihil Sacris Literis contrarium dica-tur: intempeſtiuum tamen exiſtimo, eam controuerſiam hîc mouere, prius atque ſolliciter. Illud in genere pro-mitto, nihil me dicturum, quod in Sacras Literas iniuriū ſit, & ſi cuius Copernicus mecum conuincatur, pro nul-lo habiturum. Atque ea mens mihi ſemper fuit, inde à quo Copernici Reuolutionum libros cognoſcere cœpi.

Cum igitur hac in parte nulla religione impedirer, quo minus Co-pernicum, ſi conſentanea diceret, audirem: primam fidem mihi fecit illa pulcherrima omnium, quæ in cœlo apparent, cū placitis Copernici con-ſenſio: vt qui non ſolum motus præteritos ex vltima antiquitate repetitos demonſtraret, ſed etiam futuros antea, non quidem certiſſime, ſed tamē longe certius, quam Ptolemæus, Alphonſus, & cæteri, diceret. Illud au-tem longe maius, quod quæ ex alijs mirari diſcimus, eorum ſolus Coper-nicus pulcherrime rationem reddit, cauſamque admirationis, quæ eſt i-gnoratio cauſarum, tollit. Nunquam id facilius docuero Lectorem, quam ſi ad Narrationem Rhetici legendam illi auctor & perſuaſor exiſtam. Nam ipſos Copernici libros Reuolutionum legere non omnibus va-cat.

(2) Atq; hoc loco nunquam aſſentiri potui illis, qui freti exēplo ac-cidentariæ demonſtrationis, quæ ex falſis præmiſſis neceſſitate Syllogi-ſtica verum aliquid infert. Qui, inquā, hoc exemplo freti contendebant, fieri poſſe, vt falſæ ſint, quæ Copernico placent hypotheſes, & tamen ex illis vera Φαινόμϟα tanquam ex genuinis principijs ſequantur.

Exemplum enim non quadrat. Nam iſta ſequela ex falſis præmiſſis fortuita eſt, & quæ falſi natura eſt; primum atque alii rei cognatæ accō-modatur, ſeipſam prodit: niſi ſponte concedas argumentatori illi, vt in-finitas alias falſas propoſitiones aſſumat, nec vnquā in progreſſu, regreſ-ſuque ſibiipſi conſtet. Aliter ſe res habet cum eo, qui Solem in cētro col-locat. Nam iube quidlibet eorum, quæ reuera in Cœlo apparent, ex ſemel poſita hypotheſi demonſtrare, regredi, progredi, vnum ex alio colligere, & quiduis agere, quæ veritas rerum patitur: neq; ille hæſitabit in vllo, ſi genuinum ſit, & vel ex intricatiſſimis demonſtrationum anfractibus in ſe vnum conſtantiſſime reuertetur. Quod ſi obijcias, idem partim adhuc poſſe, partim olim potuiſſe dici de tabulis & hypotheſibus antiquis, quod nempe Φαινομϟοις ſatisfaciant: Atque illas tamen à Copernico, vt falſas reijci: Poſte igitur eadem ratione & Copernico reſponderi: nempe quā-uis egregie eorum, quæ apparent rationem reddat, tamen in hypotheſi errare. Reſpondeo, primum, antiquas hypotheſes præcipuorum aliquot capitum, nullam plane rationem reddere. Cuiuſmodi eſt, quod ignorant,

B 3 numeri,

CHAPTER I.
THE REASONING WITH WHICH THE HYPOTHESES OF COPERNICUS AGREE, AND EXPOSITION OF THE HYPOTHESES OF COPERNICUS.[1]

Although it is proper to consider right from the start of this dissertation on Nature whether anything contrary to Holy Scripture is being said, nevertheless I judge that it is (1) premature to enter into a dispute on that point now, before I am criticized. I promise generally that I shall say nothing which would be an affront to Holy Scripture, and that if Copernicus is convicted of anything along with me, I shall dismiss him as worthless.[2] That has always been my intention, since I first made the acquaintance of Copernicus's *On the Revolutions*.

In this area, then, I should not be prevented by any religious scruple from listening to Copernicus, if what he said was consistent. My confidence was first established by the magnificent agreement of everything that is observed in the heavens with Copernicus's theories; since he not only derived the past motions which have been recapitulated from the remotest antiquity, but also predicted future motions, not indeed with great certainty, but far more certainly than Ptolemy, Alfonso, and the rest.[3] However what is far more important is that, for the things at which from others we learn to wonder, only Copernicus magnificently gives the explanation, and removes the cause of wonder, which is not knowing causes.[4] The easiest way for me to show the reader that would be for me to incite and persuade him to read Rheticus's *Narratio*, for not everyone has the leisure to read Copernicus's *On the Revolutions* itself.

(2) On this point I have never been able to agree with those who rely on the model of accidental proof, which infers a true conclusion from false premises by the logic of the syllogism. Relying, as I say, on this model they argued that it was possible for the hypotheses of Copernicus to be false and yet for the true phenomena to follow from them as if from authentic postulates.

For the model does not fit. The conclusion from false premises is accidental, and the nature of the fallacy betrays itself as soon as it is applied to another related topic—unless you gratuitously allow the exponent of that argument to adopt an infinite number of other false propositions, and never in arguing forwards and backwards to reach consistency. That is not the case with someone who places the Sun at the center. For if you tell him to derive from the hypothesis, once it has been stated, any of the phenomena which are actually observed in the heavens, to argue backwards, to argue forwards, to infer from one motion to another, and to perform anything whatever that the true state of affairs permits, he will have no difficulty with any point, if it is authentic, and even from the most intricate twistings of the argument he will return with complete consistency to the same assumptions. But you may object that it can to some extent still be said, and to some extent could once have been said about the old tables and hypotheses, that they satisfy the appearances, yet they are rejected by Copernicus as false; and that by the same logic the reply could be made to Copernicus that although he gives an excellent explanation for what is observed, yet he is wrong in his hypothesis. I reply first that the old hypotheses simply do not account at all for a number of outstanding features. For instance, they do not give the reasons for the number, extent, and time of the retrogressions, and why they

numeri,quantitatis,temporifque retrogradationum caufas: & quare illæ ad amuſſim ita (3) cum loco & motu Solis medio conueniant. (4)
Quibus omnibus in rebus, cum apud Copernicum ordo pulcherrimus
appareat,cauſam etiam ineſſe neceſſe eſt.Deinde earum etiam hypotheſium,quæ conſtantem apparentiarum cauſam reddunt,& cum viſu con-
ſentiunt,nihil negat Copernicus,potius omnia ſumit & explicat. Nam
quod multa in hypotheſibus vſitatis immutaſſe videtur, id reuera nō ita
ſe habet. Fieri namque poteſt,vt idem contingat duobus ſpecie differē-
tibus præſuppoſitis,propterea quod illa duo ſub eodem genere ſunt,cu-
ius gratia generis primo id contingit,de quo agitur.Sic Ptolemæus Stellarum ortus & obitus demonſtrauit, non hoc medio termino proximo,
& coæquato;Quia terra ſit in medio immobilis. Neq; Copernicus idem
hoc medio demonſtrat,quia terra à medio diſtans voluatur.Vtriq; enim
ſuffecit dicere(quod &vterque dixit)ideo hæc ita fieri,quia inter cœlum
& terram intercedat aliqua motuum ſeparatio, & quia nulla inter fixas
ſentiatur telluris à medio diſtantia.Igitur Ptolemæus non demonſtrauit
falſo & accidentario medio,ſi quæ demonſtrauit Φαινόμυα.Hoc tantum
in legem κατ᾽ αὐτὸ peccauit , quod exiſtimauit, hæc ita propter ſpeciem
euenire,quæ propter genus eueniunt. Vnde apparet ex eo,quod Ptole-
inæus ex falſa mundi diſpoſitione,vera tamen, & Cœlo, noſtriſq; oculis
conſona demonſtrauit,ex eo inquam,nullam eſſe cauſam,ſimile quid etiam de Copernicanis hypotheſibus ſuſpicādi. Quin potius manet,quod
initio dictum eſt:non poſſe falſa eſſe Copernici principia, ex quibus tam
cōſtans plurimorū Φαινομξ́ων ratio,ignota veteribus,reddatur, (5) quatenus ex illis redditur. Vidit hoc feliciſſimus ille Tycho Brahe,Aſtrono-
mus omni celebratione maior, qui quamuis omnino de loco terræ à Copernico diſſentiret, tamen ex eo retinuit id, cuius gratia rerum hactenus
incognitarum cauſas habemus: Solem nempe eſſe centrum quinq; pla-
netarum.Nam & hoc anguſtius eſt mediū ad demonſtrandas repedatio-
nes:τὸ Sol in centro immobilis. Sufficit enim generale illud, Sol in cētro
Planetarum quinque. Cur autem ſpeciē pro genere ſumeret Copernicus,& Solem inſuper in centro mundi, terram circa eum mobilem face-
ret:aliæ cauſæ fuerunt. Nā vt ex Aſtronomia ad Phyſicam,ſiue Coſmo-
graphiam deueniam,hæ Copernici hypotheſes non ſolum in Naturam
rerum non peccant,ſed illam multo magis iuuant. Amat illa ſimplicita-
tem,amat vnitatem. Nunquam in ipſa quicquam otioſum aut ſuperfluū
extitit:at ſæpius vna res multis ab illa deſtinatur effectibus. Atqui penes
vſitatas hypotheſes orbium fingendorum finis nullus eſt: penes Copernicum plurimi motus ex pauciſſimis ſequuntur orbibus. Vt interim ta-
ceam penetrationem orbiū Veneris & Mercurij,& alia, quibus antiqua
Aſtronomia in tanta orbium fingendorum libertate etiamnum laborat.
Atq; ſic Vir iſte nō tantum naturam oneroſa illa & inutili ſupellectili tot
immenſorum orbium liberauit : ſed inſuper etiam inexhauſtum nobis
theſaurum aperuit diuiniſſimorum ratiociniorum, de totius Mundi,ʼo-
mniumq;corporū pulcherrima aptitudine.Neq; dubito affirmare,quic-
quid à poſteriori Copernicus collegit, & viſu demōſtrauit,mediantibus
Geometricis axiomatis,id omne vel ipſo Ariſtot. teſte, ſi viueret (quod
frequenter optat Rheticus) à priori nullis ambagibus demonſtrari poſ-
ſe.Verum

Chapter I

agree precisely, as they do, (3) with the positions and mean motion of the Sun. (4) On all these points, as a magnificent order is shown by Copernicus, the cause must necessarily be found in it. Second, of those hypotheses which give a reliable reason for the appearances, and agree with observation, Copernicus denies nothing, but rather adopts and expounds them. For although he seems to have altered a great deal in the customary hypotheses, in fact that is not the case. For it can happen that the same conclusion follows from two suppositions which are different in species, because they are both included in the same genus, and the point in question is a consequence of the genus. Thus Ptolemy did not derive the risings and settings of the stars from the proximate intermediate premise of the same logical status, "Because the Earth is motionless at the midpoint." Nor did Copernicus derive the same conclusion from the intermediate premise, "Because the Earth revolves at a distance from the midpoint." For it was sufficient for each of them to say (as both did) that those phenomena follow from the propositions that there is a difference between the motions of the heavens and the Earth, and that there is no sensible distance between the Earth and the midpoint in comparison with the fixed stars. Therefore what appearances Ptolemy did derive, he did not derive from a false and accidental intermediate premise. His only breach of the rules as such was that he believed the consequences which follow from the genus to follow from the species.[5] Hence it is evident that Ptolemy derived from a false arrangement of the universe what was nevertheless true, and in agreement both with the heaven, and our own eyes, and that there is in that no ground for suspecting anything of the same sort of the Copernican hypotheses. Rather the point stands which was made at the outset, that Copernicus's postulate cannot be false, when so reliable an explanation of the appearances—an explanation unknown to the ancients—is given by them, (5) insofar as it is given by them.[6] This was seen by the highly successful Tycho Brahe, an astronomer beyond all praise, who although he entirely disagreed with Copernicus on the position of the Earth, yet retained from him the point which gives us the reasons for matters hitherto not understood, that is, that the Sun is the center of the five planets. For the proposition that the Sun is motionless at the center is a more restricted intermediate premise for the derivation of retrogressions, and the general proposition that the Sun is in the center of the five planets is sufficient.

Yet for Copernicus's taking the species as the genus, and in addition setting the Sun at the center of the universe, and the Earth in motion round it, there were other reasons. For, to turn from astronomy to physics or cosmography, these hypotheses of Copernicus not only do not offend against the Nature of things, but do much more to assist her. She loves simplicity, she loves unity. Nothing ever exists in her which is useless or superfluous, but more often she uses one cause for many effects. Now under the customary hypotheses there is no end to the invention of circles, but under Copernicus's a great many motions follow from a few circles. For the moment I will not mention the interpenetration of the spheres of Venus and Mercury and other points on which the ancient astronomy with its extreme freedom to invent circles is still in difficulty. And so this great man has not only freed Nature from the burdensome and useless paraphernalia of all those immense circles; but in addition he has opened to us an inexhaustible treasury of calculations on the fitting together of the whole universe and of all the bodies in it. Nor do I hesitate to affirm that everything which Copernicus inferred *a posteriori* and derived from observations, on the basis of geometrical axioms, could be derived to the satisfaction of Aristotle,

77

fe. Verum de his omnibus fusius & pro dignitate pridem egit Rhetici narratio , & Copernicus ipse : & si quid copiosius explicari potest,
(6) alius id loci & temporis erit , nunc attigisse sufficit: vt ea mentio-
ne constaret lectori altera caussa , quæ me in Copernici partes per-
traxerit.

Neque tamen temere, & sine grauissima præceptoris mei Mæstlini clarissimi Mathematici auctoritate, hanc sectam amplexus sum. Nam
is, etsi primus mihi dux & præmostrator fuit, cum ad alia , tum præcipue
ad hæc philosophemata, atque ideo iure primo loco recenseri debuisset:
tamen alia quadam peculiari ratione (7) tertiam mihi causam præbuit
ita sentiendi: dum Cometam anni 77. deprehendit, constantissime ad
motum Veneris à Copernico proditum moueri , & capta ex altitudine
superlunari coniectura , in ipso orbe Venerio Copernicano curriculum
suum absoluere. Quod si quis secum perpendat, quam facile falsum à seipso dissentiat,& econtra, quam constanter verum vero consonet: non
iniuria maximum argumentum dispositionis orbium Copernicanæ vel
ex hoc solo cœperit.

Vt autem ea omnia , quæ de hypothesibus vtrisque dixi, verissime ita se habere deprehendas:
accipe hanc breuem explicationem hypothesium Copernici, duasque tabulas ad hoc facientes.

Pro cognoscendo ordine Sphærarum Mundi secundum Copernici sententiam , intuere Tabulam primam in fine huius capitis , & quæ
ei adscripta sunt. (8) Terræ pro diuerso respectu tribuuntur à Copernico motus quatuor (Copernicus breuitati intentus tres dicit, qui reuera quatuor sunt) qui omnes reliquorum Planetarum motibus aliquam
apparentem varietatem conciliant.

Primus est ipsius Sphæræ seu Orbis, qui tellurem ceu stellam circa
Solem annuatim circumagit. Atq; is orbis, cum sit eccentricus, (9) eccentricitate insuper mutabili, (10) tripliciter nobis considerandus est.
(11) Initio remota eccentricitate ; Orbis igitur hic, motusque Terræ
has commoditates præstat : quod non indigemus tribus eccentricis in
vsitatis hypothesibus , scilicet Solis, Veneris & Mercurij. Nam pro eo,
quod terra circa hos tres planetas circumuehitur, Terricolæ existimant
tres illos circa se immobiles circumuehi. Sic ex vno motu tres faciunt.
Quod si plures essent stellæ intra orbem terræ , pluribus etiam hunc motum ascriberent. Cadunt etiam hoc orbe posito tres magni epicycli,
Saturni, Iouis, & Martis, cum eorum motibus. Id quomodo accidat,
in adiunctis parallelis, schematibus videri potest : Rursum enim, quia
Terra in conspectu Saturni (quasi quiescentis, quia tardior est) in orbe
suo circumit, à Saturno recedens & accedens: existimant incolæ, Saturnum in epicyclo suo circumire, accedere, recedere, se vero in centro
orbis sui quiescere. Circulum igitur A B putant esse epicyclos g, i, l. Item
propter telluris hunc eundem accessum ad Planetas & recessum in orbe
suo , videntur nobis ipsæ quinque planetarum latitudines aliquam varietatem accipere; quam librationem vt saluaret Ptolemæus, necesse ipsi fuit quinque alios motus statuere : qui omnes, posito vnico telluris
motu cadunt.

Et quamuis hi omnes motus, vndecim numero, è mundo extermi-
nati

if he were alive (which Rheticus repeatedly wishes for), *a priori* without any evasions.[7] However, Rheticus's *Narratio* and Copernicus himself have long since dealt with all this on a broader scale and as its importance deserves. If a more extensive exposition of any point is possible, (6) another place and time will do; here it is sufficient to have touched on it, so that this mention will make clear to the reader another reason for my having been completely converted to Copernicus's side.

Yet I did not embrace this cause rashly, and without the very weighty support of that famous mathematician, my teacher Maestlin. For, although as my first director and guide, both generally and in these philosophical questions especially, he should rightly have been placed at the head of this list, nevertheless by another particular argument (7) he furnished me with a third reason for accepting the theory when he realized that the comet of the year '77 moved in complete conformity with the motion of Venus stated by Copernicus, and, by a conjecture drawn from its altitude's being greater than the Moon's, that its whole path was in the actual sphere of Venus.[8] Now on careful consideration of how easily the false disagrees with itself, and on the other hand how reliably truth is consistent with truth, a very strong argument for the Copernican arrangement of the spheres will quite rightly be drawn from that fact alone.

To realize that everything which I have said about both hypotheses is absolutely true, here are a brief exposition of the hypotheses of Copernicus, and two plates to assist you.

To find the order of the spheres of the universe according to Copernicus's theory, look at the first plate at the end of this chapter, and what is written on it. (8) To the Earth according to its various aspects four motions are attributed by Copernicus (he himself, intent on brevity, says three, though in actual fact there are four) which in combination reconcile an apparent variation with the motions of the remaining planets.

The first is that of the sphere or circle itself, which carries the Earth like a star round the Sun annually. Now this circle, being eccentric, and (9) furthermore with a variable eccentricity, (10) has to be considered by us in three ways. (11) To start with disregard the eccentricity. Then this circle, and the motion of the Earth, produce the following advantages, that we do not require three eccentric circles as in the customary hypotheses, namely those of the Sun, Venus, and Mercury. For instead of the Earth being carried round those three planets, the Earthdwellers believe that those three are carried round themselves when they are motionless. Thus out of one motion they make three. If there were more stars within the Earth's circle, they would also ascribe this motion to more. There also disappear, if this circle is assumed, three large epicycles, those of Saturn, Jupiter and Mars, together with their motions. How that comes about can be seen in the attached parallel diagrams; for again, because the Earth as seen from Saturn (taken as at rest, because it is the slower) goes round in its circle, moving further from and nearer to Saturn, its inhabitants believe that Saturn in its epicycle goes round, comes nearer, goes further away, but they themselves are at rest at the center of its circle. Therefore they think that the circle AB is the epicycles g, i, l. Further, on account of the Earth's coming nearer to the planets and going further away from them, as has been mentioned, in its circle, the actual latitudes of the five planets seem to admit a certain variation; and for Ptolemy to save this oscillation, it was necessary for him to establish five other motions, all of which disappear if we assume a single motion of the Earth.

And although all these motions, eleven in number, are banished from the universe

nati sint, substituto hoc vnico terræ motu: nihilominus adhuc aliarum plurimarum rerum causæ redduntur,quas Ptolemæus ex tam multis motibus reddere non potuit.

Nam primo à Ptolomæo quæri potuit,qui fiat,quod Eccentrici tres Solis,Veneris & Mercurij habeant æquales reuolutiones? Respondetur enim,quod non vere reuoluantur ipsi,sed pro ipsis vnica terra. 2. Quare quinque Planetæ fiunt retrogradi. Luminaria non item? Respondetur primo de Sole,quia is quiescit: vnde fit, vt motus terræ,qui semper directus est, ipsi Soli mere & imperturbate inesse videatur, tantum per partem oppositam cœli.De Luna vero,quia motus Terræ annuus, ipsius cœlo vere communis est cum terra. (12) Duo autem quæ habent eundem motum per omnia,videntur inter se quiescere. Vnde motus Terræ in Luna non sentitur,vt in cæteris planetis.De superioribus Saturno,Ioue & Marte respondetur: Quia ipsi sunt tardiores terra: & quia circulus & motus iste Terræ putatur ipsis inesse. Quare sicut illis,qui ex L Saturni globo prospicerent, Terra interdum progredi videretur; dum iret per medietatem P B N supra Solem:interdum regredi,dum iret per N A P, stare vero in N & P: sic necesse est,vt nobis ex terra prospicientibus Saturnus volui videatur in partes oppositas. Vt dum est terra in B N A, Saturnus videtur in b n a alterius tabulæ. Inferiores Venus & Mercurius ideo regredi videntur, quia sunt velociores terra; vnde perinde ac si terra staret immota, Venus,currens in parte circuli remotiori, contrariam plane describit viam illi,quam conficit in parte circuli sui vicina terræ.

3. Ita quæri potuit(sed nihil respondente Ptolemæo) quare in magnis orbibus sint tam exigui epicycli, & quare in paruis orbibus tam immanes: hoc est,quare περιφάρεσις Martis sit maior Iouiâ, & huius maior quam Saturni? Et cur non Mercurius etiam maiorem, quam Venus,habeat,cum sit inferior Venere;siquidem quatuor reliquorum semper inferior maiorem habet? Hic facilis est responsio. Mercurij enim & Veneris veros orbes,veteres epicyclos esse putarunt. Mercurij autem,vt velocissimi,minimus etiam orbis est. Superiorum vero vt cuique Telluris orbis propior est,sic maiorem ad eum proportionem habet, & maior apparet. Mars igitur proximus habet maximam æquationem,Saturnus altissimus minimam. Nam si oculus in G constitueretur, ei orbis P N videretur sub angulo T A V. At si in L esset, idem orbis videretur sub angulo R L S.

4. Pariter non iniuria mirati sunt veteres, cur tres superiores semper in oppositione cum Sole sint humilimi in suo epicyclo, in coniunctione altissimi: vt si Terra, Sol & g sint in eadem linea,quare Mars tum non possit in alio loco epicycli esse,quam in γ. In Copernico causa facile redditur;Non enim Mars in epicyclo, sed terra in orbe suo hanc varietatem causatur;Hinc si terra ex A in B discesserit,Sol erit inter G Martem & B Terram.Et tum Mars videbitur in Epicyclo ex δ in γ ascendisse. At Terra in A existente,quod est punctum ipsi G proximum: G Mars & Sol videbuntur ex A inuicem oppositi. Atq; hæc sunt, quæ ex tabula ad oculum demonstrari possunt.

Iam deinceps consideremus etiam eccentricitatem huius orbis. (13) Copernicus facit Apogæum Solis(vel Terræ)vt & cæterorum moueri,

by the substitution of this single motion of the Earth, nevertheless reasons are supplied for a great many other matters for which Ptolemy for all his many motions could give no reason.

For in the first place one might ask of Ptolemy how it comes about that the three eccentrics of the Sun, Venus, and Mercury have equal times of revolution? For the answer is, that they themselves do not really revolve, but instead of them the Earth does on its own. 2. Why do the five planets make retrogressions, whereas the luminous stars do not? The answer is first, in the Sun's case, that it is because it is at rest; and the result is that the motion of the Earth, which is always in the same direction, seems straighforwardly and uninterruptedly to belong to the Sun itself, though in the opposite direction with respect to the heaven. In the case of the Moon, however, as the Earth's motion is annual, its own motion with respect to the heaven is indeed shared with the Earth: (12) two bodies which have the same motion in every way seem to be at rest with respect to each other. Hence the motion of the Earth is not observed in the Moon, as it is in the other planets. In the case of the superior planets, Saturn, Jupiter, and Mars, the answer is: because they are slower than the Earth, and the circle and motion of the Earth are imputed to them. Consequently, just as to anyone looking from L (the globe of Saturn), the Earth would sometimes seem to be moving forwards, so long as it was going by way of PBN above the Sun, and sometimes backwards, while it was going along NAP, but to stand still at N and P, in the same way to us, looking from the Earth, Saturn must necessarily seem to be turning in the opposite directions. Thus while the Earth is on BNA, Saturn seems to be on bna in the other plate. The inferior planets Venus and Mercury seem to move backwards because they are faster than the Earth. Hence exactly as if the Earth were stationary, Venus, passing along the further part of its circle, clearly describes a path in the opposite direction to that which it traces in the part of its circle which is next to the Earth.

3. Similarly one could ask (but with no answer from Ptolemy) why in the large circles the epicycles are so tiny, and why in the small circles they are so huge; that is, why the correction[9] for Mars is larger than that for Jupiter, and for Jupiter larger than for Saturn? And why Mercury does not have an even larger correction than Venus, since it is lower, seeing that among the other four planets the lower one always has the larger correction? Here the answer is easy. For in the case of Mercury and Venus the ancients thought that the true circles were epicycles. But Mercury's circle, although it is the fastest planet, is also the smallest. However, in the case of the superior planets, the nearer the Earth's circle is to each of them, the greater the ratio of it to the Earth's circle, and the larger it appears. Consequently Mars, the nearest, has the largest correcting factor, and Saturn, the highest, has the smallest. For if the eye were situated at G, the circle PN would appear to it to be subtended by the angle TGV. But if it were at L, the same circle would then appear to be subtended by the angle RLS.

4. Similarly the ancients rightly wondered why the three superior planets are always in opposition to the Sun when they are at the bottom of their epicycles, but in conjunction when they are at the top; for example if the Earth, the Sun, and g are in the same line, why Mars cannot be at any other point in its epicycle but at γ. In Copernicus's theory the reason is easily supplied. For it is not Mars on an epicycle but the Earth on its own circle which causes this variation. Thus if the Earth moves from A to B, the Sun will be between Mars at G and the Earth at B. And at that point Mars will seem to have climbed up on its epicycle from δ to γ. But when the Earth is at A, which is the point nearest to G, Mars at G and the Sun will seem from A to be in opposition to each other. These are the points which can be demonstrated from the diagram at sight.

ueri,nõ per deferentes, ſed per epicyclium paulo tardius orbe ſuo ad initium rediens. Hic motus Apogæi etiam aliquid inſert in motibus cæterorum Planetarum. Nam Ptolemæus cæterorum eccentricitates computat à centro terræ; quod ſi centrum Eccentrici Telluris & Apogæum per conſequentiam ſignorum diſceſſerint in aliam partem Zodiaci, rélictis poſt ſe aliorum Apogæis tardioribus; accidet aliqua mutatio eccétricitatum in planetis cęteris. Hoc valde rurſum mirabitur Ptolemæi Aſtronomia, atque ad confingendos nouos orbes confugiet; quibus demonſtret, hæc ita fieri poſſe, cum tamé ex motu Telluris vnico ſecutura ſint. Atq; hoc quidem multa poſt ſecula vix demũ fiet, ſed tertio (15) mutatio eccentricitatis terrenæ, qua centrum eccentrici ad Solem accedit, & ab eo recedit, inde à Ptolemæo ad nos vſque magnum quid in Marte & Venere intulit: quorum eccentricitates cum mutatæ videantur, quid Ptolemæum dicturum putas? Nunquid rurſum nouos circulos in cęterorum infinitam turbam aſciſceret, ſi viueret? quibus omnibus in Copernico opus minime eſt. Hæc tot & tanta Copernicus per vnius circuli **A B** poſitionem & motum præſtitit: vnde merito, quamuis exiguus eſſet, **MA-GNO** cognomen dedit. Hic primus motus cœlo Lunæ cum Tellure communis fuit.

Iam porro videamus, quid reliqui motus telluris efficiant; qui accidunt intra illum Lunæ orbiculum ad **A**.

Secundus igitur motus non integri orbis, ſed *(16)* qrbiculi cœleſtis, terræ globum proxime ceu nucleum includentis, tendit in oppoſitũ ab ortu in occaſum, perinde vt epicyclia ſuperiorum, quibus eorum eccentricitas ſaluatur à Copernico. Huius annua cõſtitutione fit, vt æquinoctialis ſemper in eandem mundi partem declinet. Poli enim Æquinoctialis ſiue corporis ab huius polis per 23. gradu cum dimidio, diſtant. Qui motus cum pauxillo velocior ſit motu annuo orbis magni, facit ſectiones circulorum, ſiue (17) æquinoctiorum loca paulatim in præcedentia moueri. Quare per hunc exiguum globulum cadit illa monſtroſa, ingens, ἄναϛρος Nona Sphæra Alphonſinorum, vt cuius officium in illum orbiculum antea neceſſarium tranſlatum eſt. Cadit etiam motus deferentium Apogæum Veneris, vt quod non aliter mouetur, niſi ſi fixæ moueri ſtatuantur.

(18) Tertius motus eſt Polorum globi terreni, conſtans duabus librationibus, quarum vna eſt altera duplo celerior, & ad rectos angulos. Is adminiſtratur per quatuor circulos, ſic vt bini circuli ſingulas librationes faciant, & librationes ipſæ permixtæ corollæ intortæ ſpeciem præbeant, in hunc modum: Vna libratio in Coluro ſolſtitiorum fit, & ſaluat variationem declinationis Zodiaci, ſero poſt Ptolemęi tępora animaduerſam: tale quid & Ptolemæo opus fuiſſet confingere, & nonnulli moderni, vndecimo Mundi orbe iam conficto, præſtare conati ſunt.

Altera libratio, quæ fit in coluro Æquinoctiorum, ſaluat inæqualem præceſſionem Æquinoctiorum, & eliminat octauæ fixarum Sphæræ, quæ vltima eſt apud Copernicum, motum trepidationis, illique quietem ſuam reſtituit. Atque ne non & hic motus aliquid in cæteris motibus fœneretur: tollit irregularitatem motus, quem

C   omnium

Chapter I

Next let us take into account also the eccentricity of this circle. (13) Copernicus makes the apogee of the Sun (or of the Earth) move, like those of the other planets, not along deferents, but along a small epicycle which returns to its starting point a little more slowly than on its own deferent circle. This motion of the apogee also has some effect on the motions of other planets. For Ptolemy (14) computes the eccentricities of the others from the center of the Earth; but if the center of the Earth's eccentric and its apogee shifts eastwards to another part of the zodiac, leaving behind them the slower apogees of the others, some change in the eccentricities of the other planets will result. Again the astronomy of Ptolemy will greatly wonder at this and will take refuge in inventing new circles by which to demonstrate that it is possible, whereas it will follow from a single motion of the Earth. That indeed will only just come about after many centuries; but thirdly, (15) a change in the Earth's eccentricity, by which the center of its eccentric moves closer to the Sun and moves further away from it, between Ptolemy's time and our own has had a great effect on Mars and Venus; and when their eccentricities seem to be changed, what do you think Ptolemy would say? Would he again admit new circles to the infinite crowd of others, if he was alive? All of which are scarcely needed in Copernicus. All these great and numerous phenomena Copernicus accounted for by the location and motion of the single circle AB, so that it was proper that he gave it, although it was tiny, the title of Great.[10] This first motion with respect to the heaven was common to the Moon and the Earth.

Now let us go on to see the effects of the remaining motions of the Earth, which take place within the little circle of the Moon at A.

The second motion, then, which is not a motion of the entire circle but only of (16) the little heavenly circle which closely enfolds the Earth's globe like a kernel, is in the opposite direction, from east to west, like the epicycles of the superior planets, by which their eccentricity is saved by Copernicus.[11] The result of its annual occurrence is that the equator always slopes towards the same point in the universe. For the poles of the equator or of the actual globe are 23½° from the poles of this circle. This motion, being very slightly faster than the annual motion of the Earth's orbital circle, makes the intersections of the circles, that is, (17) the positions of the equinoxes, move gradually westwards. Hence this tiny little globe does away with that vast, monstrous, starless ninth sphere of the Alfonsine compilers, as what used to be its essential function has been transferred to this little orbit. The motion of the circles which carry round the apogee of Venus also disappears, unless the fixed stars are deemed to move.

(18) The third motion is that of the poles of the terrestrial globe, consisting of two oscillations, one of which is twice as rapid as the other, and at right angles to it.[12] It is accomplished by means of four circles, in such a way that each oscillation is produced by two circles, and the oscillations themselves combine together to form the shape of an interwoven garland, in the following manner: one oscillation is on the colure of the solstices, and saves the variation in the declination of the zodiac, which was noticed late after the time of Ptolemy, and is something which Ptolemy would have needed to invent, and several moderns have tried to represent by inventing an eleventh circle of the universe.

The other oscillation, which is on the colure of the equinoxes, saves the irregular precession of the equinoxes, and eliminates the motion of trepidation of the eighth sphere, that of the fixed stars, which is the last according to Copernicus, and restores it to rest. And to make sure this motion also contributes

omī·un, septem Planetarum, & Apogæorum motus habere debuiſ-
ſent(nō ſinę miniſterio aliquot nouorum circulorum) quia compertum
eſt omnes motus æqualiter per fixas incedere.

Quartus denique motus eſt ipſius globi terreni & circumfuſi aeris
proprius,cuius periodus eſt 24.horarum in eandem mundi plagam cum
cæteris,nempe ab occaſu in ortum : propter quē totus mundus reliquus
ab ortu in occaſum, imperturbatis magno miraculo motibus ſecundis
ferri putatur.Cadit igitur illa incredibiliter alta & pernix decima Sphæra ἄναϛρος,cuius & totius mūdi tanta eſſet in Ptolemæo pernicitas, vt vno
nictu oculi aliquot millia milliarium tranſirent. Ac quæſo te,ad tabellam reſpicias, & cogites,quod tellus hæc noſtra, de cuius motu diſputa-
tur,exigui circelli lunaris ad A , ſeptuageſimam vix demum partem diametri æquet: Ab hoc circello dein ad S   turni amplitudinem, & ab hac
ad fixarum inæſtimabilem altitudinem e ulos intende, & denique conclude, vtrum factu crediturque facilius, punctulum illud intra A circel-
lum,& ſic tellurem in vnam plagam rotari, an vero totum mundum decem diſtinctis motibus(quia decem ab inuicem ſoluti orbes) infanda ra-
piditate ire in plagam alteram, nec quoquam, niſi ad illud punctulum,
telluris imagunculam,eamque ſolam immobilem, reſpicere, quia extra
nihil eſt.

Huc pertinet Tabella prima & ſecunda.

IN CAPVT PRIMVM
Notæ Auctoris.

(1) **I**Ntempeſtiuum.] *Occurrit huic ſcrupulo Copernicus ipſe ,in præfatione ad Paulum Tertium Pontif. Maxim. ſed paulo rigidiuſcule: cuius orationis pœnas luit denique, plus quam
70.annis ab editione libri , áque morte ſua elapſis :* ſuſpenſus enim eſt, *inquit cenſura, donec
corrigatur, opinor autem, etiam hoc ſubintelligi,* donec explicetur. *Quomodo enim non ſit
ſcripturæ contrarius, quippe in propoſito longiſſime diuerſo, conatus ſum oſtendere rationibus & ex-
emplis , in* Introductione in Commentaria de motibus Martis. *Ipſius etiam Copernici verba explicaui dilucidius in fine libri I.* Epitomes Aſtronomiæ: *quibus locis ſpero religioſis ſatisfactum iri:
dummodo & ingenium & cognitionem Aſtronomiæ talem ad hoc iudicium afferant, vt gloria diuinorum operum viſibilium , ipſorum patrocinio tuto credi poſſit. Eſt ſane aliqua lingua Dei, ſed eſt,
etiam aliquis digitus Dei. Et quis neget linguam Dei eſſe, attemperatam & propoſito ſuo , & ob id,
lingua populari, hominum? In rebus igitur euidētiſſimis torquere Dei linguam, vt illa digitum Dei in
natura refutet, id religioſiſſimus quiſque maxime cauebit. Legat, cui curæ ſunt laudes Creatoris &
Domini noſtri,legat,inquam,librum meum quintum* Harmonicorum: *& percepta motuum politia
exquiſitiſſime* Harmonica, *deliberet ſecum, ſatin' iuſta, ſatin' prægnantes cauſſæ fuerint quæſitæ
conciliationis inter linguam & digitum Dei: anne expediat, ea conciliatione repudiata, famam
hanc Operum diuinorum pulchritudinis immenſæ,cenſuris opprimere; quæ fama vt ad rudis popu-
li: quinimo , vt ad vulgi literatorum notitiam vel leuem percuniat, nullis vnquam imperiis effici
poſſit. Renuit inſcitia reſpicere in auctoritatem,ad pugnam vltro prouolat , freta multitudine , &
ſcuto conſuetudinis,telis veritatis impenetrabili.*
    Acies vero dolabræ in ferrum illiſa, poſtea nec in lignum valet amplius. Capiat hoc cuius in-
tereſt.

(2) **Atque hoc loco.]** *Eandem inſtantiam: in particulari etiam hypotheſi eccentricitatis,diſcuſſi in Commentariis Martis cap.21. Oſtendique , qua de cauſa & quatenus falſa hypotheſis
interdum verum prodat.*

(3) **Cum loco & motu Solis medio.]** *Nondum ſciebam, quod poſtea in Comment.*
                                                                    Martis

84

something to the other motions, it removes an unevenness of movement which the motions of all the seven planets, and of their apogees as well, would have had to have (not without the services of some new circles) because it has been found that all their motions are regular relative to the fixed stars.

Lastly, the fourth motion is that of the terrestrial globe itself and of the atmosphere which immediately surrounds it. Its period is twenty-four hours and its direction with respect to the universe is the same as the rest, that is from west to east, and because of it the whole of the remainder of the universe is thought to be carried along from east to west, its secondary motions being by a great miracle undisturbed. Consequently there disappears that incredibly lofty and swift tenth and starless sphere, the swiftness of which and of the whole universe would be so great according to Ptolemy that it would traverse several thousand miles in the blinking of an eye. I ask you to look at the plate and consider that this Earth of ours, the motion of which is in dispute, scarcely equals a seventieth part of the tiny little lunar circle at A. Next turn your eyes from that little circle to the spaciousness of Saturn's, and from that to the incalculable loftiness of the fixed stars; and finally decide whether it is easier for it to happen and to be believed that that small point within the little circle A, and hence the Earth, rotate in one direction, or that the complete universe goes with ten distinct motions (as there are ten mutually independent circles) with inconceivable rapidity, and is subject to nothing but that small point, which alone is motionless, because there is nothing outside.

Here belong Plates I and II.

AUTHOR'S NOTES ON CHAPTER ONE

(1) *Premature.*] Copernicus himself faces this doubt in his preface to Pope Paul III, but with a little too much inflexibility. His discourse finally paid the penalty, after more than seventy years had elapsed since the publication of his book, and his own death; for "it is suspended," said the censorship, "until it is corrected"; though I think "until it is explained" should also be read between the lines. For I have tried to show with arguments and examples the way in which it is not contrary to Scripture, admittedly with a greatly different intention, in the introduction to the *Commentaries on the Motions of Mars.* Also I have explained Copernicus's own words more clearly at the end of Book I of the *Epitome of Astronomy.* In those passages I hope to have satisfied those with religious scruples, provided that they approach their decision on this point with sufficient intelligence and knowledge of astronomy for the glory of God's works, which are themselves visible, to be safely entrusted to their protection. Certainly God has a tongue, but he also has a finger. And who would deny that the tongue of God is adjusted both to his intention, and on that account to the common tongue of men? Therefore in matters which are quite plain everyone with strong religious scruples will take the greatest care not to twist the tongue of God so that it refutes the finger of God in Nature. Let him read, if any man is concerned for the praises of our Creator and Lord, let him read, I say, the fifth book of my *Harmonice*; and when he has perceived the most skillfully Harmonized Republic[13] of the motions, let him debate with himself whether sufficiently sound, sufficiently prolific reasons have been discovered for reconciling the tongue and the finger of God; or whether he will repudiate that reconciliation and hasten to suppress with censorship the renown of the immeasurable splendor of the works of God. That this renown should come to be known to the common people, nay rather, to the generality of the even superficially educated, could never be brought about by order. Ignorance refuses to respect authority, it resorts spontaneously to combat, relying on numbers and on the shield of habit, which is impenetrable to the weapons of truth.

Truly once the edge of an axe has been blunted on iron, it can no longer cut wood.

Let those who care understand.

(2) *On this point.*] I have also discussed this instance, with reference to the particular hypothesis of eccentricity, in the *Commentaries on Mars,* Chapter 21, and have shown for what reason and to what extent a false hypothesis sometimes reveals the truth.[14]

(3) *With the position and mean motion of the Sun.*] I did not yet know what I afterwards

TABELLA I. EXHIBENS ORDINEM SPHÆRARVM

Cœlestium mobilium: simulque veram proportionem magnitudinis earum iuxta medias suas distantias: item angulos prosthaphæreseon earundem in orbe Magno Telluris, secundum sententiam Copernici.

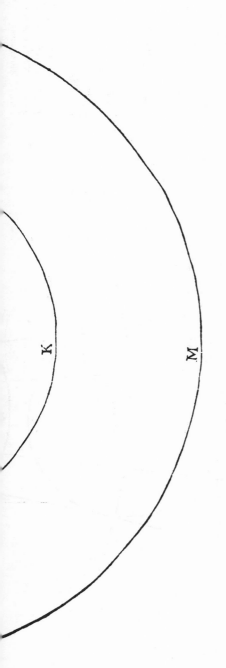

K

M

In centro, vel prope est SOL immobilis.

E F minimus circa Solem circulus est MERCVRII, qui restituitur diebus 88. fere.

Hunc sequitur VENERIS CD, cuius revolutio circa eundem Solem est dierum 224. cum besse.

Quibus sequitur A B, TELLVRIS est, cuius revolutio dierum 365. & quadrantis. Dicitur ORBIS MAGNVS, propter usum multiplicem.

Circa Tellurem est orbiculus velut epicyclus, SPHAERAE LVNARIS, ad A. eodem motu per annos ipsaciem cum tellure ad eandem stellam fixam rediens. Sed eam propria revolutio ad Solem habet die 29. cum dimidio.

Post hunc est Orbis MARTIS GH, qui cursum suum sub fixis stellis, sive ad Solem absoluit diebus 687.

Hunc excipit post magnum intervallum, Sphaera IOVIS IK, habens ambitum dierum 4332. cum quinq. octaui fere.

LM Vltimus & maximus, est SATVRNI, eius tempus periodicum dierum 10759. cum quinta.

FIXAE vero STELLAE adhuc tam inaestimabili interuallo altiores sunt, ut adeam, quae est inter Solem & Terram intercapedo sensibilis non sit. Et eae sunt in extremo sicut Sol in centro, penitus immobiles.

Angulus T G V, vel Arcus T V, prosthaphaeresis est, siue parallaxis, quam Orbis Magnus Telluris ad Sphaeram Martis habet.

Sic P I H est iusdam Orbis Magni parallaxis ad Sphaeram Iouis: & P L H, siue R L S, vel R S arcus ad sphaeram Saturni.

Ita X A Y, vel X Y arcus est parallaxis sphaera Veneris: ut & Z A E, vel Z E sphaera Mercurij parallaxis, ad Orient Magnum.

See p. 227.

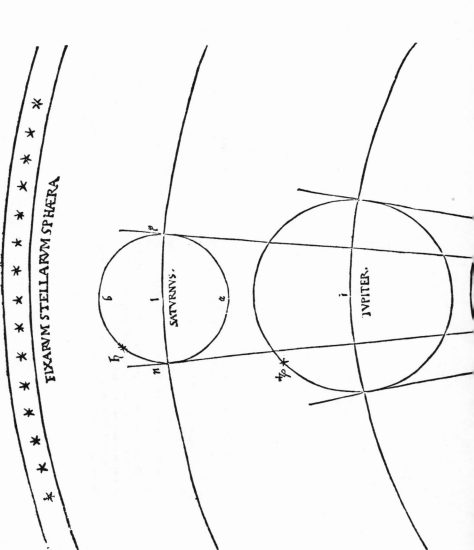

TABELLA II.

Exhibens ordinem sphærarum cœlestium, & vtcunq; proportionem orbium & epicyclorum, atque angulos vel arcus prosthaphæreseon eorúdem, iuxta medias distantias, secundū veterum sententiam.

In Centro TERRÆ est, sola immobilis.

Intimus circa Terram orbiculus LVNÆ Sphæram repræsentat, cuius motus menstruus est.

Hunc proxime MERCVRII orbis circumdat: quem sequitur VENERIS, & postea

FIXARVM STELLARVM SPHÆRA

SATVRNVS.

IVPITER.

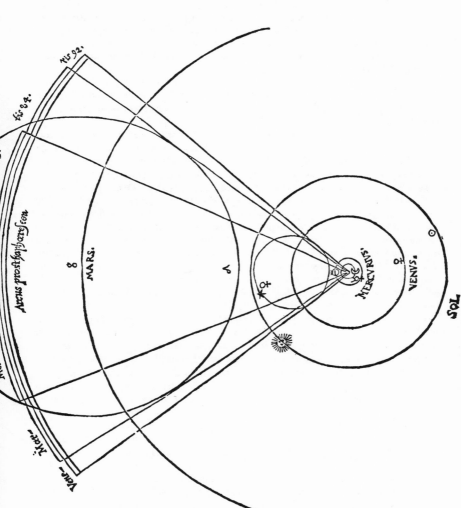

rum MARTIS, IOVIS &
-quoque STELLARVM Sphæ-
ram, arcus, quos circa terram,
ceu centrum integros deſcribe-
re, & complere quiſq, poteſt, indicant. Martis orbis biennio cõ-
uertitur. Iovis 12. annos quam
proxime, requirit, & Saturni
fere 30. ann. Fixæ Stellæ 49000.
annis, iuxta Alphonſinorum
placita, periodum reſtituunt.

Quantas ſingulorum (præ-
ter D) epicycli in concentrico
circulo proſthaphareſes, in me-
diis diſtantiis faciant, arcus, re-
Elis ex terra ductis, & epicyclos
ſingulos tangentibus intercepti,
additis graduum numeris, mon-
ſtrant.

Poſt cap. 1.
pag. 18.

See pp. 227-28.

Martis demonstrati, Anomaliam orbis Magni f.u Commutationis, qua Retrogradationes caus̄entur, restitui ad ipsum verum Solis motum & locum. Id vero in veteri Astronomiæ forma unitos adhuc magis mirari coguntur, qui eam retinent. Adeoque ex hoc ipso, demonstratiua nascuntur argumenta, retrogradationes non oriri ab aliquo motu reali, vel planetarum, vel totius Systematis cœlestis, sed ex motu Telluris vnicæ per imaginationem in Planetas omnes transferri.

(4) *Porro sententia sequens, Quibus omnib. & c. oscitanter est concepta; hoc enim dicere olebam; cum in Copernico appareat ordo pulcherrimus, qualis est inter causam & suos effectus; necesse esse, vt hæc ipsa sit vera causa retrogradationum, quam Copernicus dicit; vt sc. Hypothesis ista non sit fictitia tantum.*

(5) **Quatenus ex illis redditur.**] *Quia, vt iam sequitur, non vt ex speciali earum conformatione, sed vt ex generali, quæ est Copernico cum Braheo communis, quorundam caussa redditur; at nonnullorum tamen caussa ex ijs vt ex speciali Copernici redditur: item, Quia si particularissimas Hypothesium Copernici conditiones dimensionesque respiciamus; caussa minutiarum nonnullarum in Observationibus reddi non potuit: eaque de caussa Copernicana Hypotheses circa particularia tam quoad formam, quam quoad dimensiones, à me corrigi ad præscriptum Observatiorum debuerunt. Etsi quæ in forma dixi emendanda fuisse, illa talia sunt, vt ad perfectionem potius Hypothesium Copernici, hoc est, ad longiorem earum deductionem ab vsitata via spectent, quam ad nouam aliquam conformationem: qua vt in Commentarijs Martis alicubi dixi, Copernicus suarum ipse diuitiarum fuit ignarus.*

(6) **Alius id loci & temporis erit.**] *Potissima huius operæ moles, quod Astronomica attinet, in Commentaria Martis incubuit; in Physicis vero seu Metaphysicis argumentis corrogandis fusior sum in Epitomes Astr.lib.IV. qui liber ipsum τὸ ἔρχον continet, quod hoc loco sum pollicitus. Vide totum.*

(7) **Tertiam mihi causam ita sentiendi.**] *Idem tamen me vltro admonuit postea, non esse necessariam hanc collectionem. Nam cum Cometa motum non in multos dies continuet, & cum habeamus libertatem intendendi remittendive eius motum in Hypothesi suscepta, vbi observationes (quæ plerumque crassæ sunt) id requirere videntur, hinc adeo sit, vt diuersæ in particularibus Hypotheses, easdem Cometæ observationes repræsentent. Et Braheus libro de Cometis fol.282. Mæstlini Hypothesin examinat, cum sua comparat, redarguitque. Ipse vero, fol.206. Hypothesin talem proponit, in qua motus Cometæ proprius circularis initio tardus, mox intensus, in fine rursum tardus exhibetur. Itaque cedo hoc genere argumenti, sic quidem informati, vt ex eo quod potuerunt artifices præstare, nuda credulitate, & generali illa coniectura, quod verum vero consonet, de veritate suppositionum quid præsumatur. At vicissim alia via eandem arcem rursum occupo. Si enim motus telluris ad hoc vtilis est, vt Cometarum motus rectilinei, eorumque perpetua, vel æquabilitas, vel augmentatio, aut contraria diminutio perpetua, satisfaciant observatis; tunc sane, quantum est verisimilitudinis in motu rectilineo æquabili, corporum vanescentium; tantum fidei accedet motui Telluris, præsertim, si flexus itinerum apparentium irregulares occasione motus Telluris prouenisse constet, aliarumque, quæ in Cometis apparent, ratio reddatur. Verbi caussa ille ipse Cometa anni 1577. ortus ex vltimis partibus sagittarij, maximum ibi motum diurnum, caput 7. minutorum, caudam 22.gr. longam exhibuit; hæc omnia fuerunt diminuta versus finem, adeo vt in signo Piscium, quod quadrante distat à sagittario, stationem peracturus videretur, nisi euanuisset. Quæritur quæ caussa, cur Cometa circa quadratum eius loci in quo maximi apparuerunt & velocissimi, appropinquent stationi, cur stationi vicini, alij occultentur sub Solem, vt istæ, alij etiam in opposito solis paulatim enanescant, vt ille anni 1618. faciunt enim ista plerique. Quod si potiaris libertate circularis motus Cometæ tribuendi, caussam per omnes Cometas eandem dicere non poteris. At si teipsum redigas ad angustias traiectionis rectilineæ, statim apparet necessitas phænomeni. Itaque planum traiectorium Cometæ anni 1577. ego ordinassem in ea linea, per quam paucis post disparitionem diebus, videndus fuisset caussa longitudinis, si superfuisset: traiectionem ipsam primum velocem, in subeuntes dies tardiorem fecissem, idque pro ratione propinquitatis partium, traiectoriæ lineæ ad Solem, quia Cometa via obliqua fugiebat à Sole, Tellusque simul à Cometa. Qua ratione efficiebatur, vt Cometa initio quidem dimidiam solis altitudinem haberet; ex eo sphæras Veneris, Telluris, Martis traiiceret, & in fine plus quam triplo altior Sole euaderet. Non mirum igitur, quod parallaxis in eo nulla deprehendi potuit. Sed de hac re plus satis: hoc loco plura si petit lector, adeat meum de Cometis libellum, quem nundinis Autumnalibus anni 1619. emisi.*

C 2 (8) Ter-

demonstrated in the *Commentaries on Mars*, that the anomaly of the Great Orbit or of the parallax, which causes the retrogressions, is restored to the actual true motion and position of the Sun. Those who keep to the old form of astronomy must find this fact much more surprising in that form. Consequently this fact in itself gives birth to compelling arguments that the retrogressions do not arise from some real motion either of the planets or of the whole system of the heavens, but that from the motion of the Earth alone they are transferred by imagination to all the planets.

(4) Furthermore I was nodding when I composed the following sentence, *On all these points, etc.,* for I meant to say, that as a magnificent order is shown by Copernicus, as there is in the relationship between a cause and its effects, the true cause of the retrogressions must necessarily be this very fact, as Copernicus says, that is, this hypothesis is not merely a fiction.

(5) *Insofar as it is given by them.*] Because, as now follows, the explanation of certain points is given, not from this special conformation of the planets, but from the general one, which is common to Copernicus and Brahe; but nevertheless the explanation of some points is given by them from Copernicus's special one. Further, because if we consider the particular detailed conditions and dimensions of Copernicus's hypotheses, no explanation could be given of some minor points in the observations; and for the sake of that explanation the Copernican hypotheses had to be corrected by me on particular points as dictated by the observations, both with respect to the arrangement and with respect to the dimensions. However, the points which I have said had to be emended in the arrangement were such that they contribute rather to the perfection of the hypotheses of Copernicus, that is, to drawing them further from the traditional path, than to some new conformation, since as I have said elsewhere in the *Commentaries on Mars*, Copernicus was himself unaware of his own riches.[15]

(6) *Another place and time will do.*] The chief burden of this task, as far as it concerns astronomy, fell on the *Commentaries on Mars*: but I have assembled a more copious collection of the physical or metaphysical arguments in Book IV of the *Epitome of Astronomy*. That book contains the actual *oeuvre* which I promised in the present passage. See the whole.

(7) *He furnished me with a third reason for accepting the theory.*] However the same person later informed me of his own accord that this argument is not necessary. For since a comet does not persist in its motion for many days, and since we are at liberty to intensify or relax its motion in the hypothesis adopted, when the observations (which are frequently rough) seem to require it, it therefore follows that hypotheses which differ in detail represent the same observations of a comet. And Brahe in his book about comets,[16] page 282, examines Maestlin's hypothesis, compares it with his own, and refutes it. On the other hand on page 206 he proposes a hypothesis such that according to it the proper circular motion of the comet is shown as slow to start with, then intensified, and finally slow again. I therefore abandon this kind of argument, which indeed is so constructed that it leads to the presumption, by sheer credulity and the general conjecture that truth is consistent with truth, that what the practitioners have been able to adduce has some bearing on the truth of the suppositions. However I again capture the same citadel by another route instead. For if the motion of the Earth has the useful result that taking the motions of the comets as rectilinear, and either constantly regular or increasing, or on the contrary constantly diminishing, will satisfy the observations, then plainly just as much credence will be given to the motion of the Earth as there is probability in attributing regular rectilinear motion to bodies which disappear, especially if it is accepted that the irregular shifts in their apparent paths are the effect of the Earth's motion, and an explanation is given of the other appearances observed in comets. For example, the comet of the year 1577 itself when it rose in the furthest region of Sagittarius showed its greatest daily motion there, a head of 7' and a tail 22° long. All these were reduced towards the end, so much so that in the sign of Pisces, which is a quadrant away from Sagittarius, it would have seemed almost to have reached a stationary point, if it had not disappeared. The reason is therefore required why comets, at about a quadrant from the place where they have appeared to be largest and fastest, approach a stationary point, why when they are close to a stationary point some are hidden by nearness to the Sun, as that one was; while others disappear little by little in opposition to the Sun, as did that of the year 1618. For this is what they mostly do. But if you grant yourself the liberty of ascribing circular motions to a comet, you cannot speak of the same explanation for all comets. But if you confine yourself to the restrictions of a rectilinear path, the necessity of the phenomenon is at once apparent. Thus I would have placed the rectilinear path of the comet of the year 1577 on the line on which it would have been visible a few days after its disappearance, if it had survived. Its actual passage I would have made first rapid, for the succeeding days slower, and that in proportion to the nearness of the regions where the line of its path lay to the Sun, because the comet was moving away from the Sun at an angle, and the Earth at the same time from the comet. For that reason the result was that the comet, to start with, had half the altitude of the Sun; after that it traversed the spheres of Venus, the Earth, and Mars; and in the end it finished by being more than three times higher than the Sun. It is therefore not surprising that no parallax could be detected in its case. But that is more than enough on this point: if the reader wants more on the subject, he should go to my little book on the comets,[17] which I published at Michaelmas 1619.

(8) Terræ tribuuntur motus quatuor.] *Scribendo ego id temporis adhuc didici; ne perturberis igitur multitudine istâ motuum: proprie duo tantùm sunt, vnus ab interno pendens principio, convolutionis diurnæ, circa proprium centrum, alter extrinsecus à Sole Telluri illatus, annuus circa Solem; etsi moderatur illum formatque vis magnetica, fibris Telluris insita, qui verò tertius hic censebatur, ille quies est potius axis Telluris in situ parallelo, dum centrum circa Solem fertur, & qui quartus hic ventitatur, is est leuicula perturbatio huius quietis, orta ex aberratione duorum primorum & propriorum. Sed de his infra plura.*

(9) Eccentricitate insuper mutabili.] *Hoc coacti sunt statuere auctores, cæteri de Sole, Copernicus de Tellure, quia nimium tribuunt Obseruationibus Hipparchi & Ptolemæi: sed quæ non sunt tantæ subtilitatis, vt dogma tanti momenti possit iis superædificari. Itaque in Commentariis Martialium speculationum, & lib. V I. Epitomes parte I. opinionem istam, vt Physicæ cœlesti inimicam admodum, fidenter reieci, nec dum cedo sententia: euidentem imbecillitatem opinationis huius, alibi demonstrabo.*

(10) Tripliciter nobis considerandus.] *Non quod triplex ipse sit; sed quia vnus & idem existens, tria distincta habet, quæ singula suos multiplices vsus & munia habent in Astronomia reformata.*

(11) Initio remota eccentricitate.] *Id est, seposita consideratione Eccentricitatis. Quædam enim præstat orbis iste Eccentricus, non ipsa sua Eccentricitate, sed illa sola re, quod circa Solem vertitur.*

(12) Duo autem.] *Cœlum puta Lunæ (non Luna per se) & Tellus, habent eundem motum translationis de loco in locum, per Orbem magnum, ergo cum semper Terra loco eodem sit, quo loco inuenitur & cœlum Lunæ; Cœlum igitur hoc Lunæ, & per id, Luna ipsa, causa quidem cœli sui, nullam talem ex motu Telluris apparentiam suscipit motus sui, quatem ex Terra translatione, Sol suscipit, ipse verè quiescens. Id secus esset, si Terra promota, Cœlum Lunæ quiesceret, aut moueretur de loco in locum, motu alio distincto: tunc enim motus centri Telluris per imaginationem etiam in cœlum Lunæ transcriberetur; & sic etiam totum cœlum Lunæ, pro ratione situs sui posset retrogradum videri, non minus, quam Planetæ quinque.*

(13) Copernicus facit Apogæum Solis.] *Duo hic innuuntur, alterum Solem ipsum attinet, alterum ex Sole redundat in Planetas. Ptolemæus Solem collocat in Eccentrico, Eccentricum includit duobus deferentibus: Copernicus Epicyclo affigit planetam, Epicyclum concentrico. Ptolemæus igitur, vt Apogæa promoueat, Deferentibus suis attribuit motum peculiarem tardissimum; Copernicus idem præstat, per aberrationem restitutionis Epicycli à restitutione Concentrici, cum sit vtraque annua ferè. Verisimilius autem est, motus illos tardos, ex aberratione esse, quam ex motu positiuo. Præsertim cum epicyclo motus annuus, tantum respectu Eccentrici sui insit, à quo circumacto Epicyclus se euoluit in plagam contrariam; at respectu fixarum, quietis potius speciem præ se fert; quia in hac euolutione sit vt eædem Epicycli partes iisdem fixarum plagis semper obuertantur, nisi quantulum turbat aberratio. Ego verò in Commentariis Martis, & in Epit. Astr. libro I V. caussam trado physicam, tam Eccentricitatis, quam transpositionis Apogæorum, quæ caussa insita est in fibris corporis planetæ, nec indiget, vel deferentibus, vel Epicyclis. Sed hoc membrum, Solem ipsum (seu Terram) attinens, intellige obiter saltem inculcatum; vt ex eo iam ostendatur, quid ex Apogæi Solis transpositione redundet in Planetas cæteros.*

(14) Eccentricitates computat à centro Terræ] *Hæc dilucidiora fiunt per intuitum Tabulæ V. Est quidem hoc ratiocinari, dicere quid post multa sæcula sit futurum; cum iste scrupulus de præsenti nondum vrgeat Astronomiam veterem. Sed sic comparatum est cum transsumptione placitorum Ptolemæi particularium in Hypothesin Copernici; vt non potuerit à me omitti mentio ista. Nam etiam Copernicus Eccentricitates quinque planetarum computauit velut à Centro Orbis magni: quasi illud (non verò ipsum Solis centrum vicinissimum) sit genuina basis Systematis planetarii. Per hos verò 25. annos, ex quo libellum hunc edidi, sic est à me constituta Astronomia, vt Eccentricitates omnes (primariorum Planetarum) ad ipsissimum Solis centrum, ceu reram Mundi basin referantur. Itaque manere possunt Eccentricitates Planetarum omnium, quorsumcunque se recipiat Apogæum Solis. Vide in Martialibus Commentariis partem primam de æquipollentia Hypothesium, præsertim Caput VI.*

(15 Mutatio Eccentricitatis terrenæ.] *Hæc ex admonitione ipsius Copernici transcripta sunt. Et verum est, qui Centrum Orbis Solis à Tellure (vel Telluris à Sole) nimium dimouet, vt fecisse contendo Ptolemæum & Hipparchum; is si Planetarum Eccentricitates ad hoc punctum refert,*

alias

Chapter I

(8) *To the Earth...four motions are attributed.*] From writing this at that time I have continued to learn. Do not, therefore, be perplexed by this multiplicity of motions. Properly speaking there are only two, one depending on the internal origin of the daily revolution about its own center, the other the annual motion round the Sun conveyed from outside by the Sun to the Earth, though the latter is controlled and shaped by the magnetic power residing in the bowels of the Earth. The motion which was here reckoned as the third is rather the immobility of the Earth's axis in a parallel position, while its center is carried round the Sun; and the one which is here held out as the fourth is the slight disturbance of this immobility which arises from the aberration of the two first and primary motions. But more on these points below.

(9) *Furthermore with a variable eccentricity.*] The authorities were forced to this conclusion, the rest with reference to the Sun, Copernicus with reference to the Earth, because they rely too much on the observations of Hipparchus and Ptolemy; but they are not of such precision that a doctrine of such importance can be based on them. Consequently in my *Commentaries on the Observations of Mars*, and Book VI of my *Epitome*, Part 1, I rejected that opinion very confidently as repugnant to celestial physics, and I do not abandon that conclusion. The obvious stupidity of this supposition I shall demonstrate elsewhere.

(10) *Has to be considered by us in three ways.*] Not because it is itself triple, but because though it is one and the same it has three distinct aspects, which each have their own multiple applications and functions in the reorganized astronomy.

(11) *To start with disregard the eccentricity.*] That is, consider the eccentricity separately. For this eccentric orbit produces certain effects not by its own eccentricity but solely because it revolves round the Sun.

(12) *Two bodies.*] The heaven of the Moon that is to say (not the Moon as such), and the Earth have the same motion of translation from place to place on the Great Orbit. Consequently since the Earth is always in the same place as that in which the heaven of the Moon is also found, then the heaven of the Moon, and by its means the Moon itself, does not on account of its own heaven receive any such appearances of motion as the Sun does from the displacement of the Earth, though it is itself at rest. It would be different if the Earth moved and the heaven of the Moon were at rest, or moved from place to place with another distinct motion; for in that case the motion of the center of the Earth would also be transferred by the imagination to the Moon's heaven, and thus the whole of the Moon's heaven could, according to its location, seem to retrogress just as much as the five planets.

(13) *Copernicus makes the apogee of the Sun.*] Two points are hinted at here. One applies to the Sun itself, the other extends from the Sun to the planets. Ptolemy places the Sun on an eccentric, and encloses the eccentric between two deferents: Copernicus locates the planet on an epicycle, and the epicycle on a concentric circle. Consequently Ptolemy, in order to make the apogees move forward, attributes a special very slow motion to his deferents. Copernicus produces the same effect by a divergence of the period of revolution of the epicycle from that of the concentric, each being almost annual. Now it is more probable that those motions are slow on account of a divergence than on account of an additional motion, particularly since the epicycle possesses an annual motion only with respect to the eccentric, as the epicycle rolls in the opposite direction from the rotation of the eccentric; but with respect to the fixed stars it presents rather a kind of immobility, because in this rolling motion it comes about that the same parts of the epicycle are always turned towards the same regions of the fixed stars, unless an aberration disturbs them slightly. Now in my *Commentaries on Mars*, and in Book IV of my *Epitome of Astronomy*, I relate the physical cause both of the eccentricity, and of the shift of the apogees, a cause which resides in the bowels of the planet, and does not need either deferents or epicycles. But take this clause, referring to the Sun itself (or to the Earth) as an interpolation inserted merely in passing; so that the effect which extends from the shift of the Sun's apogee to the other planets is now made clear from it.

(14) *Computes the eccentricities...from the center of the Earth.*] These points are much more readily apparent from a glance at Plate V. To say what is going to happen many centuries later is indeed prophesying, though such a compunction about the present does not yet affect ancient astronomy. But the comparison between this extrapolation of Ptolemy's particular conclusions and the hypothesis of Copernicus was such that I could not omit mention of the point. For Copernicus also computed the eccentricities of the five planets as if from the center of the Great Orbit, as if that (and not the actual center of the Sun which is very near) were the true basis of the planetary system. During these twenty-five years since I first published this little book, I have so established astronomy that all the eccentricities (of the primary planets) are referred to the actual center of the Sun, as the true basis of the universe. Hence the eccentricities of all the planets can remain unchanged, wherever the Sun's apogee shifts to. See in my *Commentaries on Mars* the first part on the equivalence of hypotheses, especially Chapter 6.

(15) *A change in the Earth's eccentricity.*] These words were included in accordance with the advice of Copernicus himself. And it is true that anyone who moves the center of the Sun's orbit too far from the Earth (or the Earth's too far from the Sun), as I contend that Ptolemy and Hipparchus did, if he refers the eccentricities of the planets to that point must necessarily apportion different quantities to them than

aliasiis quantitates largiatur neceſſe eſt, quam qui hodie Solis Eccentricitatem emendatam habet. At ſi Eccentricitates computentur ab ipſo centro Solis, vt ego ſtio, tunc nihil illas attinet hac muta-tio Eccentricitatis Solis ſeu Terræ, ſeu vera illa ſit, vt credidit Copernicus, ſeu, vt ego, falſa & perſua-ſione nuda nixa. Inſpice ſuper hac re tabulam V. & narrationem Rhetici: vt & Martialium meo-rum caput vltimum.

(16 Orbiculi cœleſtis, Terræ globum ceu nucleum.] *Imaginationi huic anſam præbuit Copernicus: ſeu ſeruire voluerit captui, ſiue reuera & ipſe hæſerit in perplexitate rei, quæ ſchematibus planis ſublenari nequit, ſolidis poſſet quidem, ſed illa difficillime apparantur. Vtut ſe res habeat, motus iſte reuera motus non eſt, quies potius dicenda: nec melius vlla re poteſt repræſenta-ri, quam ipſiſſima ſua cauſſa phyſica, quæ ex Martialibus, & Epitomes Aſtr.lib.I. II. III. & VI. eſt iſta. Terræ globus dum animo motu circumfertur circa Solem, tenet interim axem conuolutionis ſuæ ſibiiſ ſemper parallelum in diuerſis ſitibus, propter fibrarum naturalem & magneticam in-clinationem ad quieſcendum: vel etiam propter continuatem diuinæ conuolutionis circa hunc a-xem, quæ illum tenet erectum, vt ſit in turbine incitato & diſcurſitante. Quare ſicut motus iſte re-uera non eſt, ſed quies potius, ſic etiam orbiculo commentitio nihil eſt opus: & iure hic me antiqua & erronea perſuaſionis de ſoliditate Orbium reum egit Tycho Braheus, qui lecto libello literas hac de cauſa ad me dedit.*

(17 Æquinoctiorum loca paulatim in præcedentia.] *Omnis doctrina præceſſio-nis æquinoctiorum, contemplatione axis & Polorum Telluris abſoluitur: vt nec Nona Sphæra, nec orbiculo illo circa terram ſit opus. Vide Comment. Martis partem V. Et Epit. Aſtr. lib. II. III. & VII.*

(18) Tertius motus eſt Polorum.] *Secundum motum in meram axis quietem rede-gimus, tertius iam ad ſecundum eſt reducendus, & cum eo in vnum conſtandus. Si enim cauſſarum phyſicarũ obſeruatione axis Telluris poſt vnam reuolutionem annuã inuenitur inſenſibili aliquo retror-ſum inclinatus à ſitu priſtino; & ſituetur nihilominus conſtantem inclinatione ad latera mundi, ſeu polos via regia; ſi tertio etiam Ecliptica, quippe Orbita Telluris, vt reliquorum Planetarum orbitæ, la-titudines ſuas habet à via regia, eaſque per ſimilem præuentionem tranſlocabiles de loco in locum ſub fixis: ex his obtentis ſequitur vltro ſine vlla Polorum libratione; & declinationem Ecliptiæ mutari, & æquinoctia nonnihil nunc incitari, nunc retardari; quin imo ſequitur hoc etiã amplius, quod Co-pernico inanimaduerſum, Tycho Braheus & Landgrauius Haſſiæ detexerunt, fixarum mutari la-titudines. Etſi vero libratio æquinoctiorum non tanta nec tam celer tunc elicitur, quanta ex libra-tionibus Copernici: at de illa quantitate non tantum nondum liquet, ſed conſtans æqualitas ante & poſt Ptolemæum deprehenſa, totum negotium, vna cum obſeruationibus Ptolemæi propemodum in dubium vocat. Sola enim ætas Ptolemæi eſt, quæ exorbitat: reliquarum ætatum obſeruationes con-gruunt ad æquabilem regulam; Copernicum enim, qui ſuæ ætatis aſſociatione librationem hanc enixus eſt, proximi ætate obſeruatores fide digniſſimi reſutant. Vide hac de re mea Commentaria de Marte Capitibus vltimis, & Epi-tom. Aſtron.libr. VII.*

if he knew the present-day corrected eccentricity of the Sun. But if the eccentricities are computed from the center of the Sun itself, as I compute them, then this change in the eccentricity of the Sun or the Earth does not affect them at all, whether it is genuine, as Copernicus believed, or as I believe, false and based on mere opinion. On this matter see Plate V and the *Narratio* of Rheticus, also the last chapter of my *Commentaries on Mars*.

(16) *The little heavenly circle which...enfolds the Earth's globe like a kernel.*] Copernicus provided a handle for this simile; whether he wished to accommodate our capacity for understanding, or whether he himself was truly caught in perplexity over the point, which could not be relieved by flat diagrams, though it could be by solid models, which are extremely difficult to supply. However that may be, the motion in question is not truly a motion, but should rather be spoken of as rest; and it cannot be better represented by anything than by its actual physical cause, which from the *Commentaries on Mars*, and the *Epitome of Astronomy*, Books I, II, III, and VI, is the following. While the Earth's globe travels round in its annual motion about the Sun, all the time it keeps its axis of revolution always parallel to itself in its various positions, on account of the natural and magnetic tendency in its inner parts towards staying at rest, or even on account of the continuity of the diurnal rotation about this axis, which holds it upright, as happens with a top which has been set in motion and is spinning. Consequently just as this is not truly a motion, but is rather rest, similarly there is no need of an imaginary little circle; and Tycho Brahe rightly accused me of this ancient and erroneous belief about the solidity of the spheres, and when he had read my little book wrote to me on this topic.[18]

(17) *The positions of the equinoxes...gradually westwards.*] The whole theory of the precession of the equinoxes is disposed of by consideration of the axis and poles of the Earth, and there is no need either of a ninth sphere or of the little circle round the Earth. See the *Commentaries on Mars*, Part V, and the *Epitome of Astronomy*, Books II, III, VII.

(18) *The third motion is that of the poles.*] We have reduced the second motion merely to the axis's staying at rest. The third must now be assimilated to the second, and combined with it. For if by the intervention of physical causes it is found that the Earth's axis after a single annual revolution is inclined by an insensible amount backwards from its original position; and if nevertheless it maintains a constant inclination towards the edges of the universe, or the poles of the Royal Way;[19] and thirdly, if the ecliptic (that is to say the orbit of the Earth), like the orbits of the other planets, takes its latitudes from the Royal Way, and they are by a similar forward motion capable of moving from one position to another as against the fixed stars; then from these premises it follows that of their own accord without any libration of the poles not only does the declination of the ecliptic alter, but also the equinoxes sometimes move faster, and sometimes more slowly to a certain extent. Furthermore, it follows even more strongly, as Copernicus failed to notice and Tycho Brahe and the Landgrave of Hesse revealed, that the latitudes of the fixed stars alter. Even though the libration of the equinoxes which then emerges is not as large or as rapid as from the librations of Copernicus, yet not only is there still no agreement on that quantity, but the fact that it has been discovered to be constant and regular before and after Ptolemy calls almost the whole affair along with Ptolemy's observations into doubt. For it is only Ptolemy's age which is out of keeping. The observations of other ages are consistent with a regular law. For Copernicus, who originated this libration by comparison with his own age, is refuted by thoroughly reliable observers who are very close to his age. See on this point my *Commentaries on Mars*, in the final chapters, and the *Epitome of Astronomy*, Book VII.

CAPVT II.

Primariæ demonstrationis delineatio.

QVIBVS ita præmissis, vt ad propositum veniam; atque modo recensitas Copernici hypotheses de mundo nouo, nouo argumento probem:rem à primo,quod aiunt,ouo, nouo qua breuitate fieri poterit,repetam.

Corpus erat id, quod initio Deus creauit;cuius definitionem si habeamus,existimo mediocriter clarum fore,cur initio corpus non aliam rem Deus creauerit. Dico quantitatem Deo fuisse propositam : ad quam obtinendam omnibus opus fuit, quæ ad corporis essentiam pertinent: vt ita quātitas corporis, quatenus corpus, quædam forma, Definitionisque origo sit. Quantitatem autem Deus ideo ante omnia existere voluit ; vt esset curui ad Rectum comparatio. Hac enim vna re diuinus mihi Cusanus, alijque videntur: quod Recti, Curuique ad inuicem habitudinem tanti fecerunt, & Curuum Deo,Rectum creaturis ausi sint comparare : vt haud multo vtiliore operam præstiterint,qui Creatorem creaturis, Deum homini,iudicia diuina humanis;quam qui curuum recto,circulum quadrato æquiparare conati sunt.

Cumque vel in hoc solo satis constitisset penes DEVM quantitatum aptitudo,& curui nobilitas: accessit tamen & alterum longe maius: Dei trinuni imago in Sphærica superficie, Patris scilicet in centro, Filij in superficie, Spiritus in æqualitate σχέσεως inter punctum & ambitum. Nam quæ Cusanus circulo, alij forte globo tribuerent:ea ego soli Sphæricæ superficiei arrogo. Nec persuaderi possum, Curuorum quicquam nobilius esse,aut perfectius ipsa Sphærica superficie. Globus enim plus est Sphærica superficie , & mixtus rectitudini, qua sola impletur intus. Circulus vero nisi in plano recto existat, hoc est, nisi Sphærica superficies,aut globus plano recto secetur ; circulus nullus erit. Vnde videre est, multas illic à Cubo in globum , hîc à quadrato in circulum secundario defluere proprietates,propter diametri rectitudinem.

Sed cur denique Curui & Recti discrimina , curuique nobilitas Deo fuerunt proposita in exornando mundo? Cur enim? nisi quia à Cōditore perfectissimo necesse omnino fuit,vt pulcherrimum opus constitueretur, *Fas enim nec est,nec vnquam fuit* (vt loquitur ex Timæo Platonis Cicero in libro de vniuersitate) *quicquam nisi pulcherrimum facere eum , qui esset optimus.* Cum igitur Idæam mundi Conditor animo præconceperit (loquimur humano more, vt homines intelligamus) atque Idæa sit rei prioris, sit vero, vt modo dictum est, rei optimæ, vt forma futuri operis & ipsa fiat optima; Patet quod his legibus quas Deus ipse sua bonitate sibi præscribit,nullius rei Idæam pro constituendo mundo suscipere potuerit,quam suæ ipsius essentiæ: quæ bifariam, quam præstans atqᵉ diuina sit considerari potest,primo in se,quatenus est vna in essentia,trina in personis,deinde collatione facta cum creaturis.

Hanc

CHAPTER II.
OUTLINES OF THE PRIMARY DERIVATION

After these preliminaries, to come to the point, and to demonstrate by new evidence Copernicus's hypotheses about a new universe, which have just been reviewed, I shall repeat the argument, as they say, from scratch, with as much brevity as possible.

It was matter which God created in the beginning; and if we know the definition of matter, I think it will be fairly clear why God created matter and not any other thing in the beginning. I say that what God intended was quantity. To achieve it he needed everything which pertains to the essence of matter; and quantity is a form of matter, in virtue of its being matter, and the source of its definition. Now God decided that quantity should exist before all other things so that there should be a means of comparing a curved with a straight line. For in this one respect Nicholas of Cusa and others seem to me divine, that they attached so much importance to the relationship between a straight and a curved line and dared to liken a curve to God, a straight line to his creatures;[1] and those who tried to compare the Creator to his creatures, God to Man, and divine judgments to human judgments did not perform much more valuable a service than those who tried to compare a curve with a straight line, a circle with a square.

And although under the power of God this alone would have been enough to constitute the appropriateness of quantities, and the nobility of a curve, yet to this was also added something else which is far greater: the image of God the Three in One in a spherical surface, that is of the Father in the center, the Son in the surface, and the Spirit in the regularity of the relationship between the point and the circumference. For what Nicholas of Cusa attributed to the circle, others as it happens have attributed to the globe; but I reserve it solely for a spherical surface. Nor can I be persuaded that any kind of curve is more noble than a spherical surface, or more perfect. For a globe is more than a spherical surface, and mingled with straightness, by which alone its interior is filled. Furthermore a circle exists only on a flat plane; that is, only if a spherical surface or a globe is cut by a flat plane, can a circle exist. Hence it may be seen that many properties are imparted both to the globe by the cube, and to the circle by the square, that is from an inferior source, on account of the straightness of the diameter.

But after all why were the distinctions between curved and straight, and the nobility of a curve, among God's intentions when he displayed the universe? Why indeed? Unless because by a most perfect Creator it was absolutely necessary that a most beautiful work should be produced. "For it neither is nor was right" (as Cicero in his book on the universe quotes from Plato's *Timaeus*) "that he who is the best should make anything except the most beautiful."[2] Since, then, the Creator conceived the Idea of the universe in his mind (we speak in human fashion, so that being men we may understand), and it is the Idea of that which is prior, indeed, as has just been said, of that which is best, so that the Form of the future creation may itself be the best: it is evident that by those laws which God himself in his goodness prescribes for himself, the only thing of which he could adopt the Idea for establishing the universe is his own essence, which can be considered as twofold, inasmuch as it is excellent and divine: first in itself, being one in essence but three in person, and secondly by comparison with created things.

Hanc imaginem, hanc Idæam mundo imprimere voluit, vt is fieret optimus atque pulcherrimus, vtque is eam fufcipere poffet; Quantum condidit, quantitatefque Sapientiffimus côditor excogitauit, quarum omnis, vt ita dicam, effentia in hæc duo difcrimina caderet, Rectû
& Curuum, ex quibus Curuum nobis duobus illis modo dictis modis
Deum repræfentaret ; Neque enim exiftimandum eft, temere extitiffe
tam apta præfigurando Deo difcrimina, vt Deus non de his ipfis cogita-
uerit, fed quantum corpus propter alias caufas, alioque confilio condiderit;atque poftea Recti & Curui comparatio,& hæc cum Deo fimilitudo,fuapte fponte, quafi fortuito extiterit.

Quin potius verifimile eft, initio omnium certo confilio Curuum
& Rectum à Deo electa, ad adumbrandam in mundo diuinitatem Conditoris;atque vt hæc exifterent, quantitates fuiffe, atque vt quantitas haberetur, conditum effe primo omnium Corpus.

Videamus modo, ecquomodo Creator Optimus has quantitates
in mundi fabrica adhibuerit : & quid verifimile fit noftris ratiocinationi-
bus à Conditore factum effe: vt illud poftea, cum in Antiquis, tum in nouis hypothefibus quæramus, eique palmam tribuamus, penes quem illud
reperietur.

Mundum igitur totum figura claudi fphærica, abunde fatis difpu-
tauit Ariftoteles, ductis inter cætera ex nobilitate fphæricæ fuperficiei
argumentis: quibus etiamnum vltima Copernici fixarum fphæra quamuis motu carens, eandem figuram tuetur, recipitque Solem tanquam cê-
trum in intimum finum. Orbes vero cæteros rotûdos effe circularis ftellarum motus arguit. Curuum igitur ad mundi ornatum adhibitum effe,
vlteriore probatione non eget. Cum autem tria quantitatum genera videamus in mundo, figuram, numerum & amplitudinem corporû : Curuum quidem adhuc in fola figura reperimus. Neque enim amplitudinis
vlla ratio ex eo eft, quod infcriptum fimili (fphæra fphæræ, circulus circulo)ex eodem Centro, aut vndiquaque tangit, aut nullibi : & Sphæricum
ipfum, cum folum & vnicum fit in fuo quantitatis genere; non poteft a-
lius numeri, quam ternarij fubiectum effe. Quod fi igitur folum Curuum
Deus in conditu refpexiffet, præter Solem in centro, qui patris fphæram
fixarum, vel aquas Mofaicas in ambitu, quæ filij ; auram cœleftem omnia
replentem, fiue extenfionem & firmamentum illud, quod Spiritus imago effet ; præter hæc, inquam, nihil exifteret in hoc ædificio mundano.
Nunc vero cum & fixæ fint innumerabiles, & mobilium non incertiffi-
mus catalogus, & cœlorum magnitudines inæquales inuicem ; neceffe
eft caufas eorum omnium ex rectitudine petamus. Nifi forte Deum pu-
tabimus quicquam in mundo temere feciffe, dum rationes optimæ fup-
peterent: id quod nemo mihi perfuadebit, vt vel de fixis fentiam: quarum
tamen fitus maxime omnium confufus, quafi fortuitus fementis iactus
nobis videtur.

Veniamus igitur ad Rectas quantitates. Sicut autem antea Sphærica fuperficies ideo affumpta eft, quia perfectiffima fuit quantitas: ita iam
vno faltu ad corpora tranfeamus, vt quæ ex Rectis perfectæ funt quantitates, & tribus dimenfionibus conftant : nam Idæam mundi perfectam
effe conuenit. (1) Lineas vero & fuperficies rectas, vt infinitas, & proin
ordi-

This pattern, this Idea, he wished to imprint on the universe, so that it should become as good and as fine as possible; and so that it might become capable of accepting this Idea, he created quantity; and the wisest of Creators devised quantities so that their whole essence, so to speak, depended on these two characteristics, straightness and curvedness, of which curvedness was to represent God for us in the two aspects which have just been stated. For it must not be supposed that these characteristics which are so appropriate for the portrayal of God came into existence randomly, or that God did not have precisely that in mind but created quantity in matter for different reasons and with a different intention, and that the contrast between straight and curved, and the resemblance to God, came into existence subsequently of their own accord, as if by accident.

It is more probable that at the beginning of all things it was with a definite intention that the straight and the curved were chosen by God to delineate the divinity of the Creator in the universe;[3] and that it was in order that those should come into being that quantities existed, and that it was in order that quantity should have its place that first of all matter was created.

Now let us see in what way the best of Creators used these quantities in the structure of the universe; and what is likely, by our reckoning, to have been made by the Creator;[4] so that thereafter we may search for it, both in the ancient and in the new hypotheses, and award the palm to the one within which it is found.

That the whole universe is enclosed by a spherical shape has been thoroughly well argued by Aristotle,[5] drawing arguments among others from the nobility of a spherical surface; and by these arguments even now Copernicus's outermost sphere, that of the fixed stars, although it is without motion, preserves the same shape, and takes the Sun, as its center, into its innermost recess. On the other hand the circular motion of the stars is evidence that the other orbits are round. Yet there is no lack of further proof that curvature was used in the pattern of the universe. Although we see three kinds of quantity in the universe, the shape, number, and extension of objects, so far we find the curved only in shape. For there is no measure of extension, from the fact that like is inscribed within like (sphere within sphere, circle within circle) about the same center, or touches it at all points, or at none; and the spherical itself, since it is alone and unique in its own kind of quantity, cannot be subject to any other number but three. But yet if at the Creation God had taken cognizance only of the curved, except for the Sun in the center, which was the image of the Father, the Sphere of the Fixed Stars, or the Mosaic waters,[6] at the circumference, which was the image of the Son, and the heavenly air which fills all parts, or the space and firmament, which was the image of the Spirit—then, except for these, I say, nothing would exist in this cosmic structure. But in fact as there are innumerable fixed stars, and the well established tally of planets, and the irregular sizes of the heavens, we must of necessity seek the causes of them all in straightness, unless perhaps we suppose that God has made anything in the universe at random, even though excellent reasons were available. Of that nobody will persuade me; and that is my opinion even on the fixed stars, although their position is the most disordered of all, and looks to us like seed scattered indiscriminately.

Let us come then to straight quantities. However in the same way as previously a spherical surface was assumed, because it was the most perfect quantity, let us now pass at a bound to solid bodies, as the quantities which are perfectly formed from the straight, and are made up of three dimensions; for it is accepted that the Idea of the universe is perfect. Nevertheless, let us reject (1) straight lines and

ordinis minime capaces, è mundo finito, ordinatiſſimo, pulcherrimo eijciamus. Rurſum ex corporibus, quorum infinities infinita ſunt genera,
ſeligamus aliqua cēſu habito per certas notas : puta, quæ aut latera aut
angulos, aut plana, ſingula vel alterna, vel quouis conſtanti modo mixta
habeant inuicem æqualia : vt ita bona cum ratione ad finitum aliquid
veniatur. Quod ſi quod genus corporum per certas conditiones deſcri-
ptum, intra ſpecies quidem numero finitas conſiſtit ; ſed tamen in ingen-
tem numerorum copiam multiplicatur: eorum corporū angulos & centra planorum (2) pro fixarum multitudine, magnitudine, ſituque de-
monſtrando, ſi poſſumus, adhibeamus : ſin autem is labor non eſt hominis, ergo tantiſper differamus numeri, ac ſitus earum rationem quærere;
dum quis nobis ad vnum omnes, quot quantæue ſint, deſcripſerit. Miſſis
igitur fixis, atq; ei permiſſis, qui ſolus numerat multitudinem ſtellarum,
& ſingulas nomine vocat, (Pſ. 147.) ſapientiſſimo Artifici ; nos oculos ad
propinquas, paucas & mobiles conuertamus.

    Denique igitur delectum corporum ſi habuerimus, atque omnem
mixtorum turbam eiecerimus, retineamus vero ſola illa, quorum omnia
plana & æquilatera, & æquiangula fuerint; reſtabunt nobis hæc quinque
Corpora Regularia, quibus Græci hæc aſcripſère nomina, Cubus ſeu
Hexaedrum, Pyramis ſeu Tetraedrum, Dodecaedrum, Icoſaedrum, O-
ctaedrum. Quodque his quinque plura eſſe non poſſint, vide Euclid. lib.
13. poſt prop. 18. ſcholion.

    Quare ſicut horum definitus & exiguus admodum eſt numerus, cæ-
terorum aut innumerabiles, aut infinitæ ſpecies, ita decuit in mūdo duo
eſſe ſtellarum genera, euidenti diſcrimine ab ſe inuicem diſtincta (cuiuſmodi motus & quies eſt) quorum vnum genus infinito ſimile, vt fixæ, alterum anguſtum vt Planetæ. Non eſt huius loci diſputare de cauſis, cur
hæc moueantur, illa non. Sed poſito, quod Planetæ motu indigue-
rint, ſequitur, (3) vt hunc obtinerent, rotundos orbes accipere de-
buiſſe.

    Habemus orbem propter motum, (4) & corpora propter nu-
merum & magnitudines; quid reſtat amplius, quin dicamus cum Platone, θεὸν ἀεὶ γεωμετρεῖν, atq; in hac mobilium fabrica corpora orbibus, & orbes corporibus inſcripſiſſe tantiſper, dum nullum amplius corpus reſta-
ret, quod non intra & extra mobilibus orbibus veſtitum eſſet. Nā ex 13.
14. 15. 16. 17. lib. 13. Euclidis videre eſt: quā hæc corpora natura ſua ſint apta
ad hanc inſcriptionem & circumſcriptionem. Quare ſi quinq; corpora
mediantibus & claudentibus orbibus, inferantur ſibi mutuo: habebimus
numerum ſex orbium.

    Quod ſi aliqua mundi ætas hoc pacto de mundi diſpoſitione diſpu-
tauit, vt ſex orbes poneret mobiles circa Solem immobilem ; illa vtique
veram Aſtronomiam tradidit. *Atqui eiuſmodi ſex orbes habet Copernicus, eoſ-*
Propoſ. *que binos in eiuſmodi ad inuicem, proportione: vt hæc quinque corpora omnia aptiſ-*
ſime interlocari poſſint : quæ ſumma erit eorum quæ ſequuntur. Quare tantiſper
audiendus eſt, dum quis aut aptiores ad hæc Philoſophemata protulerit
hypotheſes; aut docuerit, fortuito in numeros atque in mentem hominis
irrepere poſſe, quod optima ratione ex ipſis naturæ principijs deductum
eſt. Nam quid admirabilius, quid ad perſuadendum accommodatius dici aut

surfaces, as they are infinite, and consequently scarcely admit of order, from this complete, thoroughly ordered, and most splendid universe. On the other hand let us select from among solid bodies, the varieties of which are infinitely infinite, by picking out a few in accordance with particular distinguishing features — such as those which have edges, or vertices, or faces, singly or in pairs, or combined in some regular way, respectively equal, so that in this way we may arrive at some finite number on a logical basis. But if a class of bodies defined by definite characteristics, though it falls among those species which are finite in number, nevertheless proliferates to a vast number, let us use the angles and centers of the plane faces of those bodies, if we can, (2) to derive the myriad number, the size, and the position of the fixed stars. Yet if that task is not humanly possible, then let us put off seeking the logic of their number and position until such time as someone has given us an account of the number and size of every last one of them. So let us pass over the fixed stars, leave them to that most wise Craftsman who alone numbers the multitude of stars and calls them each by name (Psalm 147),[7] and turn our eyes to those which are near, few, and moving.

So if in the end we make a selection of bodies, and reject the whole crowd of hybrids, but retain only those which have all their faces equilateral and equiangular, we shall be left with those five regular solids to which the Greeks allotted the following names: cube or hexahedron, pyramid or tetrahedron, dodecahedron, icosahedron, octahedron. There cannot be more than these five — see Euclid, Book XIII, scholium after Proposition 18.

*

**

Therefore, just as the number of the latter is limited and very small, while the species of the rest are either innumerable or infinite, so it was proper that there should be in the universe two kinds of stars, distinguished from each other by an obvious criterion (such as motion and rest), of which one kind is apparently infinite,[8] like the fixed stars, and the other is restricted, like the planets. This is not the place to discuss the reasons why the latter move and the former do not. But if it is assumed that the planets required motion, it follows that (3) they had to receive round orbits in order to acquire it.

We know the orbit by the motion, (4) and the solid bodies by their number and sizes. What else remains except to say with Plato, "God is always a geometer," and in this structure of moving stars he has inscribed solids within spheres, and spheres within solids, until no further solid was left which was not robed outside and inside with moving spheres. For by Propositions 14, 15, 16, and 17 of Euclid's thirteenth Book, it is clear how these solids are suited by their nature for this inscription and circumscription. So if the five solids are fitted one inside another, with spheres between them and inclosing them, we shall have a total of six spheres.

Now if any age of the universe has discussed the arrangement of the universe on the basis of the assumption that there are six spheres moving round a motionless Sun, it has undoubtedly given a true account of astronomy. *But Copernicus has six spheres of that sort, and each pair of spheres in such proportion to each other that all these five solids can very readily be fitted in between them, which is the essence of what follows.* So we must concur with him, until someone has either put forward hypotheses which give a better solution to these problems, or asserted that a system which has been deduced by excellent reasoning from the very principles of Nature can creep accidentally into the numbers and into the mind of Man. For what could be said or imagined which would be more remarkable, or more convincing, than that what Copernicus established by observation, from the effects, *a posteriori*, by a lucky rather than a confident guess,

Proposition

dici aut fingi poteft;quam,quod ea quæ Copernicus ex φαινομένοις,ex effectibus,ex pofteriorib. quafi cœcus baculo greffum firmãs)vt ipfe Rhetico dicere folitus eft(felici magis quam confidenti coniectura conftitu-
it,atque ita fefe habere credidit)ea inquam omnia rationibus à priori, à
caufis, à Creationis idæa deductis rectiffime conftituta effe deprehendantur.

Nam fi quis philofophicas iftas rationes,fine rationibus, & folo ri-
fu excipere atque eludere voluerit: propterea,quod nouus homo fub finem feculorum,tacentibus illis philofophiæ luminibus antiquis, philo-
fophica ifta proferam;illi ego ducem, auctorem & præmonftratorem ex
antiquiffimo feculo proferam *Pythagoram*:cuius multa in fcholis mentio,
quod cum præftantiam videret quinque Corporum,fimili plane ratione ante bis mille annos,qua nunc ego, Creatoris cura non indignũ cen-
fuerit ad illa refpicere;atque rebus mathematicis phyfice, & ex fua qualibet proprietate accidentaria cenfitis , res non mathematicas accom-
modauerit. (5) Terrã enim Cubo æquiparauit, quia ftabilis vterque,
quod tamen de cubo non proprie dicitur. Cœlo Icofaedrum dedit,quia
vtrumque volubile: Igni Pyramida,quia hæc volantis igniculi forma;re-
liqua duo corpora inter aerem & aquam diftribuit , propter fimilem v-
trinque cum vicinis cognationem.Sed enim Copernicus illi viro defu-
it,qui prius,quid effet in mundo, diceret: abfque eo non fuiffet, dubium
non eft,quin quare effet,inueniffet, atque hæc cœlorum proportio tam
nota nunc effet, quam ipfa quinque corpora;tam item recepta, quam
hoc temporum decurfu inualuit illa de Solis motu, deque quiete Telluris opinio.

Verum age vel tandem experiamur, vtrum inter orbes Copernici
fint iftæ corporum proportiones. Ac initio rem craffiufcule cenfeamus.
Maxima diftantiarum differentia in Copernico eft inter Iouem & Martem: Vt vides in explicatione hypothefium Tab. 1. & infra cap.14. & 15.
Martis enim diftantia à Sole non æquat tertiam partem Iouiæ. Quæra-
tur igitur corpus,quod maximam facit differentiam inter orbem circũ-
fcriptum & infcriptum (6) (concedatur nobis hæc καταχρησις cauum
pro folido cenfendi)quod eft Tetraedrum fiue Pyramis. Eft igitur inter
Iouem & Martem Pyramis.Poft hos maximam faciunt differentiam diftantiæ Iupiter & Saturnus.Huius enim ille paulo plus dimidium æquat.
Similis apparet in cubi intimo & extimo orbe differentia. Cubum igitur
Saturnus ambit,cubus Iouem.

Æqualis fere proportio eft inter Venerem & Mercurium, nec ab-
fimilis inter orbes Octaedri. Venus igitur hoc corpus ambit, Mercurius
induit.

Reliquæ duæ proportiones inter Venerem & Terram, inter hanc
& Martem minimæ funt, & fere æquales,nempe interior exterioris do-
drans aut bes. In Icofaedro & Dodecaedro funt etiam æquales diftantiæ
binorum orbium: Et proportione vtuntur minima inter reliqua regularia corpora. Quare verifimile eft, Martem ambire terrã mediante alter-
utro horum corporum:Terram autem à Venere fummotam, mediante
reliquo.Quare fi quis ex me quærat,cur fint tãtum fex orbes mobiles, re-
fpondebo,quia non oporteat plures quinque proportiones effe, totidem

D   nem-

like a blind man, leaning on a stick as he walks (as Rheticus himself used to say) and believed to be the case, all that, I say, is discovered to have been quite correctly established by reasoning derived *a priori*,[9] from the causes, from the idea of the Creation?

For if anyone who listens to this philosophical reasoning wants to evade it without reasoning and merely with a laugh, because I am putting forward this piece of philosophy almost at the end of the ages as a newcomer, though the ancient luminaries of philosophy say nothing of it, then I will offer him as a guide, authority, and demonstrator from the earliest age Pythagoras, who is much spoken of in the lecture rooms because seeing the pre-eminence of the five solids,[10] by plainly similar reasoning, two thousand years before I now do so, he judged it not unworthy of the Creator's concern to take account of them, and he made things which were not mathematical fit mathematical things physically, and by classifying them according to some accidental property of their own. (5) For he compared the Earth with the cube, because each is stable, although that is not an essential property of the cube. To the heaven he gave the icosahedron,[11] because they both rotate; to fire the pyramid, because that is the shape of a rising flame. The other two bodies he divided between air and water, as each has a similar affinity with its neighbors on either side. Yet Pythagoras did not have Copernicus to state beforehand what there was in the universe. If he had not been without him, there is no doubt that he would have discovered the reason why it was; and this proportion in the heavens would now be as well known as the five solids themselves, and also accepted to the same extent as during this lapse of time the belief in the motion of the Sun and the immobility of the Earth has weakened.

But come, let us at length test whether the proportions of the five solids are found between the spheres of Copernicus; and to start with let us assess them rather roughly. The greatest difference of the distances in Copernicus is between Jupiter and Mars, as you may see in the explanation of the hypotheses, Plate I, and below in Chapters XIV and XV. For the distance of Mars from the Sun is not as much as a third of that of Jupiter. A solid is therefore required which makes the difference between the circumscribed and inscribed sphere a maximum (6) (if we may be allowed the solecism of counting it as hollow instead of solid), and that is the tetrahedron or pyramid. Therefore there is a pyramid between Jupiter and Mars. After them Jupiter and Saturn yield the greatest difference in distance. For the former amounts to a little more than half the latter. A similar difference is found in the interior and exterior sphere of a cube. Therefore Saturn goes round a cube, and a cube round Jupiter.[12]

The proportion between Venus and Mercury is almost equal, and that between the spheres of an octahedron is not greatly different. Therefore Venus goes round outside that body, Mercury inside.

The remaining two proportions, between Venus and the Earth, and between the latter and Mars, are very small, and almost equal, that is, the interior sphere is three quarters or two thirds of the exterior. In the icosahedron and the dodecahedron also the radii of the two spheres are equal, and stand in the smallest proportion to each other, compared with the rest of the regular solids. Therefore it is probable that Mars goes round the Earth with one or other of these solids in between; whereas the Earth is separated from Venus by the interposition of the other. Therefore if I am asked why there are only six moving spheres, I shall answer that it is because there ought not to be more than five proportions,

nempe,quot regularia funt in mathefi corpora. Sex autem termini con-
fummant hunc proportionum numerum.

Huc pertinet Tabula tertia.

Annotatio in Caput fecundum,antiqua.

☞ *fol.præced.*   *Quodque his quinque)* Corporum nobilitas eft ex fimplicitate, & ex
æqualitate diftantiæ planorum à centro figuræ. Sicut enim norma & regula creaturarum Deus eft;fic Sphæra corporum. Atqui ea habet dictas
proprietates.1.Eft fimpliciffima,quia vno clauditur termino,feipfa fcilic.
2. Omnia eius puncta æqualiffime à centro diftant. Ex corporibus igitur
proxime accedunt regularia ad Sphæræ perfectionem. Eorum definitio
hæc eft, vt habeant, 1.omnia latera, 2. plana , & 3.angulos,fingula æquales & fpecie & magnitudine, quod eft fimplicitatis ; quam pofitam definitionem fequitur illud vitro,quod 4.omnium planorum centra æqualiter à medio diftent, 5. quod infcripta globo omnibus angulis tangant fu-
perficiem, 6. quod in ea hæreant, 7. quod infcriptum globum omnibus
planorum centris tangant,8.quod proinde infcriptus globus hæreat im-
motus,9.& quod idem centrum habeat cum figura. Quibus rebus effi-
citur altera fimilitudo cum Sphæra, quæ eft ex æqualitate diftantiæ planorum.

•   (7) Scholion autem illud ita fonat: Aio vero.præter dictas quinq;
Supr.ibid. figuras non poffe aliam conftitui figuram folidam, quæ planis & æquilateris & æquiangulis contineatur,inter fe æqualibus. Non enim ex duobus triangulis , fed neque ex aliis duabus figuris folidus conftituetur angulus.

    Sed ex tribus triangulis,conftat Pyramidis angulus.

    Ex quatuor autem, Octaedri.

    Ex quinque vero,Icofaedri.

    Nam ex triangulis fex & æquilateris , & æquiangulis ad idem pun-
ctum coeuntibus,non fiet angulus folidus. Cum enim trianguli æquilateri angulus, recti vnius beffem contineat, erunt eiufmodi fex anguli rectis quatuor æquales. Quod fieri non poteft. Nam folidus omnis angulus,minoribus quam rectis quatuor angulis continetur,per 21.11.

    Ob eafdem fane caufas,neque ex pluribus quam planis fex eiufmo-
di angulis folidus conftat.

    Sed ex tribus quadratis Cubi angulus continetur.

    Ex quatuor nullus poteft.Rurfus enim recti quatuor erunt.

    Ex tribus autem pentagonis æquilateris, & æquiangulis Dodeca-
edri angulus continetur. Sed ex quatuor nullus poteft. Cum enim Pen-
tagoni æquilateri angulus rectus fit , & quinta recti pars, erunt quatuor
anguli rectis quatuor maiores. Quod fieri nequit. Nec fane ex alijs po-
lygonis figuris folidus angulus continebitur, quod hinc quoque abfurdū
fequatur.Quamobrem perfpicuum eft,præter dictas quinque figuras a-
liam figuram folidam non poffe conftitui , quæ fub planis æquilateris &
æquiangulis contineatur.

Planum

that is the same number as there are regular solids in mathematics. But six boundaries make up this number of proportions.

Here belongs Plate III.[13]

Original note to Chapter II.[14]

There cannot be] The nobility of solids depends on their simplicity, and on the equality of the distance of the faces from the center of the figure. For just as God is the model and rule for living creatures, so the sphere is for solids. Now the sphere has the following properties: 1. It is extremely simple, because it is enclosed by a single boundary, namely itself. 2. All its points are at a precisely equal distance from the center. Therefore among bodies the regular solids approach most closely to the perfection of the sphere. Their definition is that they have: 1. all their edges, 2. their faces, and 3. their vertices respectively equal both in kind and in size, which is a sign of simplicity. From the adoption of this definition it. follows automatically that 4. the centers of all the faces are equally distant from the midpoint, 5. that if they are inscribed in a globe all their vertices touch its surface, 6. that they are fixed within it, 7. that they touch the inscribed globe at all the centers of their faces, 8. that consequently the inscribed globe is fixed and immobile, 9. and that it has the same center as the solid. These properties yield another resemblance to the sphere, which results from the equality of the distances between the faces.

(7) Now the scholium expresses it in this way: I say that apart from the five stated figures no other solid figure can be formed which is bounded by equilateral and equiangular faces which are equal to each other. For a solid angle cannot be formed from two triangles, nor from any other two figures.

But from three triangles the vertex of a pyramid is formed.

From four, that of an octahedron.

From five, that of an icosahedron.

Now from six equilateral and equiangular triangles meeting at the same point, a solid angle cannot be produced. For since the vertex of an equilateral triangle contains two thirds of a right angle, six angles of that type will be equal to four right angles. Which is impossible. For every solid angle is bounded by four angles less than right angles, by Euclid Book XI, Theorem 21.

For the same reason, neither can a solid angle be formed from more than six plane angles of that type.

But the vertex of a cube is bounded by three squares.

From four such, no solid angle can be formed, for again there will be four right angles.

However, the vertex of a dodecahedron is bounded by three equilateral and equiangular pentagons. But from four such, no solid angle can be formed. For since the angle of an equiangular pentagon is one and one-fifth right angles, four such angles will be greater than four right angles. Which cannot be the case. Nor, plainly, can a solid angle be bounded by other polygons, because from that also an absurdity would follow. Consequently it is evident that apart from the five stated figures no other solid figure can be formed which is bounded by equilateral and equiangular faces.

* page 97

** Above, at the same place.

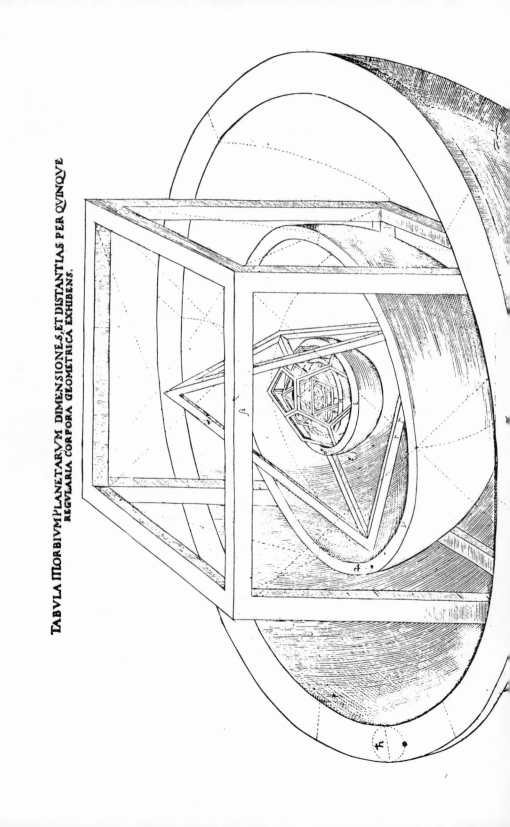

TABVLA IIIORBIVM PLANETARVM DIMENSIONES, ET DISTANTIAS PER QVINQVE
REGVLARIA CORPORA GEOMETRICA EXHIBENS.

See p. 228.

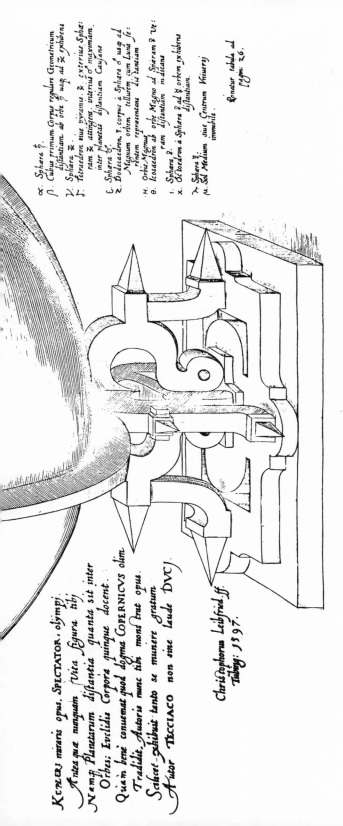

Kepleri miraris opus, Spectator olympi,
Antea quæ nunquam Visa figura tibi.
Namq Planetarum distantia quanta sit inter
Orbes; Euclidis Corpora quinque docent.
Quam bene conuenat quod dogma Copernicus olim.
Tradidit, Autoris nunc tibi mons erat opus.
Scilicet exhibuit tanto se munere gratum
Autor Tecciaco non sine laude Duci.

Christophorus Leibfrid ff.
Tubing: 1597.

α. Sphæra ♄.
β. Cubus primum Corpus regulare Geometricum distantiam ab orbe ♄ usq ad ♃ exhibens.
γ. Sphæra ♃.
δ. Tetraedron siue pyramis ♃. exterius Sphæram ♃ attingens, interius ♂ maximam inter planetas distantiam Causans.
ε. Sphæra ♂.
ζ. Dodecaedron, ♂ corpus à Sphæra ♂ usq ad Magnum orbem tellurem cum Luna feręntem repræsentans distantiam.
H. Orbis Magnus.
θ. Icosaedron ab orbe Magno ad Sphæram ♀ ueram distantiam indicans.
ι. Sphæra ♀.
χ. Octaedron à Sphæra ♀ ad ☿ orbem exhibens distantiam.
λ. Sphæra ☿.
M. Sol Medium siue Centrum Vniuersi immobile.

Renatur tabula ad pagm: 26.

		Planum		Plana	Latera	Angul.		Orbem inscriptū.
Cubus		quadrangulum		6	12	8		mediocrem.
Octaedron	habet	triangulum		8	12	6		cubo æqualem.
Dodecaed.	habet	quinquangulum		12	30	20		maximum.
Icosaedron		triangulum		20	30	12		dodec. æqualem.
Tetraed.			triangulum		4	6	4	minimum.

IN CAPVT SECVNDVM
Notæ Auctoris.

(1) Lineas vero & superficies.] *O male factum.E mundone eijciamus? Imo postliminio re-uocaui in Harmonicis.Cur autem eijciamus? An quia infinitæ,& proin ordinis minime capaces? Atqui non ipsæ,sed mea illius temporis inscitia,communis mihi cum plerisque , ordinis illarum minime capax erat.Itaque lib.I.Harmonicorum,& delectum aliquem inter infinitas docui,& ordinem in ijs pulcherrimum in lucem protuli.Nam cur lineas nos ex archetypo mundi eliminemus; cum lineas Deus opere ipso expresserit,motus sc.Planetarum?Lingua igitur corrigenda,mens tenenda. In corporum numero,sphærarum amplitudine constituenda primitus,eliminentur sane lineæ: at in motibus,qui lineis perficiuntur, exornandis,ne contemnamus lineas & superficies, quæ solæ proportionum Harmonicarum sunt origo.*

(2) Pro fixarum multitudine.] *Ingens discrimen argumento nominum , est inter fixa & mobilia; cur non sit aliquod etiam in vtriusque generis exornatione? Quis Ordinis pulchritudinem intelligeret,si non iuxta cerneret fixarum exercitum ordinis expertem? Quis Astronomiam disceret,si perpetua esset similitudo schematismorum, seu constellationum?Est suus formis ornatus,est & Materiæ.Sit igitur propria , materia & pulchra exornatio, quæ facta est per infinitam & molem & multitudinem, & varietatem,tam situ, quam magnitudinis claritatisque.*

(3) Vt hunc,motum,obtinerent, rotundos orbes accipere debuisse.] *Non illos solidos,male hic sum intellectus à Tychone Braheo, sed spacia, prorsum quidem circularia, vt reuolutiones syderum in seipsas redire & perpetuæ esse possent;versus polos vero itidē circularia,id est superficies sphæricas,propter motus latitudinum;non quod polis opus habuerint,à quibus,vt sphæra materialis,affigerentur.*

(4) Et corpora propter numerum.] *Corpora intellige Geometrica regularia solida quinque;hæc vt archetypum,Orbes vero,vt opus exstruendum.*

(5) Terram enim Cubo æquiparauit.] *Vide lib.I.Harmon.in proœm.fol.4.& lib. II.prop.XXV.& lib.V.cap.I.Et Epit.Astr.lib.IV.fol.456.*

(6) Concedatur nobis hæc.] *Vere quidem aut sphæricum etiam inter solida censendum,quod globum dicimus;aut hæc corpora solida dici non merentur?nec erant à soliditate, hoc est à perfectione trium dimensionum argumenta texenda,pro Orbium exornatione per ea.Nam & ipsi orbes(seu spacia) caui sunt, & figuræ hæ ob id nobiles sunt , quia sphærici perfectionem omnimoda conclusione spacij, quod amplexæ sunt , quam proxime æmulantur. Soliditas vero tam in globo,quam in his figuris,est genuina materiæ idea vt superficies formæ.*

(7) Scholion autem illud.] *Hoc est dimidium libri mei II. Harmon.de Congruentia planorum in solido.*

D 2 CA-

Chapter II

	Type of face	No. of faces	No. of edges	No. of vertices	Inscribed sphere
Cube	Quadrilateral	6	12	8	Medium
Octahedron	Triangle	8	12	6	Equal to cube
Dodecahedron	Pentagon	12	30	20	Largest
Icosahedron	Triangle	20	30	12	Equal to dodeca-hedron
Tetrahedron	Triangle	4	6	4	Smallest

has (bracket linking Type of face column)

AUTHOR'S NOTES ON CHAPTER TWO

(1) *Straight lines and surfaces.*] What a mistake! Are we to reject them from the universe? Instead, I reinstated them, as was their right, in the *Harmonice*. But why are we to reject them? Is it because they are infinite, and so scarcely admit of order? But it was not they themselves which scarcely admitted the possibility of their being ordered, but my own ignorance at that time, held in common with most people. So in Book I of the *Harmonice* I not only explained a certain principle of selection among these infinities, but also brought to light a most splendid order in them. For why should we eliminate lines from the archetype of the universe, seeing that God represented lines in his own work, that is, the motions of the planets? Therefore my language should be corrected, my intention retained. In establishing the number of the bodies, and the size of the spheres, lines should indeed be eliminated in the first place; but in displaying the motions, which are accomplished by lines, let us not despise lines and surfaces, which alone are the origin of the harmonic proportions.

(2) *To derive the myriad number...of the fixed stars.*] There is a vast difference as far as names are concerned between the fixed and moving stars: why should there not also be some difference in the displaying of the two kinds? Who would understand the splendor of order, if he could not perceive alongside it the host of the fixed stars without order? Who would learn astronomy, if there was an endless similarity between the patterns or constellations? Their display is provided by their forms, and by their matter. Then let them be displayed in a way which is appropriate for their matter, and splendid, as it is the result of both their infinite bulk and their infinite number, and of the variety in both their position and their size and brightness.

(3) *They had to receive round orbits, in order to acquire (this motion).*] Not solid spheres (I was misunderstood here by Tycho Brahe), but spaces, which are indeed completely circular, so that the revolutions of the stars could return to the same positions and be perpetual. In the direction of the poles they are likewise circular, that is, their surfaces are spherical, on account of the motions in latitude, not because they needed poles to be fixed to, like a material sphere.

(4) *And the solid bodies by their number.*] Understand this to mean the five regular geometrical solids: they are, so to speak, the archetype, and the spheres the work to be constructed.

(5) *For he compared the Earth with the cube.*] See Book I of the *Harmonice*, page 4 of the preface, Book II, Proposition 25, and Book V, Chapter 1; and the *Epitome of Astronomy*, Book IV, page 456.

(6) *If we may be allowed.*] In fact, either the spherical shape, which we call the globe, should be counted among the solids, or else these bodies do not deserve to be called solid, and arguments that the spheres were displayed in accordance with them should not have been drawn from their solidity, that is from the perfection of their three dimensions. For the spheres (or spaces) themselves are hollow, and also the reason for the nobility of these figures is that they emulate as closely as possible the perfection of the spherical by their complete enclosure of the space which they have surrounded. Solidity indeed both in the globe and in these figures is the true Idea of their matter, as surface is of their form.

(7) *Now the commentary.*] That is the half of Book II of my *Harmonice*, which is about the concurrence of planes in a solid figure.

CAPVT III.

Quod hæc quinque corpora in duos ordines distinguantur; & quod terra recte locata sit.

ORRo autem fortuitum hoc videri posset, atque à nulla fluens causa, quod sex orbes Copernici recipiant intra suas vnius ab alio distantias hæc quinq; corpora, nisi is ipse ordo esset inter illa, quo ordine ego singula interlocaui. Nam si Saturnus Ioui, tam propinquus esset, quã est Venus Telluri, vicissim si hæ duæ ab inuicem tanto interuallo distinguerentur in Copernico, quanto distinguuntur Iupiter & Mars: alio ordine vtendum fuisset in inserendis corporibus. Foret enim inter duos primos orbes primo loco Dodecaedron vel Icosaedron, quarto vero loco Tetraedrum. Qui ordo cum non possit admitti rationibus Mathematicis, facile foret futilitatem concepti Theorematis patefacere. Nunc autem videamus nos, ecquibus rationibus probetur, debuisse hoc ipso ordine disponi corpora inter orbes. Initio distinguuntur hæc corpora in tria primaria, Cubum, Tetraedrum, Dodecaedrum, & duo secundaria, Octaedrum & Icosaedrum. Quodque verissimum hoc sit discrimen, nota vtriusque generis proprietates. 1. Primaria plano inter se differunt: secundaria vtuntur eodem triãgulari. 2. primariorum quodlibet proprium habet planum: cubus quadratum, Pyramis triangulum, Dodecaedron quinquangulum. secundaria planum triangulum à Pyramide mutuantur. 3. primaria omnia simplici vtuntur angulo, nempe tribus planis comprehenso: secundaria quatuor aut quinque planis in vnũ solidum adscicunt. 4. Primaria nemini suam debent orig nem & proprietates: secundaria pleraque ex primariis, facta commutatione, adepta sunt, & quasi genita ex illis. 5. Primaria non moueri cõcinne possunt, nisi acta diametro per centra vnius aut oppositorum planorum: secundaria vero acta per oppositos angulos diametro. 6. Primariorum est proprium stare: secundariorum pendere. Siue enim hæc in basin prouoluas, siue illa in angulum erigas: visus vtrinque deformitatem aspectus refugiet. 7. Adde denique quod primaria perfecto numero tria sunt: secundaria imperfecto duo; quodque illa omnes anguli species habent, Cubus rectum, Pyramis acutum, Dodecaedrum obtusum; hæc vero ambo in obtusi solius genere versantur. Et Octaedri quidẽ angulus per omnes tres species vagatur, in iunctura laterum obtusus; inter coeuntia duo latera ex opposito, rectus; ipse vero solidus, acutus. Cum igitur manifestũ esset discrimen inter corpora, conuenientius fieri nihil potuit, quam vt Tellus nostra, totius mundi summa & compendium, atque adeo dignissima stellarum mobilium, orbe suo inter dictos ordines distingueret, locumque eum sortiretur, quem ipsi superius attribuimus.

CA-

CHAPTER III.
THAT THESE FIVE SOLIDS ARE CLASSIFIED INTO TWO TYPES; AND THAT THE EARTH HAS BEEN CORRECTLY LOCATED

Now, it might seem fortuitous, and not the result of any cause, that the six spheres of Copernicus accept these five solids into the spaces between them, if their actual pattern were not the pattern in which I have placed them. For if Saturn were as close to Jupiter as Venus is to Earth, or on the other hand if the last two were separated from each other in Copernicus by a gap of the same size as are Jupiter and Mars, it would have been necessary to use a different pattern for interpolating the solids. For between the first two spheres in the first position would be a dodecahedron or icosahedron, but in the fourth position a tetrahedron. Since this pattern could not be acceptable to mathematical reasoning, the futility of the theorem which I have adopted would easily be exposed. As things are, however, let us see the reasoning by which it is confirmed that the solids had to be arranged in this precise pattern among the spheres. To start with, these solids are classified into three primaries, the cube, tetrahedron and dodecahedron, and two secondaries, the octahedron and icosahedron. For the correctness of this distinction, note the properties of each class. 1. The primaries differ from each other in shape of face; the secondaries both have triangular faces. 2. Every one of the primaries has its particular type of face: the cube has the square, the pyramid the triangle, the dodecahedron the pentagon; the secondaries borrow the triangular face from the pyramid. 3. All the primaries have a simple vertex,[1] that is, one which is included between three faces; the secondaries combine four or five faces in one solid angle. 4. The primaries owe their origin and properties to no one; the secondaries have got several things from the primaries by borrowing, and are so to speak generated by them. 5. The primaries cannot move appropriately except on a diameter drawn through the centers of a single or of opposite faces; but the secondaries on a diameter drawn through opposite vertices. 6. It is characteristic of the primaries to stand upright, of the secondaries to balance on a vertex. For if you roll the latter onto their base, or stand the former on a vertex, in either case the onlooker will avert his eyes at the awkwardness of the spectacle. 7. Add finally that the primaries are three, the perfect number, the secondaries two, an imperfect number; and that the former have all types of vertex, the cube a right angle, the pyramid acute, and the dodecahedron obtuse, but the latter both employ a single type of angle, the obtuse. In fact, in the case of the octahedron all three types of angle occur: the obtuse at the junction of the faces; a right angle between two edges running from opposite vertices; whereas the actual solid angle is acute. Therefore, since there was an obvious distinction between the solids, nothing could be more appropriate than that our Earth, the pinnacle and pattern of the whole universe, and therefore the most important of the moving stars, should by its orbit differentiate between the two classes stated, and should be allotted the position which we have attributed to it above.[2]

CAPVT IV.

Quare tria corpora terram ambiant, duo reliqua induant?

PATERE nunc, Lector æquanime, vt ludam aliquantif-
per in re feria, & nonnihil Allegoriis indulgeam. Etenim
exiftimo ex amore Dei in hominem caufas rerum in mũ-
do plurimas deduci poffe. Certe equidem nemo negabit,
in domicilio mundi exornando Deum ad incolam futurum identidem refpexiffe. Finis enim & mundi & omnis
creationis homo eft. Terram igitur;quæ genuinam Creatoris imaginem
datura & alitura effet,exiftimo dignam à Deo cenfitam, quæ circumiret
inter medios planetas fic, vt totidem illa haberet intra orbis fui comple-
xum,quot extra habitura effet. Vt hoc Deus obtineret, Solem reliquis
quinque Stellis accenfuit, quamuis ille toto genere difcreparet. Idque
eo magis confonum videtur, quod cum fupra Sol Dei patris imago fuerit, credibile eft, hac affociatione cum reliquis Stellis argumenta ventu-
ro colono præbere debuifle φιλανθρωπίας, & ἀιθρωπωπαθείας, quam Deus v-
furpaturus erat erga homines, ad domefticam familiaritatem vfque fefe
demittens. Nam in Veteri Teftamento, frequenter in numerum hominum venit, & Abrahami amicus audire voluit ; ficuti Solem videmus in
numerum mobilium venire. Cum autem Sol à terra ambiretur : pofitis,
quæ dicta funt, neceffario ille ordo corporum intra terram includendus
fuit, qui duo faltem complectitur : nempe vt mobilia duo cum immobili
Sole eundem efficerent numerum ternarium, qui eft in exclufis ab orbe
terre. Sic igitur Luna præfertim terram ambeunte, domicilium noftrum
optimus Creator in medio feptem Planetarum collocauit. Nam fi trium
reliquorum ordo ad Solem acceffiffet; fuiffent igitur intra terram cum
Sole quatuor Stellæ,duæ vero tantum extra. Quæ numeri ἀταξία cum ratione careat, omiffa eft à Creatore: Cum item continere fit perfectioris,
vt actio, contineri vt paffio imperfectioris ; primaria vero perfectiora fint
cæteris; conuenit, vt trium ordo contineret terram, reliqua contineren-
tur intra orbis terreni ambitum. Atque fic habemus obiter caufam, cur
extra terram tres moueatur Planetæ, intra duo; quæ fi minus Lectori probatur, cogitet, honorarium hoc effe, non præcipuum. Nam etfi nefcire-
mus caufam ob quam fupra terram (vel Solem Ptolemæi) tres irent Stellæ, tamen fequentia ftarent cum præcedentibus ; quia nobis de RE conftat. Nec quifquam vnquam dubitauit, quin ♄ ♃ ♂ fuperiores fint. Tãtum illud teneamus ; cum tres in Copernico Planetæ fint fupra terram,
oportere nos ordinem trium primariorum corporum Cubum, Pyramida, Dodecaedron extra orbem telluris collocare , Octaedrum vero & Icofaedron intra; fi palmam in hoc negotio
velimus obtinere.

CA-

CHAPTER IV.
WHY SHOULD THREE BODIES GO ROUND THE EARTH, THE REMAINING TWO INSIDE IT?

Bear with me now, patient reader, if I trifle for a moment with a serious subject, and indulge in allegories a little. For I think that from the love of God for Man a great many of the causes of the features in the universe can be deduced. Certainly at least nobody will deny that in fitting out the dwelling place of the universe God considered its future inhabitant again and again. For the end both of the universe and of the whole creation is Man. Therefore in my opinion it was deemed by God fitting for the Earth, which was to provide and nourish a true image of the Creator, to go round in the midst of the planets in such a way that it would have the same number of them within the embrace of its orbit as outside it. To achieve that, God added the Sun to the other five stars, although it was totally different in kind. And that seems all the more appropriate because, the Sun above being the image of God the Father, we may believe that by this association with the other stars it was bound to provide evidence for the future tenant of the loving kindness and sympathy which God was to practice towards men, even as far as bringing himself down into their intimate friendship. For in the Old Testament he frequently came among their number, and was willing to be known as a friend of Abraham; just as we see the Sun is numbered among the moving stars. However, since the Sun was encircled by the Earth, granted what has been said, that class of bodies which in fact includes two had necessarily to be contained within the Earth's orbit, that is, in order that those two moving stars along with the unmoving Sun should make up the number of three, which is the number of those outside the Earth. Thus with the Moon as a special case encircling the Earth, the best of Creators placed our domicile in the middle of the seven planets. For if the class of the other three had been added to the Sun, there would have been four stars including the Sun inside the Earth, but only two outside. Since this irregularity of number lacks order, it was dismissed by the Creator. Also, since containing is proper for the more perfect, as it is active, but being contained for the more imperfect, as it is passive, but the primaries are more perfect than the rest, it is fitting that the class of three should contain the Earth, but the rest should be contained within the circuit of the terrestrial orbit. And thus we have in passing the reason why three planets move outside the Earth, two inside; and if this meets with less approval from the reader, let him reflect, that this is a by-product, and not the main point. For even if we did not know the reason why three stars moved above the Earth (or the Sun in Ptolemy), nevertheless what follows would be consistent with what precedes, because we are certain of the facts. Nor has anyone ever doubted that Saturn, Jupiter, and Mars are superior. Let us just hold on to this point: since in Copernicus the three planets are above the Earth, we should locate the class of the three primary solids, the cube, pyramid, and dodecahedron, outside the Earth's orbit, but the octahedron and the icosahedron inside it, if we wish to win the palm in this affair.

CAPVT V.

Quod cubus primum corporum, & inter altißimos planetas.

VENIAMVS modo ad primaria tria, suaque singulis spacia tribuamus. Et Cubus quidem ad fixas appropinquare debuit, primamque proportionem, quæ inter Saturnum & Iouem est, constituere; quia dignißima mundi pars extra terram sunt fixæ: vt circuli (post centrum) circumferentia: Cubus vero primum corpus in suo ordine. 1. Solus enim à sua basi generatur, cum reliqua quatuor non generétur faciebus suis, sed aut secta sint è Cubo, vt Pyramis, reiectis 4 pyramidibus rectangulis: aut aucta, vt Dedocaedron, appositis sex pentaedris. 2. Solus in homogeneos cubos sine prismate resolui potest. 3. Solus est quaqua versum, & in tres directas dimensiones porrigitur. Nam reliquorum faciés inclines sunt, & alicubi, cum se duabus directis sectionibus præbeāt, in reliqua sectorem frustrantur. 4. Hinc est, quod solus habet tot facies, quot habet ternaria dimésio terminos, nempe sex, & duplum numerum laterum, scilicet duodecim. 5. Solus vndiquaq; habet æqualem angulum, scilicet rectum. At in Pyramide regula, quæ sedet adhibita medijs planis, discrepat, si eā versus angulum intorqueas; nec solidi anguli ad eam normam quadrant, quæ interiectum longum lateralem angulum metitur. 6. Hinc etiam soli cōpetit, quod ex μονοβίβλω Ptolemæi citat Simplicius super Arist. lib. 1. de cœlo cap. 1. pro causa perfectionis in ternario; quod scilicet non plures tribus rectis perpendicularibus ad locum solidum in solidos rectos diuidendum concurrere possint. 7. Est solidorum rectilineorum omnium simplicissimum corpus. Quod etsi in Pyramide ambigitur, tamen ex eo facile euincitur, quod pyramidis mensura Cubus est, mensuram autem priorem esse conuenit. Mensura vero est non tantum ex instituto hominú, qui quicquid solidorum metiuntur, eius quantitatem in paruis cubiscis cōcipiunt animo: sed multo magis natura. Rectus enim angulus æqualis est alteri, quo cum in planum extenditur. Est igitur perpetuo sibi æqualis ipsi, atque adeo vnus, cæterorum vtrinque infiniti sunt. Mensuram autem decet vnam & eandem, atque etiam finitam esse. 8. Hinc (1) tam fœcunda est recti in circulum inscriptio, sine quo mediante, nec triangulum, nec quinquangulum, nec ab eis deriuata inscribi possunt. 9. Sed neque illud prætereundum quod perfectißimo animali solers natura sex easdem διασέσις perfectißime attribuit: non obscuro argumento, quam hoc corpus penes illam sit in pretio.

Nam homo ipse quidam quasi cubus est, in quo
sex quasi plagæ sunt, supera, infera,
antica, postica, dextra,
sinistra.

CA-

CHAPTER V.
THAT THE CUBE IS THE FIRST OF THE SOLIDS, AND BETWEEN THE HIGHEST PLANETS

Let us now come to the three primaries, and allot to each its own space. Now the cube should be close to the fixed stars, and establish the first proportion, that between Saturn and Jupiter, because the fixed stars are the most important part of the universe outside the Earth, just as that of a circle (after its center) is the circumference; and the cube is the first solid in its class. 1. For it alone is generated by its base, whereas the other four are not generated by their faces, but are either parts cut from the cube, as is the pyramid, which is derived by cutting off four rectangular pyramids, or compound, as is the dodecahedron, which is derived by the addition of six pentahedra. 2. It alone can be resolved into homogeneous cubes with no prism. 3. It alone faces in all directions, and extends in three directions at right angles. For the faces of the others are oblique, and at some point, although they allow division in two directions at right angles, frustrate it in the remaining direction. 4. It follows from this that it alone has the same number of faces as the three dimensions have directions, namely six, and twice that number of edges, that is twelve. 5. It alone has an equal angle, that is, a right angle, in every respect. But in the case of the pyramid, the formula which holds good when applied to planes of symmetry fails if you turn it on a vertex; and the solid angles do not square with the rule which governs the intervening angles between the lengths of the edges. 6. Hence also it alone agrees with what Simplicius quotes from the *Monobiblos*[1] of Ptolemy on Aristotle, *De caelo,* Book I, Chapter 1, about the reason for the perfection of the number three—that is, that not more than three straight lines perpendicular to each other can meet at a point to define a solid angle consisting of right angles. 7. It is of all rectilinear solids the most simple. Even if that is disputed with respect to the pyramid, nevertheless it is easily substantiated from the fact that the cube is the measure of the pyramid, and it is accepted that the measure is prior. Indeed it is the measure not only by the convention of men, who whenever they measure a solid conceive its quantity in their minds in terms of tiny cubes; but it is the measure much more by Nature. For one right angle is equal to any other which is spread out in the same plane. It is therefore perpetually equal to itself, and therefore one: of the rest there is an infinity on both sides. But a measure should be one and the same, and also finite. 8. (1) It is for this reason that the inscribing of a right angle in a circle is so fruitful, for without its intervention neither a triangle, nor a pentagon, nor any of the figures derived from them can be inscribed. 9. Yet further we should not pass over the fact that sagacious Nature has most perfectly allocated the same six directions to the most perfect animal—a clear sign of how she prizes this body. For a man is himself like a cube, in which there are so to speak six regions: upper, lower, fore, hind, right, left.

109

CAPVT VI.

Quod inter Iouem & Martem Pyramis.

AM cur Cubum excipiat Pyramis, nemo admodum mirabitur, cum 1. illa fere de principatu aufit cum cubo contendere. 2. Infuper vel ipfa, vel ὁμώλογα irregularia faciunt ad cæterorum compofitionem. Nam Icofaedron componunt 20. Pyramides, paulo breuiores Tetraedricis: Octaedrum octo adhuc breuiores. Dodecaedron etfi quadrato occulto conftat, tamen in pyramidas refolui necefle eft. 3. Neque contemnendum hoc, quod Tetraedrum in quatuor perfectas pyramidas & vnum Octaedron laterû dimidio minorum refolui poteft. 4. Sicut in planis omnia multangula in triangula refoluuntur, ita reliqua folida menfurandi caufa in pyramidas, quas deinde cubis, vt triangula quadratis, metimur. Eft igitur reliquorum menfura, & omnium facilime à cubo menfilis. 5. Hinc pleræque eius lineæ, vt & cubicæ tam facile quantitatê ex ratione diagonij accipiunt, non tamen aliter quam quadratis numeris. 6. Pyramidis etiam regularitas ex folis lateribus pendet: cubi etiam ex angulis. Atq; fic pyramidum inter æquilatera non plus vna eft, at in ἐξαέδρῳ, quamuis æqualibus lateribus, tamen infinita varietas eft Angulorum. Quo nomine, fi nullæ aliæ eflent rationes fitne præferêda cubo, an poftponenda, in dubio relinquo. 7. Hanc naturæ folertiam imitati homines primum materiam ad perpendiculum erigunt, rectifque angulis contignant, deinde triangulis firmant & ftabiliunt. 8. Infuper acutum angulû cum habeat pyramis, prior eft obtufangulis. Nam id femper primum eft in ordine, quod iuftam habet quâtitatem; hoc fequi videtur minus iufto, quia & longius abefle videtur ab infinitate, quâ plus iufto, & fimplicius etiam eft. Nam obtufangulum videtur quodammodo multiplex ex recto & acuto. Quo minus mirandum, cur paucitas angulorum in bafi, & ipfarum etiam bafium tetraedri non deroget cubo. Nam angulorum & bafium numerus ad fufceptam anguli fpeciem neceffario fequitur. Vnde fi rectus prior eft acuto, prius etiam ἐξαέδρον, quam Tetraedron, Tetragonoedrum quam Trigonoedrum. 9. Atque id etiam inde colligi poteft, quod perfectum vbique primum, poft, id, quod deficit, demum, quod excedit. Cum igitur Senarius facierum numerus perfectus fit, fequitur pyramidem, quæ deficit, non quidem præcedere debere cubum, at immediate fequi.

Habemus cur inter Iouem & Martem fecundo loco fit pyramis. Supra in fufpenfo fuit, quod corpus tertio loco fit inter Martem & terram. Illud vero hic facile deciditur. Cum enim è primarijs refiduum fit Dodecaedrum, erit illud ordine tertium, inter Martem & terram; de cuius proprietatibus quid fentiendum fit, collatione cum prioribus facta, facile patebit.

C A-

CHAPTER VI.
THAT THE PYRAMID IS BETWEEN JUPITER AND MARS

Nobody will now greatly wonder why the pyramid follows the cube, since 1. the former has almost dared to contend with the cube for the chief place. 2. In addition either they themselves or the irregular solids which are similar to them contribute to the composition of the rest. For an icosahedron is composed of twenty pyramids, slightly shorter than in the tetrahedron; and there are eight, which are shorter still, in the octahedron. Though a dodecahedron is based on a concealed square, yet it must necessarily be analyzed into pyramids. 3. Nor must we disregard the fact that a tetrahedron can be analyzed into four perfect pyramids and one octahedron with edges which are half as long. 4. Just as in plane surfaces all polygons can be analyzed into triangles, so the other solids are analyzed for mensuration purposes into pyramids, which we then measure by cubes, just as we measure triangles by squares. It is therefore the measure of the others, and of them all the easiest to measure by cubes. 5. Hence most of its lines, and also the cubes in it, take their magnitude as easily from the dimensions of the diagonal, though only in terms of the squares of numbers. 6. The regularity of a pyramid depends only on its edges: that of a cube also on its vertices. Thus there is only one type of pyramid among those which are equilateral, but in the case of a hexahedron, even though the edges may be equal, there is an infinite variety of angles. On this showing, if there were no other arguments, I should leave it in doubt whether the pyramid should be placed before or after the cube. 7. In imitation of this sagacity of nature, men first set up building material perpendicularly, and join it at right angles, and then fix it and strengthen it by triangles. 8. Furthermore, since a pyramid has an acute angle, it takes precedence over obtuse-angled solids. For that which has the exact measure is always first in order; and that which is less than the exact magnitude seems to come next, because it both seems to be further from infinity than that which is more than the exact, and is also simpler. For the obtuse angle seems in a sense compounded of a right angle and an acute angle. So we need not wonder why the fewness of the angles at the base of a tetrahedron, and also of the bases themselves, does not detract from the cube. For the number of the angles and bases is necessarily less important than the type of angle which is formed. Hence if the right angle takes precedence over the acute, so does the hexahedron over the tetrahedron, and solids formed from squares over solids formed from triangles. 9. And from this it can also be inferred, that the perfect everywhere has first place, the next, that which is deficient, and the last, that which is in excess. Therefore since the sixfold is the perfect number of faces, it follows that the pyramid, which is deficient, should not indeed come before the cube, but immediately after it.

We have shown why the pyramid is in the second position between Jupiter and Mars. Earlier it was undecided which solid is in the third position between Mars and the Earth. But that is now easily determined. For since the dodecahedron is the remaining one of the primaries, that will be the third in order, between Mars and the Earth. What we should conclude about its properties will easily appear from a comparison with those which come before it.

CAPVT VII.

De secundariorum ordine & proprietatibus.

ECVNDARIA quod attinet, cum Octaedron sit prius Icosaedro, mirū alicui videri possit, cur quod ordine Naturæ posterius est, in mundo præcedat? Nam quia Mars Dodecaedron sortitus est cum Tellure, sequitur ex ijs quæ diximus, inter Tellurē & Venerem interesse Icosaedron. Et prius esse Octaedron Icosaedro multa probant. Primū enim Octaedron natum est (non vere quidem, sed ita quasi natum sit) ex Cubo & pyramide primis in suo ordine; quorum illius numerum laterū, huius basin triangulam mutuatur. Icosaedron vero à pyramide, & Dodecaedro postremis in suo ordine nascitur. Rursum enim ex illa basin, ex hoc numerum laterum mutuatur. 2. Octaedron & Icosaedron si ex angulis aspicias, illud cubi basin quadratam ostēcat, hoc Dodecaedri quinquangulam. 3. Octaedrum cubo æqualtum est, vt videbimus, & Icosaedron Dodecaedro. 4. Octaedron cum cubo, Icosaedron cum Dodecaedro permutant numerum basium & angulorum. Nam Cubi bases & Octaedri anguli sunt sex, illius anguli & huius bases octo. Sic Dodecaedri bases & Icosaedri anguli sunt vtrinq; duodecim: vicissim illius anguli & huius bases sunt viginti. 5. Octaedron Cubi rectum angulum imitatur, Icosaedron Dodecaedri obtusum. Ex quibus patet Octaedron caput esse sui ordinis, sicut cubus primorum est princeps.

CAPVT VIII.

Quod Octaedron sit intra Venerem & Mercurium.

VOD autem propterea statim ad Dodecaedron in mundo sequi debeat, non sequitur. 1. Nam quia reuera duo diuersi sunt ordines, possunt etiam in diuersas mundi plagas spectare suis capitibus. 2. Atque adeo, quia Cubus dignissimæ mundi regioni extra Terram appropinquat, circumferentiæ scilicet siue fixis: par erat, vt & alterius ordinis caput digniori loco mundi intra Telluris orbem accederet. Nihil autem dignius centro & Sole. 3. Quod si etiam vtriusque ordinis situm pro vno censeamus, quid elegantius fieri poterat, quam vt ille vtrinque similibus & primis corporibus clauderetur. 4. Pulchrius etiam est, multifaria corpora adinuicem sequi in medio, & à pluralitate basium vtrinque sensim ad paucitatem discedi, si nihil aliud prohibeat: quam si ad multarum basium, corpus sequeretur, vnum paucarum basium, & denique succederet rursum aliud longe plurium, quam erat vtrumque. 5. Atque cum Dodecaedron esset in suo ordine vltimum, conueniebat, vt illi succederet ex

CHAPTER VII.
ON THE ORDER OF THE SECONDARIES AND THEIR PROPERTIES

As far as the secondaries are concerned, although the octahedron takes precedence over the icosahedron, could anyone think it puzzling that the one which comes after in the order of Nature, comes before in the universe? For because Mars together with the Earth has been allotted the dodecahedron, it follows from what we have said, that between Earth and Venus is the icosahedron. And there are many proofs that the octahedron takes precedence over the icosahedron. For first the octahedron was born (not literally born, but in a manner of speaking) from the cube and the pyramid which are first in their class: it borrows from them the number of edges of the former, and the triangular base of the latter. On the other hand the icosahedron is born from the pyramid and the dodecahedron which are the last in their class. For similarly it borrows from the former its base, and from the latter its number of edges. 2. If you look at the octahedron and the icosahedron from their vertices, the former shows the square base of the cube, the latter the five-sided base of the dodecahedron. 3. The octahedron is the same height as the cube, as we shall see, and the icosahedron as the dodecahedron. 4. The octahedron interchanges with the cube, the icosahedron with the dodecahedron, its number of bases and angles. For the cube has six bases and the icosahedron six vertices; the former eight vertices, the latter eight bases. Similarly the bases of the dodecahedron and the vertices of the icosahedron are twelve in each case: correspondingly the vertices of the former and the bases of the latter are twenty. 5. The octahedron copies the right angle of the cube, the icosahedron the obtuse angle of the dodecahedron. For these reasons it is clear that the octahedron is the chief member of its class, just as the cube is the leader of the first class.

CHAPTER VIII.
THAT THE OCTAHEDRON IS BETWEEN VENUS AND MERCURY

Nevertheless, what should come immediately after the dodecahedron in the universe does not follow from this argument. 1. For because there are in fact two different classes, it is even possible that their principal members may face towards different directions in the universe. 2. And because the cube is close to the most important region of the universe outside the Earth, that is, the circumference, or the fixed stars, it was proper that the chief of the other class should come to the more important position in the universe within the orbit of the Earth. However, nothing is more important than the center and the Sun. 3. Moreover, if we take the arrangement of both classes as the same, what could be more elegant than for it to be bounded on both sides by similar and principal solids? 4. For it is more beautiful for many-faced solids to follow one after another in the middle, and to move out bit by bit on both sides from many bases to few bases, if nothing else prevents it, than if a solid of many bases were followed by one of few bases, and then there succeeded another of far more bases than either. 5. Also since the dodecahedron was the last in its class, it was suitable for it to be succeeded by the

ret ex altero ordine, quod effet fui fimile. 6. Etiam hoc ad Telluris dignitatem pertinet, vt vtrinque fimiliter, quantum fieri poffet, ftiparetur. Cum igitur ita cecidiffet, vt exterius proxime ambiretur multifacio, par erat, vt interius etiam proxime compleĉteretur multifacium. Duo igitur hi ordines quinque horum corporum ita funt à fapientiffimo Conditore in vnum redaĉti, vt calcibus inuicem ad Tellurem, quæ maceries ipforum eft, obuerterentur, capitibus in diuerfas mundi plagas difcederent.

IN CAPVT III. IV. V. VI. VII. VIII.
Notæ Auĉtoris.

PLures corporum diftinĉtiones, & hæc ipfa fufius inuenies lib. IV. Epitomes, aliqua etiam, ortum & combinationem fpeĉtantia, lib. V. Harmon. cap. I. Et infra in hoc ipfo libello cap. XIII.

In Caput V. Notæ Auĉtoris.

(1) Hinc tam fæcunda eft, Reĉti in circulum infcriptio.] Ex anguli fcil. reĉti aptitudine, & quod omnis in femicirculo reĉtus eft angulus.

CAPVT IX.

Diftributa corpora inter Planetas, proprietates aptatæ, demonftrata ex corporibus cognatio planetarum mutua.

NON poffum præterire, quin hîc aliqua ex ea Physices parte, quæ eft de Planetarum qualitatibus, delibem ; vt appareat, etiam vires ipforum naturales hunc ordinem feruare, eamq; ad inuicem proportionem retinere. Nam fi eos planetas, qui terram ambeunt, illis etiam corporibus, quæ fibi infcripta continent accenfeas, inclufis autem Planetis à Telluris orbe illa corpora tribuas, quibus vterque circumfcribitur, quod optima ratione fieri poffe exiftimo : Saturnus habebit Cubum, Iupiter Pyramida, Mars Dodecaedron, Venerem Icofaedron, Mercurium Oĉtaedron. Terra vero cum nihil fit nifi limes, neutri accenfetur. Solem etiam & Lunam Aftrologi maximo interuallo à cæteris quinque diftinguunt, vt ita non opus fit illorum hîc meminiffe, & numerus corporum pulchre cum quinque Planetis conueniat.

Iupiter igitur (1) in medio maleficarum beneficus ipfe multos in admirationem rapuit, & Ptolemæum etiam ad caufarum inueftigationem extimulauit. Nos fimile quid videmus in pyramide, quæ inter duo corpora partim cognata, partim abhorrentia inuicem adeo ab vtroque difcrepat, vt fere de loco periclitetur in ratiocinijs fuperioribus. Trium fuperiorum quilibet cum reliquis (2) hoftilia exercet odia. Tribus etiam eorum corporibus nihil penitus conuenit eorum, quæ appatent. Mars tamen cum Saturno in fola malitia confpirat. Huic ego comparo

E   incon-

one similarly placed in the other class. 6. Also it is fitting for the importance of the Earth that as far as possible it should have similar attendants on both sides. Therefore since it had so fallen out that on the outer side it was most closely encircled by a many-faced solid, it was proper that on the inner side it should also embrace a many-faced solid most closely. Thus the two classes of these five solids have been assembled together by the wisest of Creators in such a way that they respectively turn their heels towards the Earth, which is the barrier between them, and with their heads face outwards towards different directions in the universe.

AUTHOR'S NOTES ON CHAPTERS THREE, FOUR, FIVE, SIX, SEVEN, & EIGHT.

You will find further points of distinction between the solids, and a more extensive treatment of those above, in Book IV of the *Epitome,* and some which concern their origin and combination in Book V of the *Harmonice,* Chapter 1. Also below in the present little book, Chapter 13.

AUTHOR'S NOTES ON CHAPTER FIVE.

(1) *It is for this reason that the inscribing of a right angle in a circle is so fruitful.*] That is to say, on account of the adaptability of the right angle, and because every angle in a semicircle is a right angle.

CHAPTER IX.
THE DISTRIBUTION OF THE SOLIDS AMONG THE PLANETS; THE ATTRIBUTION OF THEIR PROPERTIES; THE DERIVATION FROM THE SOLIDS OF THE MUTUAL KINSHIP OF THE PLANETS

I cannot avoid here abstracting a little from that part of physics[1] which concerns the properties of the planets, to make it apparent that their natural powers also observe this order and keep this proportion to each other. For if you allocate those planets which encircle the Earth to the solids which they contain and which are inscribed in them, but allot to those planets which are included within the Earth's orbit those solids by which they are each circumscribed, which I think could follow from the best line of reasoning, Saturn will have the cube, Jupiter the pyramid, Mars the dodecahedron, Venus the icosahedron, Mercury the octahedron. The Earth, however, since it is only the boundary, is allocated to neither. Also between the Sun and Moon and the other five the astrologers make a very great distinction, so that there is no need to mention them in this connection, and the number of the solids agrees excellently with the five planets.

Jupiter, then, benign (1) in the midst of the malevolent, has driven many to admiration, and also stimulated Ptolemy to enquiry into causes. We see something similar in the pyramid, which, between two solids which are partly akin and partly abhorrent to it, is so different from both of them that from our earlier reasoning its position is almost in peril. Everyone of the three superior planets (2) has hatred and hostility for the others. Also among their three solids absolutely none of their observable properties agrees, though Mars conspires with Saturn in malice alone. To this I relate the variability of their angles, which is peculiar to them, and

inconſtantiam angulorum, quæ illorum propria, & communis eſt vtriquc. Igitur bonitatis argumentum erit contrarium,ſc.ſtabilitas angulorum in ſolis lateribus. Argumentum cur Iupiter, Venus & Mercurius beneficiſint. Cubus, Saturni corpus, metitur omnia reliqua ſua rectitudine; Et planeta ipſe menſores efficit, eſtque quoad ingenium rigidus, recti cuſtos, ne latum vnguem cedens, inexorabilis, inflexibilis. Sic fert anguli rectitudo.

　　Cognatio euidentiſſima eſt in baſibus, qua cum Iupiter, Venus, Mercurius (planetam dico pro corpore) eadem vtantur, cauſſam habemus eorum amicitiæ, vt ſupra. Nam ſtabilitas ineſt triangulo primum.

　　Alter gradus eſt, planum apparens cum angulo ceu vmbilico. Ne miremur igitur amplius ecquid delitiarum penes durum & igneum Martem lateat, cuius cauſſa delicatula Venus mariti fruſtrata thalamum cum Marte conſpirauerit. Nam Martis quinquangulum eſt in Venere. Sic Saturni quadrangulum in Mercurio conciliat eoſdem vtrique mores. Tertius gradus eſt, cum idem eiuſdem in duobus eſt vel apparet: Et tum illis in cauſis communis amici conuenit. Igitur in rebus Iouijs conuenit Veneri cum Mercurio, quia communi Iouis vtuntur baſi. In Saturnijs conſentit Mercurius cum Marte parumper, quia in illo Saturni quadratum, in hoc tectus cubus eſt. Apparet etiam hinc cur Veneri cum Saturno nulla cognatio, & quæ potiſſima, & cur Mercurij verſatile ingenium omnibus quatuor ſeſe applicet, minimum tamen Marti.

　　Etiam Saturnus ſolitarius eſt, amanſque ſolitudinis, plane, vt eius anguli rectitudo non poteſt ferre vllam inæqualitatem vel minimam, cuius gratia multiplex fiat. Contra Iupiter è genere infinitorum acutorum vnum angulum nactus popularis ideo tactus eſt, moderate tamen & temperanter. Auctor enim eſt amicitiarum honeſtiorum. Ita Mars & Venus populares & ipſi ſunt, ſed nimium. Nam obtuſus & prodigus ipſorum angulus intemperantiam notat. Mercurius de natura Saturni & Iouis eſt ratione anguli. Et amant literati quidem ſolitudinem, ſed inhumani tamen non ſunt. Amant eos, qui ijſdem ſtudijs oblectantur: modumque ſtatuunt in conuerſationibus, plus quam Iupiter, cnius omnis actio eſt in cœtubus hominum, interque purpuratos.

　　Iupiter & Venus fœcundi ſunt. Sane quia Iupiter facit ad plerorumque compoſitionem; Venus autem Iouis quaſi ſoboles eſt, cum vna Venus viginti Ioues breuiuſculos in ſe contineat. Iupiter autem in mares æquior, Venus in fœminas; vnde ille mas dicitur, hæc fœmina. Pyramis enim efficax eſt, Icoſaedron effectum, & ſoboles. Ex his ijſdem principijs aliquanto explicatior cauſa redditur, quare Mercurius promiſcui ſexus ſit, & quare in fœcunditate mediocris.

　　Iouis primum, dein Saturni, & demũ Mercurij tranquillitas & conſtantia morum eſt à paucitate planorum: Veneris & Martis turbulentia & leuitas à multitudine. Varium & mutabile ſemper fœmina. Et figura Veneris omnium maxime varia & volubilis. Atque hi gradus ſunt: vnde medius Mercurius, media fide.

Mercurij

common to both. Therefore the contrary, that is the constancy of the angles between their edges alone, is evidence of benignity, which is evidence that Jupiter, Venus, and Mercury are benevolent. The cube, the solid of Saturn, is the measure of all the rest by its uprightness. And the planet itself produces measurers, and is rigid in temperament, a guardian of the right, not yielding a finger's breadth, inexorable, inflexible. This is the effect of the rightness of its angle.

Kinship is most evident in the bases; and as Jupiter, Venus, and Mercury have the same base (I use the name of the planet for that of the solid), we know the reason for their friendship, as above. For stability is a property of the triangle first and foremost.

The second type of kinship is in showing a plane section which is associated with its vertex as if with an umbilicus.[2] Consequently we should not wonder any longer what attraction lurks in the harsh and fiery Mars, on account of which dainty Venus betrayed her husband's bed and intrigued with Mars. For the pentagon of Mars is in Venus. Similarly the square of Saturn in Mercury assimilates the same behavior in both. The third type is when one and the same feature of a planet is found or appears in two others; and then they share with each other the characteristics of their mutual friend. Consequently Venus shares Jovial characteristics with Mercury, because they both have Jupiter's base. Mercury resembles Mars a little in Saturnine characteristics, because the square of Saturn is in the former, a concealed cube in the latter. It is also evident from this why Venus has no kinship with Saturn, and which is the strongest kinship, and why Mercury's versatile temperament is related to all four, but least to Mars.

Also Saturn is solitary, and a lover of solitude, plainly because the rightness of its angle cannot bear any irregularity, even the slightest, which might make it inconstant. On the other hand Jupiter, having taken one out of the infinite class of acute angles, has therefore become sociable, though moderately and temperately. For it is responsible for the more honorable friendships. Also Mars and Venus are themselves sociable, but too much so. For their obtuse and lavish angle betokens intemperance. Mercury partakes of the nature of Saturn and Jupiter by reason of its angle. Men of letters do indeed love solitude, but nevertheless they are not churlish. They love those who love the same studies, and set a limit in their intercourse, more than does Jupiter, all of whose activity is among assemblies of men, and among the blue-blooded.

Jupiter and Venus are prolific, plainly because Jupiter contributes to the construction of so many, and Venus is like an offspring of Jupiter, since Venus alone contains twenty tiny little Jupiters within herself. Jupiter however is more favorable to males, Venus to women; hence the former is spoken of as male, the latter as a woman. For the pyramid is the producer, the icosahedron the product and offspring. On these same principles a clearer explanation is given why Mercury is of both sexes indiscriminately, and why it is not very prolific.

In the case of Jupiter first, of Saturn next, and lastly of Mercury, their calm and the steadiness of their character are the result of the fewness of their faces; in the case of Venus and Mars their turbulence and changeability are due to their large number of faces. Woman is always fickle and capricious;[3] and the shape of Venus is the most capricious and variable of all. These, then, are the types of kinship; and hence Mercury is intermediate and of intermediate reliability.

Mercurij verſatile & celer ingenium refert Octaedri mobilitas. Nã ſi ſuper duos angulos voluas, quatuor continua latera per medium figuræ directum iter tranſeunt. Cæteras figuras, quomodocunque voluas, videbis per medium tranſuerſa & impedita incedere latera.

Mars multis lateribus pauciora plana efficit, Venus totidem lateribus plura plana; Martis etiam multi conatus irriti ſunt; Venus conatibus illi par, proſperiore tamen vtitur fortuna. Nec id mirum eſſe debet. Facilius enim choreæ inſtituũtur quam bella, & par erat, citius ad finem peruenire amores, quam iras; quia hæ perimunt homines, illi gignunt. Eodem pacto Mercurius Saturno felicior eſt.

IN CAPVT NONVM
Notæ Auctoris.

ETſi nihil eſt hoc caput, niſi luſus aſtrologicus, nec pars operis cenſeri debet, ſed excurſus: conferat tamen illud lector cum Ptolemæi rationibus, tam in Tetrabiblo, quam in Harmonicis: videbit noſtras Ptolemaicis non inferiores, ac forte meliores eſſe.

(1) In medio maleficarum.] *Loquor cum aſtrologis. Nam ſi meam ſententiam dicam, nullus in cœlo maleficus mihi cenſetur: idque cum ob alias rationes, tum maxime propter hanc, quia hominis ipſius Natura eſt, hic in terris verſans; quæ radiationibus Planetarum conciliat effectum in ſeſe; ſicut auditus, inſtructus facultate dignoſcendi concordantias vocum, conciliat Muſica hanc vim, vt illa incitet audientem ad ſaltandum. De hac re egi multis in Reſponſo ad Obiecta Doctoris Roſlini, contra librum de Stella noua, & alibi paßim, etiamq; in lib. IV. Harmonicorum paßim, præſertim cap. VII.*

(2) Hoſtilia exercet odia.] *Hoc allegorice intellectum phyſicis rationibus defendi poteſt: vt ſi ſub odij vocabulo diſcrimen qualecunque intelligatur ſitus, motus, luminis, coloris. Vide lector caput vltimum Ptolemæi Harmonicorum, vbi prodierint, quæque in id annotauerim, præſertim vltimam meam ſpeculationem, de Saturni & Martis mutuis exceßibus vel defectibus, Iouis vero mediocritate.*

CAPVT X.

(1) *De origine numerorum nobilium.*

INFINITVM eſt ſingula perſequi: neq; ſine fructu de his Aſtrologus amplius cogitet: Videamus modo Aſtronomorum Arithmeticam, ſacroſque eorum numeros, 6. 12. 60. Igitur excepto quadrante & ſextante, ſcilicet, 15. 10. omnes ſexagenarij partes multiplices reperiuntur in his quinque corporibus. (2) Viciſſim exceptis angulis planis Octaedri & cubi, quorum vterque habet 24. Cætera omnia, quæ numerantur, ſunt pars multiplex ſexagenarij: vt exiſtimem vix vlli numero poſſe ne à Pythagora quidem vllam rem naturalem aſſignari, quę illi magis ſit propria, quam hic numerus eſt dictis quinque corporibus.

Vnus eſt Cubus, Vna pyramis, Vnum Dodecaedron, Vnum Icoſaedron, Vnum Octaedron, Vnum ſolitarium ſine ſimili.

E 2 Duo

Chapter X

The quick and variable temperament of Mercury is represented by the mobility of the octahedron. For if you roll it on two vertices, four edges in continuous succession trace a path straight through the middle of the figure.[4] However you roll the other figures, you will see that the edges pass the middle obliquely and jerkily.

Mars produces fewer faces with many edges, Venus more faces with the same number of edges; also Mars makes many useless attempts, Venus makes the same number of attempts, but enjoys better fortune. Nor should that be surprising. For dances are more easily started than wars, and it was proper that lovemaking should achieve its goal more quickly than anger, because the latter destroys men, the former begets them. By the same token Mercury is more successful than Saturn.

AUTHOR'S NOTES ON CHAPTER NINE

Although this chapter is merely an astrological game, and should be considered not a part of the work but a digression, yet the reader should compare it with Ptolemy's arguments, both in the *Tetrabiblos* and in the *Harmony*.[5] He will see that our arguments are not inferior to the Ptolemaic ones, and perhaps better.

(1) *In the midst of the malevolent.*] I am addressing astrologers. For if I express my own opinion, I consider nothing in the heaven malevolent, for the following reason particularly among others, that it is the nature of Man himself, exercised here on Earth, which by the emanations of the planets gains their influence for itself; just as the hearing, which is endowed with the ability to discern the concordance of notes, gains this power of music to stir the hearer to dance. I have dealt fully with this topic[6] in my reply to the objections of Doctor Röslin against my book on the new star, and elsewhere generally, as well as in Book IV of the *Harmonice* generally, especially in Chapter 7.

(2) *Has hatred and hostility.*] If this is understood allegorically it can be defended by physical arguments, that is, if the word "hatred" is understood to refer to some difference of position, motion, brightness, or color. See, reader, the last chapter of Ptolemy's *Harmony*, and the notes I have made on it, especially my last investigation on the amounts by which Saturn and Mars mutually exceed or fall short of each other, and the way in which Jupiter falls in between.

CHAPTER X.
(1) ON THE ORIGIN OF THE NOBLE NUMBERS

It is endless to pursue details; yet it is not fruitless for an astrologer to ponder further on these topics. Let us now look at the arithmetic of the astronomers, and their sacred numbers, 6, 12, and 60. Now except for the quarter and sixth, that is 15 and 10, all the aliquot parts of sixty are found in these five solids. (2) Conversely, except for the plane angles of the octahedron and cube, of which each has 24, everything else, which is countable, is a factor of sixty. I believe therefore that there is scarcely any number to which any natural entity could be assigned even by Pythagoras which would be more appropriate to it than this number is to the aforesaid five solids.

The cube is one, the pyramid one, the dodecahedron one, the icosahedron one, the octahedron one — one solitary and unique.

Duo corpora fecundaria;Duo ordines corporum; Bina femper fibi fimilia;Duæ eiufmodi fimilitudines.

Tres anguli bafium in pyramide,Icofaedro, Octaedro,quia bafes trilatera.Tria primaria corpora.Tres angulorum differentiæ.

Quatuor anguli & latera bafis in Cubo. Quatuor folidi pyramidis anguli.Quatuor eiufdem bafes.

Quinque corpora. Quinque anguli & latera in bafi Dodecaedrica.

Sex anguli Octaedri.Sex latera pyramidis. Sex bafes cubi. Pulcher numerus.

Octo bafes Octaedri.Octo anguli cubi.

Duodecim bafes Dodecaedri. Duodecim latera Octaedri.Item & cubi.Duodecim anguli Icofaedri.Duodecim plani anguli pyramidis.

Ecce hic numerus in omnibus quinque eft.

Viginti bafes Icofaedri.Viginti anguli Dodecaedri.

Viginti quatuor anguli, plani Octaedri & cubi. Hic alienus eft numerus, fed nec præcipuæ rei, nec ita alienus; eft enim bis 12. ter 8. quater 6.qui omnes funt in 60.

Triginta latera Icofaedri & Dodecaedri.

Sexaginta plani anguli Dodecaedri & Icofaedri.

Prætereaque nihil numeratur,nifi fummas omnium laterum & angulorum inire velimus,quod alienius eft.Tum prouenient anguli denominantium bafium 18. Facies 50.Anguli totidem,latera 90. Anguli plani 180.Numeri cognati omnes.

IN CAPVT DECIMVM
Notæ Auctoris.

(1) **D**E origine Numerorum nobilium.] *Vt fupra iam dictum eft , omnis Numerorum nobilitas(quam præcipue admiratur Theologia Pythagorica,rebufque diuinis comparat)eft primitus ex Geometria. Cum vero multa fint eius partes: hæ quidem quinque figuræ folidæ non funt prima nec vnica caufa nobilitatis huius;fed accidit,vt multa in eundem numerum confpirent. Prima enim origo aptitudinis numerorum eft ex figuris planis regularibus , circulo infcriptilibus,earumque congruentia;vnde poftea folidæ oriuntur.Vide lib.I.& II.Harmonicorum. Ne vero confundaris,vbi legeris,Demonftrationes laterum, quibus vtuntur figuræ; arceffi à numeris angulorum: quafi ideo Numerus, vt numerans,prior fit & dignior. Minime,non enim ideo numerabiles fiunt anguli figuræ, quia præceffit conceptus illius numeri,fed ideo fequitur conceptus numeri,quia res Geometricæ habent illam multiplicitatem in fe,exiftentes ipfæ Numerus numeratus.*

(2) *Vicifsim exceptis,&c.& infra, Octo bafes.] Ecce manifeftam hallucinationem, Octo,non eft pars fexagenarij,fed bene pars eft numeri 120.qui eft bis 60.*

CA-

Chapter X

The secondary solids are two; the classes of solids are two; twofold and in all cases like each other; two likenesses of the same kind.

The angles of the bases in the pyramid, icosahedron, and octahedron are three, because the bases are three-sided. The primary solids are three. The different classes of angle are three.

The angles and the sides of the bases in the cube are four. The solid angles of the pyramid are four. The bases of the same are four.

The solids are five. The vertices and edges in the base of the dodecahedron are five.

The vertices of the octahedron are six. The edges of the pyramid are six. The bases of the cube are six. An excellent number.

The bases of the octahedron are eight. The vertices of the cube are eight.

The bases of the dodecahedron are twelve. The edges of the octahedron are twelve. So are those of the cube. The vertices of the icosahedron are twelve. The plane angles of the pyramid are twelve.

Behold — this number is in all five.

The bases of the icosahedron are twenty. The vertices of the dodecahedron are twenty.

The plane angles of the octahedron and cube are 24. This number is foreign, but neither in an important respect, nor altogether foreign; for it is twice 12, three times 8, and four times 6, which are all contained in 60.

The edges of the icosahedron and dodecahedron are 30.

The plane angles of the dodecahedron and icosahedron are sixty.

Apart from these there is nothing countable, unless we wish to proceed to the sums of all the edges and angles, which is more foreign. In that case the angles of the defining bases will come to 18, the faces to 50, the vertices to the same, the edges to 90, the plane angles to 180. All these numbers are akin.

AUTHOR'S NOTES ON CHAPTER TEN

(1) *On the origin of the noble numbers.*] As has been said above, the whole nobility of the numbers (which is especially a source of wonder in the Pythagorean doctrine, and is there ranked with the divine) is originally from geometry.[1] Yet there are many parts of it: indeed these five solids are not the first and unique cause of this nobility, but it happens that many features coincide in the same number. For the first origin of the aptness of the numbers is in the regular plane figures which may be inscribed in a circle, and in the way in which they fit them. It is from this that the solids subsequently derive. See Books I and II of the *Harmonice.* Yet do not be confused, when you read that the arguments for the edges which occur in the figures are derived from the numbers of the angles, as if on that account number, as a means of counting, were prior and more important. Far from it, for the angles of a figure do not become capable of being counted, because the concept of number preceded them; but the concept of the number follows, because geometrical objects have this multiplicity in themselves, and themselves constitute a number which is counted.

(2) *Conversely, except for, etc.*, and below *The bases...are eight.*] This is an obvious aberration: eight is not an aliquot part of sixty, but is in fact of 120, which is twice 60.

CAPVT XI.

(1) *De situ corporum, & origine Zodiaci.*

INFESTOS in his capitibus habebo physicos, propterea, quod naturales planetarum proprietates ex rebus immaterialibus & figuris mathematicis deduxi, porro vero etiam ex nuda imaginatione sectionum quarundam origines circulorum inuestigare audeam. His paucis respōsum volo: quod (2) Creator Deus, cum mens sit, & quæ vult faciat, non prohibeatur; quo minus in aptandis viribus & designandis circulis ad res vel sine materia, vel imaginatione cōstantes respiciat. Et cum nihil velit ille, nisi summa cum ratione, nihilque præter eius voluntatem extiterit; dicant igitur Aduersarij, quænam aliæ rationes DEO fuerint aptandarum virium, &c. cū præter q̄ titates nihil esset? Quod si, dum nihil inueniunt, ad imperscrutabil᷄ᷓᷓ᷄ditricis Sapientiæ vires confugiant: habeant sibi sane hanc inquir᷄ᷓᷓ᷄ temperantiam, illaque cum pietatis opinione fruantur: nos vero patiantur causas ex quantitatibus verisimiles reddere: dummodo nihil indignum tanto dicamus Opifice. Nulla igitur vinctus religione, pergo ad inuestigationem Zodiaci.

Ac initio existimo verisimiliorem corporum situm excogitari non posse, quam cum Cubus maxima figurarum inseratur orbi quomodocunque (nam in circulo nullum est initium. (3) Oportet autem principia sine ratione constituere, (4) ne infinitus fiat regressus; (5) & vt aliquando transitum habeamus ab infinita potentia ad finitum actum.) Iam igitur vna facierum censeatur pro basi. Pyramis igitur inserenda cubo mediante orbe Iouio, (6) debet basin basi cubi παϱάλληλον tenere: & (7) Dodecaedron pyramidis basi. Aliter ferunt secundariorum proprietates, vt vidimus. Erigendum igitur Icosaedron intra Dodecaedri, ita vt diagonius illius fiat vtrique oppositarum basium Dodecaedron perpendicularis in centris. Eodem pacto (8) suspendendum erit Octaedron minima figurarum, intra Icosaedron, ita vt acta recta veniat, 1. per centrum basis in cubo, 2. per centrum basis Tetraedricæ, 3. per centrum quinquanguli Dodecaedrici, 4. per angulum Icosaedri, 5. per angulum Octaedri, 6. per centrum mundi, & corpus solare, & porro similibus interstitijs per oppositos, 7. Octaedri, 8. Icosaedri angulos, 9. Dodecaedrici plani centrum, 10. Tetraedri angulum, 11. Cubici plani centrum. Maioris lucis caussa relego te ad tabellam capitis secundi, vbi omnia corpora ad hunc modum expressa sunt. Quibus ita constitutis, non tantum apparens in Octaedro quadratum, æqualiter à dictis duobus angulis remotum, si producatur circumcirca; omnes figuras, atque adeo totum mundum in bina diuidet æqualia; sed etiam omnium laterum, (9) quæ quis inter dictos angulos & centra, media censere potest, eorum inquam omnium (10) si regulariter ponantur, sectiones mutuæ, quæ prospicienti ex centro

E 3   appa-

CHAPTER XI.
(1) ON THE ARRANGEMENT OF THE SOLIDS, AND THE ORIGIN OF THE ZODIAC

I shall have the physicists against me in these chapters, because I have deduced the natural properties of the planets from immaterial things and mathematical figures, and futhermore because I dare to investigate the origins of the circles by frankly imaginary cross sections. I wish to respond briefly as follows: that God (2) the Creator, since he is a mind, and does what he wants, is not prohibited, in attributing powers and appointing the circles, from having regard to things which are either immaterial or based on imagination. And since he wills nothing except with absolute reason, and nothing exists except by his will, then let my adversaries say what other reasons God has for attributing powers, etc., since there was nothing except for quantities? But if, finding nothing, they take refuge in the inscrutable powers of the Founding Wisdom, let them indeed set themselves this limitation on enquiry, and enjoy it with a sense of piety; but let them allow us to draw likely explanations from quantities, provided that we say nothing unworthy of so great a Craftsman. Bound, then, by no religious scruples, I proceed to the investigation of the zodiac.

To start with, I believe no more likely arrangement can be conceived for the solids than for the cube, the largest of the figures, to be inserted into its orbit in any way whatever, since there is no beginning in a circle. (3) Now it is necessary to establish some principles without reason, (4) so as to avoid an infinite regress, (5) and to have at some point a transition from infinite potentiality to finite realization. In that case, then, let one of the faces be taken as its base. Then the pyramid which is to be inserted into the cube with the orbit of Jupiter in between (6) must have its base parallel to the base of the cube, and (7) the dodecahedron to the base of the pyramid. The properties of the secondaries are differently arranged, as we have seen. Then the icosahedron must be set up within the dodecahedron, so that one of its diagonals is perpendicular to each of the opposite bases of the dodecahedron at their centers. On the same principle (8) the octahedron, the smallest of the figures, must be suspended within the icosahedron in such a way that a straight line may be drawn to pass 1. through the center of the base of the cube, 2. through the center of the base of the tetrahedron, 3. through the center of the pentagon of the dodecahedron, 4. through a vertex of the icosahedron, 5. through a vertex of the octahedron, 6. through the center of the universe, and the body of the Sun, and then through similar points on the opposite side, 7. a vertex of the octahedron and 8. of the icosahedron, 9. the center of a surface of the dodecahedron, 10. a vertex of the tetrahedron, and 11. the center of a face of the cube. To shed more light on this I refer you to the plate in the second chapter where all the solids are drawn out in this arrangement. With things established in that way, not only will the square which is apparent in the octahedron, and is equidistant from the two vertices referred to, if it is extended all round, divide all the figures, and therefore the whole universe, into two equal parts, but also all the edges (9) which may be reckoned as intermediate between the vertices and centers referred to; all of them I mean, (10) if they are regularly placed, will have their

apparent,verſantur in eodem quadrati Octaedrici continuato plano.Id-
que præcipue in multifacijs vt cognatis apparet. (11) Nam cæterorum
latera dicta non ſimul congrue poni poſſunt. Dodecaedron igitur,de-
cem lateribus, talem deſcribit viam, per medium tranſeunte quadrato
Octaedri, in planum ex-
tenſo:

Icoſaedron vero manifeſtam Zonam hoc pacto, tranſeunte rurſum Octaedri quadrato in
rectum extenſo:

Quod ſi hæc duo cognata corpora ita applicentur per circumfe-
rentiam(nam anguli duo vnius, & centra planorum duorum alterius adhuc, vt ſupra, tanquam poli cohærere intelliguntur)vt apparentia bina
quinquangula Icoſaedri, & bina vera Dodecaedri, angulis congruant,
progignetur circularis ſe-
ctio, quæ in planum extenſa,cum Octaedri quadrato,
ſic habet.

(12) Sin angulus vnius medio lateri alterius in ſupradictis quin-
quangulis applicetur, talis
erit ſectio.

(13) Quid reſtat igitur, quin dicamus Planetas illam viam tot ma-
nifeſtis punctis notatam á Creatore iuſſos ire, præcipue cum inter ſupra
aſſumpta colligataque centra & angulos,tanquam polos media ſit.

IN CAPVT VNDECIMVM
Notæ Auctoris.

(1) DE ſitu corporum, & origine Zodiaci.] *Totum hoc caput quantum ad ſcopum omit-*
ti potuit,nullius enim momenti eſt. Neque enim hic eſt genuinus ſitus,ſeu coaptatio inter ſe,
corporum,quinque Geometricorum,vt infra patebit:neque ſi eſſet,Zodiacus inde eſſet.

(2) Creator,cum mens ſit.] *Ecce vt fœnerauerit mihi per hos 25.annos,principium iam*
tunc firmiſſime perſuaſum: ideo ſcil.Mathematica cauſas fieri naturalium; (quod dogma Ariſtoteles
tot locis vellicauit)quia Creator Deus Mathematica vt archetypos,ſecum ab æterno habuit in abſtra-
ctione ſimpliciſſima & diuina, ab ipſis etiam quantitatibus, materialiter conſideratis. Ariſtoteles
Creatorem negauit,mundum æternum ſtatuit: non mirum,ſi archetypos reiecit:fateor enim nullam
illis vim futuram fuiſſe, ſi non Deus ipſe in illos reſpexiſſet in creando. Ergo etiam Eccentricitatum
cauſæ ex hoc principio tandem inuentæ ſunt; quarum inæqualitatem vehementer neceſſe eſt admira-
ri,quicunque de iis ſerio cogitat:quicunque cum Ariſtotele de rebus cœleſtibus,ſic quærit:Quare non
quo quilibet Planeta humilior,eo pluribus orbibus vehitur? Nam qui in hoc inquiren-
dum ſibi putauit in Aſtronomia ſui temporis,inque perſuaſione illa falſa ſolidorum orbium:idem ho-
die ſi viueret,& puram atque genuinam noſtram de cœlo doctrinam cognoſceret, multo maxime ſibi
*quærendum exiſtimaret,*Quare non,quo quilibet Planeta interior,hoc minorem etiam
Eccentricitatem habet? *Itaque omnibus rationibus,quas ipſi ſua principia ſuggererent, conſum-*
ptis,illa perpetua voce, Quare non;*ſi tandem edoceretur Ariſtoteles,cauſas huius rei pulcherrimas*
& plane neceſſarias ex Harmoniis vt ex Archetypo reddi poſſe;puto illum pleniſſimo aſſenſu & Arche-
typos,& quia horum per ſe nulla efficacia eſt,Deum mundi architectum receptuum fuiſſe.Hæc igitur
de theſi ipſa: quæ tamen ad hypotheſin in hoc quidem cap. vt cœpi dicere, non fœliciter ſuit applicata.
(3) Opor-

apparent intersections, as seen by someone looking from the center, in the same extended plane of the square in the octahedron. That is particularly evident in the case of the many-faced solids, as they are akin. (11) For the edges referred to in the others cannot be placed so that they all correspond. Therefore the dodecahedron, with ten of its edges, describes a path like this as the square in the octahedron opened out into a straight line passes through it.[1]

See top diagram, opposite.

However the icosahedron describes an obvious belt like this, again as the square in the octahedron opened out in a straight line passes through it:

See second diagram, opposite.

But if these two solids, which are akin to each other, are aligned with respect to the surrounding sphere (for up to now the two vertices of one, and the centers of two faces of the other, have been understood to be related as if they were poles) in such a way that the two apparent pentagons in the icosahedron, and the two real ones in the dodecahedron, correspond at their angles, a circular cross section will be generated which, if laid out flat, along with the square in the octahedron, is like this:

See third diagram, opposite.

(12) But if the angle of one is aligned with the middle of the edge of the other, in the aforementioned pentagons, then the cross section will be of this kind:

See bottom diagram, opposite.

(13) What remains, then, but to state that the planets have been commanded to follow that path marked out by such obvious signs by the Creator, especially since it is midway between the centers and vertices which have been assumed and linked together above, as if they were its poles.

AUTHOR'S NOTES ON CHAPTER ELEVEN

(1) *On the arrangement of the solids, and the origin of the zodiac.*] The whole of this chapter, as far as its aim is concerned, could be omitted, for it carries no weight. For this is not the true arrangement, but a fitting together of the five geometrical solids among themselves, as will be apparent below; and if it were, the zodiac would not result from it.

(2) *The Creator, since he is a mind.*] Notice how the princple of which I was then already so firmly persuaded has repaid me with interest over these 25 years—that is, that the reason why the Mathematicals[2] are the cause of natural things (a theory which Aristotle carped at in so many places) is that God the Creator had the Mathematicals with him as archetypes from eternity in their simplest divine state of abstraction, even from quantities themselves, considered in their material aspect. Aristotle denied the existence of a Creator, and decided that the universe was eternal[3]—not surprisingly, if he rejected the archetypes, for I confess that they would have possessed no force, if God himself had not had regard to them in the act of Creation. Consequently the causes of the eccentricities were eventually discovered from this principle; and their irregularity must decidedly be a source of wonder for whoever seriously reflects on them, whoever with Aristotle asks concerning things in the heavens, "Why is not each planet moved by more spheres the lower it is?" For he who thought that this was the question he ought to ask in the astronomy of his own time, and in the false belief in solid spheres, would today, if he were alive and learnt of our pure and true theory of the heaven, consider that much the most important question to ask was, "Why does not each planet have a smaller eccentricity the further in it is?" Thus when all the arguments which his principles suggested were used up against that persistent phrase, "Why not?", if in the end Aristotle was persuaded that splendid and plainly necessary causes for this matter could be derived from the harmonies as if from an archetype, I think he would accept with the fullest agreement both the archetypes and, since they are ineffectual by themselves, God as the architect of the universe. These remarks, then, refer to the thesis itself; but, as I began to say, their application to the hypothesis in this particular chapter was not happy.

(3) Oportet autem principia sine ratione constituere.] Hoc de ijs dictum est, quæ in genere quantitatum, rationem habent materia. Verbi causa, sphæricum ipsum per se vnum totum sibique vndique simile est formaliter:at materialiter, vt superficies, habet partem extra partem. Hic cū ratione partium dominetur in sphærico infinitas diuisionis, sphæricum igitur ratione ea qua in partes est diuiduum, non consideratur formaliter, sed materialiter: siue quod idem est, Partes sphærici formales nullæ sunt; quæ vero in illo considerantur partes, materiales sunt, in quantum figura sphærici vtitur materia quantitatiua, diuidique potest. Iam vero actu inscribitur Cubus sphærico; si sphæricum formaliter consideratur vt figura, locus quæstioni non est, quibus nam in punctis statuendi sint anguli cubi, sin autem materialiter consideres, vt superficiem infinitorum punctorum: tunc quidem quæstioni locus est, quibus in punctis? at responderi non potest, cum ratio nulla sit, cur potius in bis punctis, quam in alijs: quippe potest in infinitis alijs atque alijs.

Huius generis sunt & istæ quæstiones; Cum singitur spatium vltramundanum infinitum, & de eo quæritur, cur potius in hac parte spacij, quam in alia collocatus sit mundus; item cum tempus æternum (oppositum in adiecto) singitur, quæriturque, Cur demum ante sex millia annorum conditus sit mundus, Deo ab omni æternitate abstinente à creando? Nam & spatiū & tempus, in genere quantitatum, rationem habent æqualem, respectu figuratarum quantitatum. Materia vero de se rationes nullas suppeditat, ipsa in se vnam & solam proprietatem habet, infinitatem partium, actualem quidem, vel numeri, vel quantitatis, si ipsum totum actu infinitum: potentialem vero numeri, si totum actu finitum, quod solum est possibile, cum quantitas est in materia corporali physica vel cœlesti. Vide lib. Epitom. 1. Astr. fol. 40. vbi de figura cœli agitur.

(4) Ne infinitus fiat regressus] Ratio Aristoteli familiaris hic impertinenter adhibetur: imo ne principium quidem datur alicuius regressus in ass. grandis rationibus, vbi ratio plane nulla est.

(5) Et vt aliquando transitum habeamus.] Si, inquam, non est initium operis faciendum sine ratione, nullum vnquam initium erit faciendum; rationes enim ad hoc vel illud initium, vbi dantur infinita, plane nullæ sunt. Quod igitur in infinitis punctis fieri æque posset, id cum sit in eorum vno aliquo, præter omnem rationem est, quod in eo potissimum sit præteritis alijs.

(6) Debet basin basi cubi parallelam.] Atqui Geometria docet locationem Pyramidis in Cubo longe concinniorem & perfectiorem: concinniorem, quia quæ ratio est inscriptionis Geometricæ illius in isto, eadem etiam in mundo concinna erit: At Geometrice Pyramis Cubo sic inscribitur, vt, quodlibet latus Pyramidis fiat diagonios vnius plani cubici: perfectiorem vero, quia si maximè basis vna Pyramidis fiat parallela basi vni Cubi: tamen adhuc incerta est locatio laterum basis triangulæ trium, respectu laterum basis quadrangulæ quatuor. Potest enim quodlibet illorum, cuilibet horum parallelum statui; potest & angulorum vni obtendi: vt perpendiculariū potius plani triangularis cum latere Cubi in idem planum competat. Denique perfecta locatio non est, vbi non omnibus planis similes situs contingunt: at cum vnum Pyramidis planum sit parallelum plano cubi reliqua illius, nulli huius erunt parallela; Idem & de lateribus & de angulis dictum esto.

(7) Dodecaedron basi Pyramidis.] Hic iam situs ab vtraque figura abhorret, & à Pyramide, & à Cubo. Nam inscriptio Geometrica docet, angulos potius quatuor Pyramidis debere iungi (vel superponi) totidem angulis de dodecaedri viginti. Sic eadem inscriptio Geometrica Cubi in Dodecaedron docet, diagonios Dodecaedri octo de duodecim, fieri octo latera Cubi: itaque si Dodecaedron vicissim sit intra Cubum; oportet de triginta lateribus Dodecaedri sena subordinari senis planis Cubi situ parallelo.

(8) Suspendendum erit Octaedron.] Hoc pacto respondebit quidem situs Octaedri intimi in Cubo extimo, inscriptioni Geometricæ eiusdem in Cubo: at Pyramidi, Dodecaedro, Icosaedro non legitime accommodabitur, nisi situs illorum in Cubo ad leges iam præscriptas emendetur. Tūc enim concurrent in vna recta linea ex centro cōmuni figurarum omnium educta, 1. angulus Octaedri, 2. laterum Icosaedri, 3. Dodecaedri, 4. Pyramidis, media puncta, 5. centrum plani cubici: eruntq́; talium linearum sex, & situs vndiq̃, sibi ipsi similis.

(9) Quæ quis inter dictos angulos & centra, media censere potest.] Quia in Pyramide per hunc vitiosum situm impedimur, vt media latera nequeamus censere.

(10) Si regulariter ponantur.] Tunc sane etiam in Pyramide inuenientur quatuor media latera; tunc etiam situs figurarum in se mutuo; respiciet leges inscriptionum Geometricarum.

(11) Nam cæterorum dicta latera non simul congrue poni possunt.] Non possunt

(3) *Now it is necessary to establish some principles without reason.*] This was said of things which are in the class of quantities and have a material aspect. For example, as form the spherical in itself is a single whole and is alike in all directions; but as matter, being a surface, it has separate parts. In this case since, in respect of its parts, an infinity of division dominates in the spherical, therefore from that aspect in which it is divisible into parts, the spherical is not considered as form, but as matter; or, which is the same thing, formally the spherical has no parts, but what are considered as parts in it are material, inasmuch as the shape of the spherical takes on quantifiable matter, and can be divided. But in actual fact the cube is inscribed in the spherical. If the spherical is considered formally, as a shape, there is no room to ask at what points the vertices of the cube are to be placed; but if you consider it materially, as a surface of infinite points, then indeed there is room to ask, "At what points?" Yet there can be no answer, since there is no reason why it should be at certain points rather than at others, for it can be at an infinity of different points.

The following questions are also of that kind. Since space outside the universe is supposed to be infinite, the question arises in its case also, why the universe has been located in this part of space, rather than in another. Further, since time is supposed to be eternal (the opposite to its use as an adjective), the question arises why the universe was established six thousand years ago, and God abstained through all eternity from creation? For both space and time in the class of quantities have a material aspect, in respect that is of quantities which have shape. Matter, however, supplies no reason for itself, but in its own self has a single unique property, the infinity of its parts, actual indeed, either in number, or in quantity, if in totality it is actually infinite, but potential in number, if in totality it is actually finite, which alone is possible, since quantity resides in corporeal matter, physical or heavenly. See the *Epitome of Astronomy*, Book I, page 40, where the shape of the heaven is discussed.[4]

(4) *So as to avoid an infinite regress.*] It is inappropriate to apply the familiar argument of Aristotle here. Rather not even a starting point is offered for a regress, in assigning reasons, where there is plainly no reason.

(5) *And to have at some point a transition.*] If, I say, the task should not be begun without a reason, it should never be begun; for plainly there are no reasons for this or that beginning, when an infinite number are offered. Therefore as it could equally well be at an infinity of points when it is at a particular one of them; there is no possible reason why it is at that one particularly to the exclusion of the others.

(6) *Must have its base parallel to the base of the cube.*] But geometry teaches us a far more appropriate and more perfect way of locating the pyramid in the cube: more appropriate, because the reason in geometry for inscribing the former in the latter will also be appropriate in the universe (and in geometry the pyramid is inscribed in the cube in such a way that each edge of the pyramid becomes a diagonal of one face of the cube); more perfect, because even if one face of the pyramid is made parallel to one face of the cube, yet the location of the three sides of the triangular face with respect to the four sides of the quadrangular base is uncertain. For any of the former can be established as parallel to any of the latter, and can also subtend one of the vertices, in such a way that it is rather a perpendicular of a triangular face which falls in with the edge of the cube in the same plane. Lastly, the location is not perfect when similar positions do not occur for all the faces; but if one face of the pyramid is made parallel to a face of the cube, the remaining faces of the former will be parallel to none of the latter's. The same applies to both the edges and the vertices.

(7) *The dodecahedron to the base of the pyramid.*] This position is now antagonistic to each of the two figures, both the pyramid and the cube. For the geometrical method of inscription teaches us that the four vertices should rather be linked with (or superimposed on) the same number of vertices among the twenty of the dodecahedron. Similarly the same geometrical method of inscription of the cube in the dodecahedron teaches us that eight of the twelve diagonals of the dodecahedron become eight edges of the cube. Therefore if the dodecahedron is in its turn inside the cube, six of the thirty edges of the dodecahedron should be arranged opposite the six faces of the cube in a parallel position.

(8) *The octahedron...must be suspended.*] On the same showing the position of the octahedron inside within the cube outside will correspond with the geometrical method of inscription of the same in the cube; but it will not correctly fit the pyramid, dodecahedron, and icosahedron unless their position within the cube is emended in accordance with the rules now prescribed. For in that case there will fall on one straight line drawn from the common center of the figures 1. a vertex of the octahedron, 2. the midpoint of edges of the icosahedron, and 3. of the dodecahedron, and 4. of the pyramid, 5. the center of a face of the cube; and there will be six of such lines, and their position will be alike in all respect.

(9) *Which may be reckoned as intermediate between the vertices and centers referred to.*] Because in the case of the pyramid, on account of its imperfect orientation we are prevented from being able to reckon the edges as intermediate.

(10) *If they are regularly placed.*] In that case indeed even in the case of the pyramid four intermediate faces will be found; and in that case the orientation of the figures towards each other will mutually respect the rules of geometrical inscriptions.

ſunt inquam congruere latera vnius omnia,lateribus alterius, minime omnium Pyramidis. Scilicet ideo congrue poni non poſſunt,quia initium poſitionis non factum eſt regulare.

(12) Sin angulus vnius medio lateri alterius.] Hic equidem legitimus duorum horum corporum ſitus eſt ad ſe mutuo:at Octaedri ſitus,qui hic adſciſcitur,illegitimus eſt.

(13) Quid reſtat igitur, quin dicamus.] *Omnino multa reſtant,quo minus hoc dicere poſsimus.Nam ſitus,qui polos hic ſignat,illegitimus eſt. Quatenus vero in duobus,Dodecaedro & Icoſaedro,ſitus eſt legitimus;totidem poſſunt eſſe poli, quot anguli huius,plana illius;duodecim ſc. quare Zonæ intermediæ ſex:Erūt igitur incerti Planetæ,quorſum eant. In genere obſtat hoc, quod figura iſta reali ſitu partium ad ſe mutuo,non ſunt expreſſæ in mundo,ſed ſolum proportio orbium figuralium ex iis deſumpta in orbes cœleſtes fuit tranſlata,numerusq̃ orbium à figuris conſtitutus.Rectius igitur hāc quæſtionem; cur hanc potius, quam aliam viam currant planetæ, vt abſurdam repellimus. Nam cum eſſet in intentione Dei circulus,motibus planetarum neceſſarius; illi Deus per intentionem conſtituto materiale & ſtellatum ſphæricum circumiecit. Nec dubitatio aliqua Deum ab opere retinuit,quo minus initium eius facere poſſet, quaſi ſine ratione:nam tunc corpus nullum præexiſtebat, cuius ille partium reſpectu dubitaret.Spatium vero ſine corpore,pura eſt negatio:ſatisq̃ rationis eſt ad faciendum initium in infinito in Nihilo, vel cogitare leuiter de aliquo:tale enim cogitatum iam ſtatim infinitis modis eſt præſtantius,reliquo infinito non actu,nec exiſtenti,nec cogitato, & ſic prius illo, & initio aptum.Neq̃ vero primus ego ſum,qui meipſum hac inutili quæſtione fatigaui;Cur ſcilic.hac traductus ſit Zodiacus,cum potuerit alia,locis infinitis:Inuenias ſimilem huius in Ariſtot. Cur hanc potius in plagam eant Planetæ,quam in eius contrariam? Nam ne hic quidem ratio eſt vlla vnius præ altero,cum omnis linea,longitudinis cōditione,duas obtineat plagas,quæ ſunt in recta verſus duos eius terminos. Fatetur quidem ibi Ariſtoteles in genere,non omnium rationes,eodem modo quæri poſſe: adoritur tamen quæſtionem hanc; Naturam ait inter poſſibilia ſemper quod optimum,eligere:melius vero eſſe vt ferantur ſidera in plagam digniorē; atqui digniorem eſſe plagam prorſum,quam retrorſum. Ridicule. Nam prius quam motus eſſet,neutra plaga, neq̃ prorſum,neque retrorſum dicebatur;principium petitur. Argutatur quidem à ſimilitudine mundi cum animalibus, Animalia cum plagis ſuis ſex,ideam mundi ſtatuens. Atqui rurſum principium petitur. Demus enim mundum eſſe factum ad ſimilitudinē animalis; dicat igitur prius de ipſo animali, cur hoc illi ſit prorſum,illud retrorſum;& non viciſſim; hoc eſt, cur oculi,auresq̃ & nares,& lingua. & os verſus imaginem in ſpeculo dirigantur, brachiorum manuum digitorumq̃ articuli illorſum flectantur , pedum palmæ illorſum extendantur; & non potius,vt imagine in ſpeculo membra eadem, retro verſus hominem:potuit enim etiam ſic fieri: hoc eſt,potuit cor,quod nunc eſt in ſiniſtra,collocari in ſede,quā nunc putamus dextram. Et vt conſtet ratio in hac Idea mundi,quid? an non æque facile contraria potuit eius ad latera mundi fieri applicatio? quid impediuit,quo minus ſiniſtram ad Meridiem tenderet, dextram ad Septentrionem,quando plagas mundi metari iuſſa eſt? ſic enim faciem vertiſſet in plagam, quæ nobis nunc occaſus dicitur;ſic contrariam ſidera plagam proſam habuiſſent, in quam motib.ſuis tenderent.Rectius itaque ſuperſediſſet Ariſtoteles ſolutione huius ineptæ quæſtionis: ſua ipſius admonitioni obtemperans. Nam inter ea,quæ omnia ex æquo contingere poſſent, natura nullam inuenit Melioris & Deterioris electionem;hoc enim inuoluit contradictionem. Quin imo ſic argumentemur: Cū Ens,non Ente,præſtet: nondum igitur exiſtente Mundo,quæcunque eius plaga proſa concepta fuit initio,illa potiores nunc ex ſua parte rationes habet,cur proſa ſit,quam eius contraria,hoc ipſo,quia contraria eius concipitur eſſe in non Ente: quæ ſi etiā proſa facta eſſet;Mundus tamen ſimilis huic præſenti factus eſſet.Comparatio locum non habet Mundorum, vbi vnus ſolus eſt. Valeant itaque quæſtiones huiuſmodi materiales,& cum ijſdem etiam metatio Zodiaci, ſeu potius, (quia hic locus ſuis excedit cui ſucceſſu)via Regia,à Solaris corporis circulo inter eius polos medio monſtrata. Nam ſi poli & axis corporis ſolaris in plagas mundi alias verſi fuiſſent,etiam via Regia alia fuiſſet tradu&a. Quod :dem & de figuris Dodecaedro & Icoſaedro dicendū.Demus enim,munus ipſarum eſſe,metari Zodiacum ſectionibus mutuis laterum,& certi quidem ordinis,ex ſex,quos diximus eſſe poſsibiles:certe tranſlato figurarum ſitu in Mundo ſenſili, ſedes etiam alia Zodiaco obtingeret.*

CA-

(11) *For the edges referred to in the others cannot be placed so that they all correspond.*] I mean that the faces of one cannot all correspond with the faces of another, least of all in the case of the pyramid. Obviously, the reason why they cannot be placed so that they correspond is that the starting point taken for their location is not governed by rule.

(12) *But if the vertex of one is aligned with the middle of the edge of the other.*] In this case indeed the mutual orientation of these two towards each other is legitimate; but the orientation of the octahedron, which is the case in point here, is not legitimate.

(13) *What remains, then, but to state.*] Decidedly, many obstacles remain to our saying this. For the position which in this case represents the poles is not legitimate. Certainly, insofar as in two cases, those of the dodecahedron and the icosahedron, the orientation is legitimate, there can be as many poles as the latter has vertices, or the former faces, that is twelve, so that there can be six pathways in between them; and consequently the planets will be uncertain where to go. In general the difficulty is that these figures are not embodied in the universe with a real mutual orientation of their parts to each other; but the proportion of the spheres based on the figures has been taken from them and applied to the celestial spheres, and the number of the spheres has been established from the figures. It is therefore more proper for us to reject as absurd the question, "Why should the planets traverse this path rather than another?" For since in God's design a circle was necessary for the motions of the planets, when he had established it in accordance with his design he surrounded it with a material and starry sphere. Nor did any doubt restrain God from the task and prevent his taking a starting point, so to speak, without a reason; for at that time there was no body previously in existence to cause him doubt over the relationship to its parts. Indeed mere space without body is a contradiction; and in an infinite Nothing it is sufficient reason for taking a starting point even to consider one fleetingly; for one which is thus fleetingly considered is at once distinguished in an infinity of ways from the infinite number remaining which has no reality either in existence or in consideration, and is consequently prior to them, and suitable for a starting point. Nor indeed am I the first to have tired myself with this useless question, "Why was the zodiac drawn round in this position, when it could have been elsewhere, in an infinite number of positions?" You will find a similar one in Aristotle:[5] "Why do the planets move in this direction, rather than the opposite?" For even here there is no argument for one rather than the other, since every line, in virtue of having length, possesses two directions, which point directly towards its two ends. In fact Aristotle there admits in general that arguments cannot be expected to account for everything in the same way. However, the following question arises: "Nature," he says, "always chooses among the possibilities that which is best. It is better that the stars should travel in the more fitting direction; and furthermore the forward direction is more fitting than the backward." This is ridiculous. For before there was motion, neither direction was spoken of as forwards or backwards. The argument is circular. Indeed he makes great play of the analogy between the universe and animals, setting up animals with their six directions as the Idea of the universe.[6] But again the argument is circular. For let us grant that the universe was made on the analogy of an animal. Then let him say first of the animal itself why this is its forwards direction, and that its backwards; that is, why the eyes, ears, and nostrils, and tongue and mouth, point towards their image in a mirror, why the joints of arms, hands, and fingers bend in that particular direction, why the soles of the feet extend in that particular direction, and not rather, as do the same members in their image in a mirror, in the opposite direction to a man. For that could have been the case: that is, the heart, which is now on the left, could have been located in the position which we now think of as the right. And to refer the argument to this Idea of the universe—after all, could it not equally easily have applied to the sides of the universe the other way round? What was there to prevent its turning the left hand side to the south, the right hand side to the north, when it was instructed to mark out the directions of the universe? For in that case it would have turned its front in the direction which we now speak of as the west; and in that case the stars would have had the opposite direction as their forward direction, towards which they would turn in their motions. It would therefore have been more sensible for Aristotle to have left off trying to answer this pointless question, conforming with his own advice. For among things which could all equally happen, Nature found no way of choosing better or worse; for that involves a contradiction. Let us rather argue as follows. Since Being has precedence over not Being, therefore when the universe did not yet exist, whatever direction in it was conceived as the forward direction at the start, now has arguments on its side, why it should be forwards, rather than the opposite direction, for this very reason, that its opposite is conceived to be in not Being; and if the latter had been made forwards, yet the universe would have become similar to what it now is. There is no room for comparison of universes when there is only one. Then let us say farewell to material questions of this sort, and with them to the marking out of the zodiac, or rather (since the zodiac departs from its position with the passing of the ages) of the Royal Path,[7] which is shown by the circle of the solar body between its poles. For if the poles and the axis of the solar body had been turned towards different directions in the universe, the Royal Path would also have been drawn in a different position. The same also applies to the figures of the dodecahedron and icosahedron. For if we grant that their function is to mark out the zodiac by the mutual intersections of their edges, and in particular order, out of the six which as we have said are possible, certainly if the position of the figures in the sensible universe were shifted, the zodiac would also be differently situated.

CAPVT XII.

Diuifio Zodiaci, & affectus.

VL TI diuifionem Zodiaci in duodecim præcifa figna pro figmento humano habuere, tali nempe, cui
nihil rei naturalis fubfit. Neque enim hæc μόρια viribus, aut affectionibus differre naturalibus arbitran-
tur ; fed affumpta propter numeri ad rationes apti-
tudinem. (1) Quibus etfi non omnino repugno, tamen ne quid temere reijciatur, ex ijfdem principijs
diuifionis huius caufam proponam, ad quam Creatorem proprietates (fi quas illæ diftinctas habent) accommodaffe vero non erit ab-
fimile.

Numerorum fubiectum quodnam fit, fupra vidimus. Et (2) certe præter quantum, aut quanto fimile, potentia qualicunque præditum,
nihil eft in toto vniuerfo numerabile, præter Deum, qui ipfiffima vene-
randa Trinitas eft. Iam igitur (3) corpora omnia diffecuimus per Zodiacum. Videamus, (4) ecquid fectione hac Zodiacus ipfe adeptus vel
paffus fit. Sectorum igitur dicto modo, Cubi facies ex fectione refultans
erit quadrata, vt & Octaedri, Pyramidis triangula, Reliquorum duorum
decangula. Quater tria decies faciunt fummam centum & viginti. Igitur infcripta circulo, quadratum, triangulum, decangulum, ad idem pū-
ctum, arcus varios in circumferentia diftinguunt, quos
omnes metitur portio non maior centefima vicefima
totius circuli. Naturalis igitur diuifio Zodiaci in 120.
ex regulari fitu corporum inter orbes. Cuius triplum
cum fit 360. videmus hanc diuifionem non omnino
nulla ratione niti. Iam fi quadratum & triangulum rurfum ex eodem puncto feparatim defcribamus, por-
tio circuli minima erit pars duodecima ambitus, nempe Signum. Vt mirum fit, (5) & motum Solis & Lunæ menftruum, & (6) coniunctiones magnas Superiorum tam apte quadrare ad portiones, quæ ab eorun-
dem corporibus per triangulum & quadratum diftinguuntur.

(7) Atque adeo quam hæc duodenaria diuifio penes naturam in
pretio fit, exemplo cape extraneo ; vt quamuis caufa non omnino cogni-
ta fit, tamen occafio pateat, fubinde præclarius de his quinq; figuris fen-
tiendi.

Efto propofita fides aliqua, eiufque fonus Γ vt. Igitur quot
occurrunt voces à Γ vfq; ad octauam confonantes cum Γ (8) toties, nec fæpius, potes fidem rationaliter diuidere, fic vt diuifæ fidis partes & inter fe & cum integra confonent. Porro quotnam illiufmodi voces occurrant aures indicant. Ego fchemate & numeris dicam.

Vide nunc & ipfas harmonias, & fidium proportiones in nu-

F   meris;

CHAPTER XII.
THE DIVISION OF THE ZODIAC, AND THE ASPECTS[1]

Many have held the division of the zodiac into twelve exactly delimited signs to be an invention of Man, that is, of a kind which has no basis in Nature. For they consider that these portions do not differ in their natural powers or influences, but are assumed on account of the ready divisibility of the number. (1) Although I do not altogether object to that, nevertheless to avoid hasty rejection of the idea I shall from the same principles propose a reason for this division, so that it will not seem unlikely that the Creator should have accommodated their properties to it (if they have any distinguishing properties.)

What is the concern of numbers we have seen above. And certainly (2) apart from quantity, or what is similar to a quantity, and endowed with a power of some kind, there is nothing in the whole universe capable of being numbered except God, who in himself is the venerable Trinity. Now (3) we have already cut through all the solids with the zodiac. Let us see (4) what the zodiac itself has acquired or experienced in this cutting. Then of the solids cut in the way stated, the figure resulting from the cutting of the cube will be square, and similarly those from the octahedron and pyramid will be triangles, those from the remaining two decagons. Four times three tens make a total of a hundred and twenty. Then if a square, a triangle, and a decagon are inscribed in a circle, from the same point, they mark off different arcs at the circumference, the common measure of which is a portion no greater than one hundred and twentieth of the whole circle. The division of the zodiac into 120 is therefore natural because of the regular alignment of the solids among the orbits. As three times that is 360, we see that this division does not wholly lack rational foundation. Now if we similarly describe a square and a triangle separately from the same point, the smallest portion of the circle will be a twelfth of the circumference, that is a sign. So it is a wonderful thing that (5) the motion of the Sun, and the monthly path of the Moon, and (6) the great conjunctions of the superior planets are so neatly fitted to the portions which are marked off from their solids by the triangle and square.

(7) Furthermore how greatly this twelvefold division is valued by Nature may be gathered from an extraneous example, so that although the reason is not completely known, yet there will be an obvious opportunity for forming at once a much clearer opinion about these five figures. Suppose there is a string, and its note is G (ut).[2] Then the number of notes from G to the octave which are concordant with G is (8) the number of times, and no fewer, you can divide the string into rational fractions so that the divided parts of the string are concordant both with each other and with the whole. Furthermore our ears tell us how many of such notes occur. I shall make the point with a diagram and numbers.

See bottom diagram, opposite.

Now look at the actual harmonies, and the proportions of the strings in numbers, where the lowest symbol represents the note of the whole string, the

meris:vbi Nota ima fignificat vocem integræ fidis; fuprema, vocem partis breuioris; media, vocem partis longioris; Numerus imus indicat in quot partes fides diuidenda fit;reliqui,longitudines partium.

(9) Atque hæ folæ voces mihi naturales videntur,propterea quod habent indubitatum numerum. Cæteræ voces non poffunt certa proportione ad iam pofitas exprimi. (10) Nam vocem F faut, aliam ex C fol fa vt, defuper, aliam ex B mi molli inferius elicies, vtcunque hæ duæ perfectæ quintæ effe videantur. Sed ad rem. Prima & fecunda concordia quodammodo fociæ funt;fic etiam quinta & fexta. (11) Cum enim imperfectæ omnes fint: binæ femper,vna dura,altera mollis,confpirant, vt fingulis perfectis quodammodo æquiparentur. Nec admodum diuerfas diuifiones habent. Nam $\frac{3}{8}$ & $\frac{2}{3}$ fefe habent ad inuicem,vt $\frac{25}{30}$ & $\frac{26}{30}$, quæ tatum vna trigefima differunt. Sic $\frac{3}{5}$ & $\frac{2}{5}$ fe habent ad inuicem, vt $\frac{15}{40}$& $\frac{16}{40}$. Differunt igitur tantum vna quadragefima particula. Atque ita proprie loquendo, tantum quinque in Mufica habemus concordias, ad numerum quinque corporum. (12) Quod fi feptem diuifionum in 6.5. 4.3.8. 5. 2. communem minimum diuiduum quæras, rurfum inuenies 120. vt fupra,cum de diuifione Zodiaci ageremus;perfectarum vero concordiarum minimum diuiduum rurfum 12. (13) plane quafi perfectæ concordiæ à quadrato & triangulo Cubi,Tetraedri & Octaedri,imperfectæ vero à decangulo reliquorum duorum corporum prouenirent. Atque hæc fecunda eft corporum cognatio cum concordiis Muficis. (14) Sed quia caufas huius cognationis ignoramus, difficile eft accommodare fingulas harmonias fingulis corporibus.

(15) Videmus quidem duos harmoniarum ordines, tres fimplices perfectas,& duas duplices imperfectas; ficut tria primaria corpora, duo fecundaria;verum cum reliqua non conueniant,deferenda eft hec conciliatio, & alia tentanda. Nempe ficut Dodecaedron & Icofaedron fuo decangulo fupra auxerunt duodenarium vfq; ad 120. ita hic imperfectæ harmoniæ idem faciunt.

Erunt igitur ad Cubum,Pyramida & Octaedron accommodandæ perfectæ harmoniæ, ad Dodecaedron & Icofaedron imperfectæ. Quo accedit & illud, atque hercle (16) indicem digitum ad caufam harum rerum occultiffimam intendit, quod proximo capite habebimus: (17) duos nempe effe Geometriæ thefauros, vnum, fubtenfæ in rectangulo rationem ad latera ; alterum , lineam extrema & media ratione fectam , quorum ex illo Cubi , Pyramidis & Octaedri conftructio fluit, ex hoc vero conftructio Dodecaedri & Icofaedri. Vnde tam facilis & regularis eft infcriptio Pyramidis in cubum, Octaedri invtrumque, ficut Dodecaedri in Icofaedron. (18) Vt autem fingulæ Harmoniæ fingulis corporibus accommodentur , non ita in promptu eft. (19) Illud folum patet , Pyramidi deberi harmoniam , quam quintam dicunt,quartam in ordine,quia in ea minor portio eft $\frac{1}{3}$ pars integræ, ficut latus trianguli (quo Pyramis vtitur)fubtendit $\frac{1}{3}$ circuli. Hoc plura infra

confir-

highest the note of the shorter part, and the middle symbol the note of the longer part, the lowest number indicates into how many parts the string is to be divided, and the others the lengths of their parts.

See diagram opposite.

(9) Now these seem to me to be the only natural notes, on account of their having an unmistakable number. The other notes cannot be expressed by a definite proportion to those already set out. (10) For the note F (fa ut) is different if you derive it by coming down from C (sol fa ut) and if you derive it by going up from B flat (mi) even though both of them seem to be perfect fifths. But back to the point. The first and second concords have some affinity, and so have the fifth and sixth. (11) For though they are all imperfect, if one of the two is major, the other minor, the two always combine together so that they are in a sense equivalent to single perfect concords.[3] Nor are the differences between them very wide. For 1/6 and 1/5 are to each other as 5/30 to 6/30, which only differ by one-thirtieth. Similarly 3/8 and 2/5 are to each other as 15/40 to 16/40. Hence they differ by only one-fortieth part. Thus properly speaking we have only five concords in Music, corresponding to the number of the five solids. (12) But if you look for the lowest common multiple of the seven divisions into 6, 5, 4, 3, 8, 5, and 2, you will again find that it is 120, as above, when we were dealing with the division of the zodiac; but of the perfect concords the lowest common multiple is again 12, (13) plainly because the perfect concords come from the square and triangle of the cube, tetrahedron, and octahedron, but the imperfect ones from the decagon of the other two solids. This is the second point of kinship of the solids with the musical concords. (14) But because we do not know the causes of this kinship, it is difficult to fit the individual harmonies to the individual solids.

(15) Indeed we see two classes of harmonies, three simple and perfect, and two double and imperfect, just as there are three primary solids and two secondaries; but as the other points do not agree, this matching must be abandoned, and another tried. Now just as the dodecahedron and the icosahedron above by their decagon increased the twelvefold to 120, so in this case the imperfect harmonies do the same.

Therefore the perfect harmonies must be fitted to the cube, pyramid, and octahedron, the imperfect to the dodecahedron and the icosahedron. From that another point follows (and heavens! (16) it points the finger at the most secret cause of these matters), which we shall include in the next chapter: (17) that is, that there are two treasure houses of geometry: one, the ratio of the hypotenuse in a right-angled triangle to the sides, and the other, the line divided in the mean and extreme ratio. From the former of these is derived the construction of the cube, pyramid, and octahedron, and from the latter the construction of the dodecahedron and icosahedron. That is why the inscription of the pyramid in the cube, and of the octahedron in either of them, is so easy and straightforward, like that of the dodecahedron in the icosahedron. (18) However, the fitting of the individual harmonies to the individual solids is less readily achieved. (19) Only one thing is obvious, that to the pyramid should be attributed the harmony which they call a fifth, the fourth in order, because in it the smaller portion is a third of the whole, just as the side of the triangle which the pyramid employs subtends

confirmabunt,vbi de afpectibus agemus,quæ vt hic etiam intelligamus,
omnino ita cogitemus,quafi fides fit non recta linea,fed circulus. Dabit
igitur diuifio harmoniæ dictæ triangulum : in quo angulus lateri oppo-
nitur,plane vt in pyramide angulus plano. Remanét igitur Cubo & Octaedro octaua & quarta dictæ, tertia & feptima in ordine. Sed vtrum eorum vtram harmoniam tenebit? vtrum di-
cemus (20) fecundaria recipere eas,quæ lineas fcribant,& primaria,quæ figuras?tum Cubo debebitur quarta dicta. Nam fi ex fide circulum facias, & ex vna quarta rectam vfque aliam ducas tamdiu,donec
in primum punctum redeas,fiet quadrangulum,quale
planum etiam Cubus obtinet. Contra Octaedro debebitur octaua,quæ eft dimidiæ fidis. Nam in circulo
ductus ad dimidiam , & ad idem punctum facit nil nifi
lineam. Sic Dodecaedro debebitur prior imperfecta
duplex. Nam ductus per quintas & per fextas circuli
faciunt quinquangulum & fexangulum. Reftabit igitur Icofaedro pofterior imperfecta duplex , quia ductus per duas quintas repetiti vfque in idem punctum,
(21) faciunt tantum lineas. †. Sic & ductus per tres
octauas.* (22) An malumus Octaedro quartam dare,
re,quia is duodecies quartâ circuli fubtendit. Id quod
nullum latus cubi facit? Sic relinquetur Cubo octaua
harmonia perfectiffima,vt ipfe perfectiffimum corpus
eft. Forfan & illud conuenientius eft, (23) relinquere Icofaedro priorem imperfectam propter fexangulum , quod bafi triangulæ cognatum magis eft, quam
quinquangulæ: Dodecaedro vero dare diuifionem o-
ctonariam propter numerum cubicum 8. quia cubus
Dodecaedro infcriptilis. Hæc fane in medio fita fint
donec caufas quis reperierit.

(24) Veniamus modo ad afpectus. Et quandoqui-
dem modo ex fide circulum fecimus ; facile eft videre, (25) quomodo,
tres perfectæ harmoniæ pulcherrime cum tribus perfectis afpectibus
comparari poffint,fcil. cum ♀,△,□. Imperfecta vero prior B. mollis ad
vnguem fimilis eft fextili,cuius hæc nota, (26) *quemque debiliffimum
effe ferunt.

Habemus caufam (27) (qualem quidem Ptolemæus non dedit)
cur planetæ diftantes vno aut quinque fignis non cenfeantur in afpectu.
Nam vt vidimus, (28) nullam talem in vocibus agnofcit Natura concordiam. Cum enim in cæteris eadem fit ratio influentiæ & harmoniarum;credibile eft & hic effe. (29) Caufa vtrinque procul dubio eadem
eft,& ex quinque corporibus,quam alijs quærendam relinquo. (30) Cû
igitur omnes quatuor harmonic confonent fuis afpectibus,& vero adhuc
tres reftent in Mufica harmoniæ ; fufpicatus aliquando fum , non negli-
gendum effe in iudicijs natiuitatum,fi Planetæ 72.aut 144.aut 135.gradi-
bus diftent, præfertim cum videam , vnam ex imperfectis habere fuum
afpectum. Quamuis cuilibet oculato Meteororum fpeculatori facile patebit,vtrum aliqua in his tribus radijs vis infit,cum cæteros afpectus aeris

F 2   muta-

one-third of a circle. This will be confirmed below in many ways, when we deal with the aspects; but to gain acceptance of it here as well let us think of the string in all respects as if it is not a straight line but a circle. Therefore the division required for the harmony named will yield a triangle, in which the angle is opposite to the side clearly in the same way as in a pyramid the vertex is opposite to the face. Therefore for the cube and octahedron there remain the harmonies named the octave and the fourth, which are seventh and third in order. But which of them will take which harmony? Shall we say that the secondaries (20) accept those which describe lines, and the primaries those which describe figures? Then the harmony named a fourth will be attributed to the cube. For if you make a circle of the string, and draw the chord of each quadrant, one after the other, until you come back to the starting point, the result will be a square, which is also the kind of surface the cube contains. On the other hand to the octahedron will be attributed the octave, which is half the string. For in a circle, drawing the chord in a semicircle and another back to the starting point produces nothing but a straight line. Thus to the dodecahedron will be attributed the former of the imperfect pairs. For lines drawn in fifths and sixths of a circle make a pentagon and a hexagon. Therefore the icosahedron will remain the second imperfect pair, because drawing a line repeatedly in two-fifths of a circle, back to the same point, (21) makes only lines. So does drawing lines in three-eighths. (22) Or do we prefer to allot the fourth to the octahedron, because it subtends a quadrant of a circle twelve times,[4] which the edge of a cube never does? In that case for the cube will be left the octave, the most perfect harmony, as it is the most perfect solid. Perhaps it will also be more appropriate (23) to leave for the icosahedron the former of the imperfect pairs, on account of the hexagon, which is more akin to the triangular base than to the pentagonal base; but to give to the dodecahedron the eightfold division on account of the number eight's being a cube, because a cube may be inscribed in a dodecahedron. These are in fact open questions, until someone finds the causes.

(24) Let us now come to the aspects. Since we have just made a circle of the string, it is easy to see (25) how the three perfect harmonies can be most beautifully related to the three perfect aspects, that is to opposition, trine, and quartile. Of the imperfect harmonies, the first, B flat[5] resembles to the last detail the sextile, of which this is the sign, ✳, (26) and which they say is the weakest.

We have the reason (27) (which Ptolemy at least did not give) why planets which are one sign or five signs apart are not reckoned under an aspect. For as we have seen, (28) Nature recognizes no such concord among musical notes. For since in the other cases the ratio is the same for the influence and the harmonies, it is easy to believe that it is in this case as well. (29) The cause in both cases is undoubtedly the same, and depends on the five solids; but I leave it for others to seek. (30) Since, then, all four harmonies agree with their aspects, and yet there still remain three musical harmonies, I suspected at one time that we should not overlook it in casting horoscopes if planets are 72° or 144° or 135° apart, especially since I see that one of the imperfect harmonies has its aspect. However, to any observer of things on high who has eyes, it will be easily apparent whether there is any power in these three aspects, since in the case of the other aspects it is verified by invariable experience of the changing atmosphere.

mutationes conftantiffima ratificent experientia; (31) Caufæ quidem
quas probabiliter quis reddat,quod ⅓ ⅖ ⅗ in fide fonent, in Zodiaco non
operentur,hæ effe poffint.

　　1. Oppofitus folus,duo quadrata,trinus cum fextili, abfoluunt fin-
guli femicirculum: at tres hi radij nullum habent focium ad hoc munus,
quem Mufica non penitus repudiet.

　　2. Reliqui radij rationem habent facilem ex diametro,latus quinquanguli,& fubtendens duo latera quinquanguli, tria octanguli,funt in
gradu remotiore & irrationales.

　　3. Caufa , quia trinus cum fextili, quadratum cum quadrato efficiunt rectum angulum,Radij reliqui nullo pacto cum vlla recepta linea.
4.Imperfecta B mollis eft quodammodo perfecta, quia vtitur eadem diuifione cum perfectis,& eft dimidia quinta. Vnde non mirum, folam ex
imperfectis refpondere afpectui alicui,fc. fextili, qui itidem eft dimidius
trinus.Cæteræ enim nec aptæ funt in duodenarium,nec perfecti alicuius
pars funt.

　　5. Denique fex trigoni anguli,quatuor quadrati,tres fexanguli, &
duobus femicirculis comprehenfa duo fpacia implent omnem in plani-
tie locum. At tres anguli in quinquangulo minores funt quatuor rectis,
quatuor funt maiores. Vnde & illud patet,quare nec octangularis, (33)
nec duodecangularis radius, nec vllus reliquorum operetur. (34) Atq;
hîc fere feparo caufas afpectuũ à caufis concordiarũ. (35) Certe enim,
quæ ex angulis fit,genuina radijs eft ratiocinatio;cum propter angulum
in puncto fuperficiei terrenæ factum,in quo mifcentur,exiftat operatio,
(36) non vero propter figuram in Zodiaco circulo defcriptam,quę imaginatione potius quam rei veritate conftat. Diuifio vero fidis nec in circulo fit , nec angulis vtitur, fed in plano per rectam lineam perficitur.
(37) Poffunt tamen nihilominus & concordantiæ & afpectus habere
commune quid,quod eadem vtrinque caufatur,vt fupra dictum. Id vero
aliorum induftriæ relinquo fcrutandum. (38) Ptolemæi Mufica , quæ
Regiomontanus cum expofitione Porphyrij,editurus erat, fed nondum
excufa Cardanus afferit,in hac materia proculdubio verfantur. Vide etiam (39) quid ex Euclidis Muficis huc referri poffit.

I N C A P V T D V O D E C I M V M
Notæ Auctoris.

(1) Q̃Vibus etfi non omnino repugno.] *Hoc thema ex profeffo tractaui in libro de ftel-*
la noua,inque refponfo ad obiecta Roflini:nempe , quatuor quidem circuli Zodiaci qua-
drantes monftrari à conditionibus duorum motuum,diurni , & Solis annui, quas fequuntur etiam
Luminis & Calefactionis metæ : at quadrantum fingulorum fubdiuifionem interna practicè figna ni-
hil tale nec ex motu,nec ex viribus habere,cuius effectus cenferi poffit:nifi tantum generaliffimam il-
lam diftinctionem,quanti vniufcuiufque in principium,Medium,& Finem:quas tamen partes nulla
neceffitas iubet æquales effe,ac ne partes quidem: fufficit enim,vt pro medio cenfeatur , tota quadran-
tis linea,pro principio & fine,duo lineæ termini feu puncta,quæ non funt pars de linea.

　　(2) Præter quantum,aut quanto fimile, potentia qualicunque præditum, nihil eft in toto vniuerfo numerabile.]*Ridicula mihi fententia excidit,vere non fententia.Quid*
enim eft, Nihil præter Omnia? Numeratio,actio Mentis , fupeuenit rebus omnibus,diuinis &
humanis: nulla ne leuiffima quidem diftinctio eft, feu realis, feu intentionalis (fit illa prima , vel fe-
cunda

(31) The reason which one may with probability suggest why 3/8, 1/5, and 2/5 produce notes on a string, but do not operate in the zodiac, may be the following:

1. A single opposition, two quartiles, or a trine combined with a sextile each make up a semicircle; but these three aspects cannot combine together for that purpose with anything which Music will not completely repudiate.

2. The other aspects are simply related to the diameter; but the side of a pentagon, and the diagonal stretching under two sides of a pentagon, or three sides of an octagon, have a more distant relationship and are irrational.

3rd reason: that a trine with a sextile, a quartile with a quartile, form a right angle; the other aspects do not by any device with the addition of any line.[c]

4. The imperfect harmony B flat[7] is in a way perfect, because it uses the same division as the perfect harmonies, and is half a fifth. So it is not surprising that alone of the imperfect harmonies it corresponds with an aspect, that is, the sextile, which in the same way is half a trine. For the rest neither fit into the twelvefold division, nor are part of a perfect chord.

5. Lastly six of the angles of a triangle, four of the angles of a square, three of those of a hexagon, and the two angular distances included in two semicircles complete the whole circuit in a plane. But three of the angles in a pentagon are less than four right angles, and four of them are greater. From that the reason is clear (32) why neither an octile (33) nor a duodecile aspect, nor any of the others, operates.[8] (34) It is precisely here that I make a distinction between the reasons for the aspects and the reasons for the concords, (35) For the argument based on the angles is sound in the case of the aspects, since their operation is due to an angle formed at a point on the Earth's surface, at which they meet, (36) and not to a figure drawn on the circle of the zodiac, which exists in imagination rather than in reality. However, the division of the string is not done on a circle, and does not use angles, but is carried out on the flat along a straight line. (37) Yet concords and aspects may nevertheless have something in common, which suggests the same reasons in each case, as has been said above. However I leave the examination of that problem to the industry of others. (38) Ptolemy's *Music*, which Regiomontanus[9] was going to publish with Porphyrius's exegesis, but which Cardanus[10] asserts has not yet been printed, undoubtedly discusses this topic. See also (39) what can be applied to it from Euclid's *Music*.

AUTHOR'S NOTES ON CHAPTER TWELVE

(1) *Although I do not altogether object to that.*] This theme I have dealt with openly in my book on the New Star, and in my Reply to the objections of Röslin.[11] That is, the four quadrants of the zodiac circle are indeed shown by the specifications of two motions, the diurnal motion and the annual motion of the Sun, which are also followed by the turning points of its light and heat-giving; but for the subdivision of each quadrant into precisely three signs, no such warrant is given either by the motion or by their powers, of which the implication could be taken into account, with the single exception of the very general distinction of any quantity whatever into beginning, middle, and end. However there is no necessity which dictates that these parts should be equal, or even parts; for it is sufficient that the whole line representing the quadrant should be taken as the middle, and for the beginning and end the two extremes of the lines, or points, which are not a part of the line.

(2) *Apart from quantity, or what is similar to quantity, and endowed with a power of some kind, there is nothing in the whole universe which is capable of being numbered.*] I let slip a ridiculous opinion, in truth not an opinion. For what is "Nothing except for everything"? Counting, an action of the mind, applies to everything, divine and human. There is no distinction, not even the slightest, either real or in intention (whether it be of first intention, or second, or third, or whichever intention you like) which does

cuxae, vel tertiæ, vel quota libet intentionis;) quæ non quandam similitudinem habeat cum diuisione rectæ in partes. Vide, quæ de numeris disputaui lib. IV. Harmonicorum Cap. I. fol. 117. Hoc autem mihi erat in Animo, cum hanc sententiam conciperem; quicquid numeratur à nobis (præter diuinas personas in SS. Trinitate) id respectum aliquem habere quantitatuum, saltem in intentione numerantis.

(3) Corpora dissecuimus per Zodiacum.] Per imaginationem plani per sectiones illas laterum & per centrum figurarum omnium traducti, & vsque sub fixas extensi, cuius sectio cum sphæra fixarum nobis peperit in conceptione illa Eclipticam.

(4) Quid sectione hac Zodiacus ipse adeptus.] Si nimirum ex centro communi figurarum, rectæ per sectiones dicti plani cum lateribus figurarum, eiiciantur vsque sub fixas: addendum autem: Si etiam omnes quinq; figuræ tali irregulari situ inuicem coaptentur, vt singularum singula latera sectionibus suis stent in vna tali recta linea: tunc enim Zodiacus distinguetur in partes tales, quas non metitur nisi centum & vicesima totius. Cum autem situs iste sit irregularis; regularis vero per angulos Dodecaedri & Icosaedri octonos vtrinque in planum dictum incidentes, distinguat Zodiacum in irrationalia; patet hanc diuisionem non esse propriam quinque figurarum. Eam igitur in Epitom. Astr. lib. II. fol. 181. demonstraui propriam esse figurarum planarum, Regularium demonstrabilium si illæ circulo inscribantur ab vno eius puncto.

(5) Motum Solis & Lunæ menstruum.] Solis intellige annuum. Nam dum Sol annum permeat: Luna duodecim menses conficit fere. Adeoque hanc distributionem anni, & accommodationem motuum Solis & Lunæ, saltem in primo proportionis illorum conceptu, Ego archetypicam statuo, exque hac ordinatione, & ex concursu naturalium causarum motricum, causas eruo quarundam inæqualitatum in Luna: vt monui in Prolegomenis Ephemeridum, & doceo plene in Epit. Astr. lib. IV. Simile, quid ibidem inuenies etiam de proportione anni ad reuolutiones diurnas 360. (in prima intentione) quibus accedunt d. inde ob concursum causarum, reuolutiones 5. & quadrans: vnde elicitur noua æquatio temporis. Etsi delibero adhuc, obseruationesque expendo.

(6) Coniunctiones magnas superiorum.] Hoc quidem accidentarium est, non archetypicum. Nam vt doceo lib. V. Harmonicorum, Periodica Planetarum tempora sunt ex Harmonicis contemperationibus motuum extremorum: in Aphelijs enim debuit esse motuum proportio quæ 2. ad 5. fere, in Perihelijs vero, quæ 5. ad 12. vt scilicet inter Saturni Aphelium & Iouis perihelium posset esse Diapente Epi Diapason, inter vero Saturni perihelium & Iouis Aphelium, perfectum Diapason, quia hæ duæ Harmoniæ Cubo cognatæ sunt. Hæc enim prima & Archetypica in motibus est causa. Quod si igitur vt Apheliorum motuum, sic totarum periodorum proportio esset quæ 2. ad 5. tunc in annis 60. contingeret præcise duæ reuolutiones Saturni, quinque vero Iouis; in annis 12. vna Iouis: & Saturnus & Iupiter coniuncti verbi causa, in principio Arietis, præcise post 20. annos in ipso principio Sagittarij coirent iterum. Iupiter enim superato Saturno, dum Zodiacum emensus Saturnum fugientem persequitur: ille interim ex Ariete abijt tantum, vt Iupiter in quinque reuolutionibus ter solummodo assequatur ipsum, quia effugit Saturnus per duas ex quinque; ita restant tres coniunctiones in quinque Iouialibus periodis perfecto triangulo distributæ. Ecce vt hic triangularis coniunctionum situs sit necessarium consequens causæ archetypicæ, ex Harmoniis desumptæ; accidat vero trisectioni Zodiaci, seu per pyramida, seu per triangulum, si quis illam, vt in hoc capite ponebam, Archetypicam esse contenderit. Vicissim si totarum periodorum ♄ & ♃ proportio esset illa, quæpropter Harmonicas contemperationes debuit esse motuum Periheliorum, sc. 5. ad 12. tunc in annis 150. Iupiter reuerteretur duodecies, semel in annis 12. semis. Ablatis igitur 5. de 12. restarent 7. toties sc. Iupiter assequeretur Saturnum. Itaque Zodiacus per has coniunctiones diuideretur in partes 7. quarum quinis, id est 257. gradibus binæ coniunctiones à se inuicem remouerentur; verbi causa, post vnam in o ♈, contingeret altera in 17. ♉ tertia in 4. ♍. Sed quia periodica tempora componuntur ex motibus tam Apheliis, quam periheliis, exque interiectis omnibus; hinc nascitur etiam intermedia periodorum proportio, coniunctionumque per Zodiacum distributio; vt prima in principio Arietis collocata, secunda neque in ipsum principium sagittarij veniat, nec etiam vsque in 17. ♐ excurrat, sed media & æquabili ratione ad tres gradus vltra triangularem locum progrediatur. Quod si ipsa Zodiaci distinctio in tres trientes, per figuras Geometricas, genuina & archetypica causa fuisset huius dispositionis coniunctionum; vtique expressisset illa perfectum triangulum; non aberrat enim diuinum opus ab archetypo suo. Non igitur amplius mirum esse debet, cur Saturni Iouisq; congressus ad triangulum alludant; quia nec perfecta & plane accidentaria est allusio.

not have some resemblance to the division of a straight line into parts. See my discussion of numbers[12] in Book IV of the *Harmonice*, Chapter 1, page 117. However, what I had in mind when I conceived this opinion was that whatever is counted by us (except the divine persons of the Holy Trinity) has some quantitative aspect, at least in the intention of the one counting.

(3) *We have already cut through all the solids with the zodiac.*] By imagining a plane drawn through these intersections of the edges and through the center of all the figures, and extended right up to the fixed stars, the intersection of which as thus conceived with the sphere of the fixed stars has produced for us the ecliptic.

(4) *What the zodiac itself has acquired. . .in this cutting.*] Certainly if straight lines are taken from the common center of the figures through the intersections of the plane mentioned with the edges of the figures, right up to the fixed stars. However, we must add the following words: also if all the five figures fitted together in such an irregular arrangement among themselves that the individual edges of individual figures at their intersections fall on a single straight line. For in that case the zodiac will be divided into parts which are measured only in units of one hundred and twentieths of the whole. As, however, that arrangement is irregular, and the regular arrangement, in which eight each of the vertices of the dodecahedron and icosahedron fall on the said plane, divides the zodiac into irrational parts, it is evident that this is not the division which is proper to the five figures. I have therefore shown in the *Epitome of Astronomy*, Book II, page 181, that it is proper to the plane, regular figures which are capable of being constructed, if they are inscribed in a circle from a single point on it.

(5) *The motion of the Sun and the monthly path of the Moon.*] Understand "annual" motion of the Sun. For while the Sun passes through a year, the Moon completes twelve months about. And so I establish this apportionment of the year and accommodation of the motions of the Sun and Moon as the archetype; and from this orderliness and from the concurrence of the natural causes of motion, I extract the causes of certain irregularities in the Moon, as I have commented in the *Prolegomena to the Ephemerides*, and report fully in the *Epitome of Astronomy*, Book IV.[13] You will also find in the same place a similar point on the ratio of the year to the 360 diurnal revolutions (in the first intention), to which are added on account of a concurrence of causes 5 1/4 revolutions. Hence a new equation of time is elicited. However I am still pondering, and weighing up the observations.

(6) *The great conjunctions of the superior planets.*] This indeed is accidental, not archetypal. For as I report in Book V of the *Harmonice*, the periodic times of the planets are derived from the harmonic consonances of the extreme motions. For at the aphelia the ratio of the motions should have been as 2 is to 5, about; and at the perihelia as 5 to 12. Thus, for instance, between the aphelion of Saturn and the perihelion of Jupiter could be the chord of the fifth above the octave and the tonic,[14] and between the perihelion of Saturn and the aphelion of Jupiter the perfect octave, as these two harmonies are akin to the cube. For this is the first and archetypal cause among the motions. Then if the ratio of the motions of the aphelia were as that of the whole periods, which is as 2 is to 5, then in 60 years there would occur precisely two revolutions of Saturn, and five of Jupiter; and in 12 years one revolution of Jupiter. If there were a conjunction of Saturn and Jupiter, for example at the beginning of Aries, after precisely 20 years they would come together again at the beginning of Sagittarius. For while Jupiter, after overtaking Saturn, has traversed the zodiac and again pursues the fleeing Saturn, it has meanwhile departed from Aries by such an amount that Jupiter in five revolutions catches up to Saturn only three times, because Saturn escapes for two revolutions out of five. So there remain three conjunctions in five of Jupiter's periods, distributed in a perfect triangle. Notice that this triangular arrangement of the conjunction is a necessary consequence of the archetypal cause, derived from the harmonies; whereas it is an accident of the trisection of the zodiac, whether in accordance with the pyramid or with the triangle, if it is maintained, as I supposed in this chapter, that it is archetypal. On the other hand, if the ratio of the total periods of Saturn and Jupiter were what the ratio of the motions of the perihelia should have been on account of the harmonic consonances, that is as 5 to 12, then in 150 years Jupiter would complete twelve revolutions, or one in 12½ years. Then on substracting 5 from 12 the remainder would be 7, that is, Jupiter would catch up to Saturn that many times. Consequently the zodiac would be divided by these conjunctions into seven parts, and five of them, that is 257°, would be the separation of two conjunctions from each other. For example, after one in 0° of Aries, another would occur in 17° of Sagittarius, a third in 4° of Virgo. But because the periodic times are compounded of both the motions at the aphelia and perihelia, and of all the motions in between, it also arises that the ratio of the periods is intermediate, and that the conjunctions are distributed round the zodiac in such a way that if the first is located at the beginning of Aries, the second neither comes to the beginning of Sagittarius nor presses on as far as 17° of Sagittarius, but in an intermediate and uniform proportion goes forward three degrees beyond the triangular position. But if the division of the zodiac itself into three thirds, in accordance with geometrical figures, had been the true and archetypal cause, it would in any case have expressed the perfect triangle; for the divine work does not deviate from its archetype. There should not therefore be any further reason to wonder why the meetings of Saturn and Jupiter make sport with a triangle; for sporting is not perfect and is plainly accidental.

(7) Atque adeo quam hæc.] Hic ſunt ipſiſſima principia mei operis Harmonici, eaque non tantum opinationum, quæ poſterioribus temporibus corrigenda fuerint, ſed etiam veriſſima rei ipſius:Omnis enim philoſophica ſpeculatio debet initium capere à ſenſuum experimentis: hic vero, quæ ſenſus auditus teſtetur de numero vocum, cum vna aliqua conſonantium; quæ item ſenſus oculorum, de longitudine chordarum conſonantium; emendatiſſime & plene expreſſam habes.

(8) Toties, nec ſæpius.] Mirum eſt equidem, cum tot ex antiquo extiterint ſcriptores Harmonicorum nuſpiam penes ipſos occurrere obſeruationem hanc, de numero ſectionum Harmonicarum plane fundamentalem, & quæ recta ad cauſas ducit; cum tam ſit obuium cuilibet, id in chorda quacunque extenſa, cuius ſpatium ſubiectum circino diuidi poſſit, ſimplici applicatione rei duræ, vt cultri aut clauis, ad chordam, manu vna, & percuſſione partium eius interſtinctaru, cum plectro in manu altera, experimentari. Itaq; ſumma fuit iſta fœlicitas in principio ſpeculationis tēdenti ad opus Harmonicum ſcribendum: quamuis tunc quidem nondum id animo deſtinaueram. Cauſa autem, cur ſeptem ordine voces, vſque ad Diapaſon cum ima ſuſcepta conſonent, eſt iſta, quia chorda ſepties Harmonice diuidi poteſt; ſingulis enim iis actibus ſinguli conſtituuntur ſoni, conſonantes cum ſono totius. Vide lib.III. Harm. cap.II.

(9) Atque hæ ſolæ.] Verum eſt, ſi Naturale id dicas, quod prima ſtatim coaptatione ſectionum, in ipſo quaſi veſtigio cauſarum prægreſſarum elicitur; vt diſtinguatur ab eo, quod ſecūdaria ratione, velut artificialiter & imitatione Naturæ conſtituitur. At ſi non ordinem ortus, ſed proportionem ipſam reſpicias, naturalia erunt & illa interualla dicenda, quæ proportiones ſic ante conſtitutas, imitatione Naturæ ſuſcipiunt. Vt in ſequela vocum Re, Mi, Fa, Sol, La Naturale eſt interuallum, Fa, Sol, Tonus maior dictus; quippe primitus conſtituitur, quando interuallum Re, Fa, adhuc nondum eſt diuiſum: ſi iam etiam inter Re, Fa, deſignetur vox Mi, tali proportione chordæ Mi, ad chordam Re, quali eſt Sol, ad Fa, tunc & ipſa vox Mi Naturalis haberi debet. Quod vero cauſam hic reddidi diſtinctionis, quaſi Fa, Sol, habeant indubitatos numeros; Mi vero, non item: id condonandum eſt tyrocinio tunc poſito. Nam lib.III.Harmon.cap.V. & VII. cauſas optimas tradidi, quibus etiam ſono Mi, & ſimilibus ſuus indubitatus numerus aſſignatur.

(10) Nam vocem F fa vt, aliam ex.] Hoc verum eſt, ſi vtrinque velles perfectum Diapente conſtituere. Atqui, quod tunc ignorabam, pars non minima eſt diſciplinæ, de Conſonantiis adulterinis, quam tradidi lib.III.Harmon.cap.XII.

(11) Cum enim imperfectæ omnes ſint.] Ita vſitate appellantur; veteres ne pro Conſonantiis quidem habuerunt. In meo Opere Harmonices, fol.83. poſteriori nec minus & cap. I. & IV. libri III. & paſſim etiam imperfectas appellaui, ſed vox iſta non æque valet adulterinæ. Deeſt enim adulterinæ minimum aliquid, quo minus ſit plena conſonantia; nihil deeſt tertiæ & ſextæ legitimæ, quo minus inter conſonantias referantur. Itaque diſtinctionis cauſa præſtat tertias & ſextas, minores dicere conſonantias, idque non quantitatis tantum reſpectu, ſed etiam ſpeciei.

(12) Quod ſi ſeptem diuiſionum.] Hunc ego neruū argumenti tunc conſtitui; Diuiditur Zodiacus in partes 12. & 120. diuiditur & chorda in totidē harmonice: ergo numeri hi ſunt apud naturam in pretio. At cum Zodiaci diuiſio ſit à quinq; corporibus (vti tunc exiſtimabam) veriſimile, indidem & Chordæ diuiſionem eſſe, & ſic quinque illas figuras etiam Harmoniarum Ideas eſſe; tunc quidem ſequi videbatur. Sed nunc ex opere Harmonico lector cauſas Harmonicorum genuinas petat: ſunt enim non illa quinque corpora Geometrica: ſed potius figura plana in circulum inſcripta, &c.

(13) Plane quaſi perfectæ concordiæ à Quadrato & Triangulo.] Iucundum eſt, primos inuentionum conatus etiam errantes intueri. Ecce cauſas genuinas & archetypicas concordantiarum, quas manibus verſabam, cæcutiens, velut abſentes, anxie quæſiui. Figuræ planæ ſunt cauſæ concordantiarum ſcipſis, non quatenus fiunt ſolidarum figurarum ſuperficies. Fruſtra ad ſolida reſpexi in conſtituendis Harmonicis motuum proportionibus.

(14) Sed quia cauſas huius cognationis ignoramus.] Atqui cauſas iam nominatas vides, figuras planas: Atqui non cognatio non conſanguinitas, ſed nuda affinitas eſt. Figura enim plana ex vna parte diuidunt circulum harmonice, ex altera parte congruunt in figuras quinque ſolidas. Ergo & Harmonica circuli diuiſio, & quinque figuræ, in vno tertio, in figuris ſcil. planis conueniunt.

(15) Vide-

Chapter XII

(7) *Furthermore how greatly.*] Here are exactly the principles of my work on harmony, and principles not of mere opinions, which will be open to correction in later times, but the genuine principles of the actual fact. For every philosophical speculation ought to take its starting point from what is experienced by the senses: in this case, what the sense of hearing testifies about the number of notes which are in harmony with a given note, and also what the sense of the eyes testifies about the length of strings which are in harmony, you know with great accuracy and in complete detail.

(8) *The number of times, and no fewer.*] It is indeed surprising, as there have been so many writers on harmony since antiquity, that nowhere in their works does there occur this observation on the number of the harmonic divisions, though it is plainly fundamental, and leads straight to the causes; and as it is so easy for anyone to make trial of it, on any stretched string, of which the length can be covered by a pair of compasses, and divided by the simple application of a hard object, such as a knife or a key, to the string with one hand, and the striking of the parts of it which are divided off with a plectrum in the other hand. Consequently this was a great piece of good fortune at the beginning of my investigation to one who was leaning towards writing a work on harmony, though I had not yet decided on it in my mind. Now the reason why seven notes in order up to the octave are in harmony with the lowest which is sounded is this, that the string can be divided seven times harmonically; for by each act of division is established a particular sound which is in harmony with the sound of the whole. See Book III of the *Harmonice*, Chapter 2.

(9) *Now these. . .the only.*] This is true, if you call natural that which is produced immediately at the first fitting together of the divisions, in the very footsteps, so to speak, of the preceding causes; so that it is distinguished from that which is set up by a secondary ratio as if artificially and in imitation of Nature. However, if you take account not of the order but of the actual proportion, those intervals should also be called natural which take the ratios thus previously set up in imitation of Nature. Thus in the sequence of notes re, mi, fa, sol, la, the natural interval is fa, sol, called a whole tone; for it is set up, first of all, when the interval re, fa has still not yet been divided off. If the note mi were now also to be designated between re and fa, with the string for mi in the same proportion to the string for re as sol to fa, then the note mi itself ought also to be taken as natural. Indeed my here giving this as the reason for the distinction, as if fa and sol have undoubted numbers, but mi has not, must be pardoned because I was then serving my apprenticeship. For in Book III of the *Harmonice*, Chapters 5 and 7, I have expounded excellent reasons for assigning their own undoubted number to mi and similar notes.

(10) *For the note F (fa ut) is different if. . . .*] This is true, if you intend to set up a perfect fifth in both cases. But, although I did not know it then, a considerable part of the discipline concerns adjusted consonances,[15] and I have expounded it in Book III of the *Harmonice*, Chapter 12.

(11) *For though they are all imperfect.*] This is what they are usually called. The ancients did not even accept them as consonances. In my work *Harmonice*, the second page 83, and equally in Chapters 1 and 4 of Book III and generally, I have also called them imperfect; but that name is not so appropriate to an adjusted concord. For an adjusted consonance is a very small amount short of being a full consonance; and the third and sixth lack nothing which would stop them being included among the consonances. Thus the distinction is best made by calling thirds and minor sixths consonances, and that not only in respect of quantity, but also of species.

(12) *But if. . .of the seven divisions.*] At that time, I made this the mainspring of the argument. The zodiac is divided into 12 parts and 120 parts; division of a string into the same number of parts is harmonic; therefore, these numbers are important in Nature. But since the division of the zodiac is based on the five solids (as I then thought), by the same argument it is probable that the division of the string is also; and so it then seemed to follow that the five solids were the Ideas of the harmonies. But the reader should now look for the true causes of the harmonies in my work on harmony; for they are not the five geometrical solids, but rather the plane figures inscribed in a circle, etc.

(13) *Plainly because the perfect concords (came from) the square and triangle.*] It is pleasant to contemplate my first efforts at my discoveries, even though they were wrong. You can see that I anxiously sought for the true and archetypal causes of concordance, as if they were not there, like a blind man, when I had them in my hands. The plane figures are the causes of concordance in themselves, not in virtue of being surfaces of solid figures. It was in vain that I turned to the solids in establishing the harmonic proportions of the motions.

(14) *But because we do not know the causes of this kinship.*] But you see that the causes have now been named: the plane figures. But it is not a kinship, not a consanguinity, but a mere affinity. For the plane figures on the one hand divide the circle harmonically, and on the other hand agree with the five solid figures. Hence both the harmonic division of the circle and the five figures meet in a single third factor, that is, the plane figures.

(15) Videmus quidem duos Harmoniarum ordines.] *Nota hoc diligenter, & cognosce vel hoc vno exemplo vim aliarum fortuitarum collisionum.* Septem concordantiarum formas, seu septem sectiones Harmonicas, in prioribus ad quinariam redegimus vtcumque, vt binæ semper imperfectæ,pro vna censerentur. Quinarius iste in duo abit membra,vt hinc stent tres,inde duæ. Atqui & Quinarius corporum ex vna parte tria habet, ex altera duo : neque tamen illis tribus est cognatio cum his tribus;nec illa duæ respondent his duobus. Nam duæ duplices imperfectarum concordantiarū formæ communicant decangulo,quod est hic cognatum rei ex primariis corporibus tribus, & vna ex secundariis duobus. Accidit ergo respectu rei alterutrius,vt altera vtatur eadem diuisione. Talia fortuita multa eueniunt in rebus Mathematicis & Naturalibus,contra quorum concursum, vt ἀναιτίον confirmanda est iudicij nostri imbecillitas,ne statim quacung, credulitate, sine duce ratione,abripiatur. Vide quæ supra de ijs disputauerim,quæ sunt numero tria,vel sex,vel septem.

(16) Indicem digitum ad causam harum rerum occultissimam intendit.] *Ecce rursum scribendo proficientem.* Hæc enim inuenta est causa ipsissima,vt lib.III.cap.I.in axiomatibus videre est. Nam figuræ quæ perfectiores habent demonstrationes, suntque effabiles (Triangulum & Quadrangulum & Sexangulum) perfectas etiam pariunt consonantias maiores ; qua vero viliorem habent demonstrationem, & latera ineffabilia (vt Octangulum, Quinquangulum, Decangulum) viliores etiam peperere concordantias maiores imperfectas vulgo dictas. Hæc autem perfectio vel contraria vilitas,insunt consonantiis,propter ipsas figuras planas,insunt & figuris solidis : rursum igitur non cognatio sed affinitas sola intercedit duplicibus illis & imperfectioribus sectionibus Harmonicis, cum Dodecaedro primario,& Icosaedro secundario.

(17) Duos nempe Geometriæ thesauros.] *Duo Theoremata infinitæ vtilitatis,eoq, pretiosissima,sed magnum discrimen tamen est inter vtrumque.* Nam prius,quod latera recti anguli possint tantum,quantum subtensa recto,hoc inquam recte comparaueris massæ auri : alterum,de sectione proportionali, Gemmam dixeris. Ipsum enim per se quidem pulchrum est, at sine priori valet nihil:ipsum tamen promonet scientiam tunc vlterius, cum prius illud nos aliquatenus prouectos, iam destituit, scilic.ad demonstrationem & inuentionem lateris Decangularis, & cognatarum quantitatum.

(18) Vt autem singulæ Harmoniæ.] *Nil mirum,accommodationem Harmoniarum ad corpora non in promptu esse ; quod enim in sinu Naturæ non est , id depromi nequit:res ista hoc quidem numero,& hac quantitate descripta,sunt insociabiles.Etsi vero & ego in Harmonicis,lib.V.cap. IX.corporibus Harmonias associo:at id non sit causa ortus vnius ex alio;sed causa vsus, in exornatione Mundi, Argumenta associationis,cap.II.multa quidem sunt etiam ex formalibus rationibus, tam corporum,quam Harmoniarum: at illa argumenta sunt multis semper Harmoniis inter se communia,singulæ Harmoniæ singulis corporibus per ea non vindicantur : accedunt igitur diuersi generis argumenta forinseca,aut à comparatione proportionum figuralium cum Harmonicis deducta ; quibus tandem Harmoniæ non istæ,sed pleræque his maiores,associantur corporibus; at neque immediata est hæc associatio : sed tribuuntur Harmoniæ motibus illorum Planetarum , quorum Orbes bini singula sortiti sunt corpora Regularia. Ita commigrant quidem Harmoniæ in quinque corporum viciniam interstinctæ suis maceriebus,& sub eadem tecta non recipiuntur.*

(19) Illud solum patet,Pyramidi deberi Quintam.] *Imo ne hoc quidem absolute verum est. Nulla quidem ex iis quæ sunt minores, quam Diapason, cognatior est Pyramidi propter Triangulum,quod Pyramidi basin,ipsi Diapente ortum præbet.* Non potest tamen ipsi Diapente locus ibi esse,vbi Pyramis interlocatur: sed aliis notis censenda est hæc Harmoniarum ad figuras aptitudo, de quo vide lib.V.Harmon.cap.II. Quinimo ne Diapente quidem Trianguli solius proxima est proles, sed antecedit illud Diapason epi diapente; vide lib.IV. Harmon. Cap.VI.fol.154. Causam quidem huius affirmati verissimam hic in ipso textu,ignarus ipse posui,tertiam sc.partem circuli.

(20) Secundaria accipere eas, quæ lineas scribunt.] *Secundariis scil. corporibus associandas esse concordantias illas,quæ sic per sectionem chordæ repræsententur,vt , si ex chorda, perfectionem signata fiat circulus,linea recta quæ signa connectit , non fiat latus alicuius figuræ perfectæ, sed vel vna linea solitaria maneat,vel latus fiat figuræ abundantis , quas lib.I. & II. Harmon.stellas à similitudine,placuit indigetare.* Pulchrum quidem commentum causæ,pulchra distributio secundū eam,Harmoniarum inter quinque corpora,si responsum Numeri spectes,at per se, neque speciem hoc habet causæ,neque Sexta supra Diapason quicquam habet cum Icosaedro commercii.

(21) Faciunt tantum lineas.] *Quasi vero stellæ non sint etiam figura? Nimirum aliquid*
erat

Chapter XII

(15) *Indeed we see two classes of harmonies.*] Take careful note of this, and learn even from this one example the force of other accidental coincidences. In previous writings we have somehow reduced the seven forms of concordance, or the seven harmonic divisions, to five, so that the imperfect pairs were always counted as one. This set of five splits into two branches, with three on one side, two on the other. But the set of five solids also has three on one side, two on the other; yet there is no kinship between the other three and these three, nor do the other two correspond to these two. For the two double forms of imperfect concordance match the decagon, which in this case is akin to one of the three primary solids, and one of the two secondaries. It is therefore an accident in respect of one of these things that the other is classified on the same basis. Many such coincidences occur in matters of mathematics or Nature, and against such combinations, since they are random, the weakness of our judgment must be fortified lest it be seduced on the spot by a piece of credulity, unguided by reason. See my arguments above on things which are three in number, or six, or seven.

(16) *It points the finger at the most secret cause of these matters.*] You can see that again I make progress in my writing. For what has been found here is the actual cause, as is evident in the *Harmonice*, Book III, Chapter 1, in the axioms. For the figures which have more perfect derivations, and are expressible[16] (the triangle and square and hexagon) also give birth to the perfect major consonances, whereas those which have a baser derivation, and inexpressible sides (such as the octagon, pentagon, decagon) have also given birth to baser major consonances commonly called imperfect. Now this perfection, or baseness on the other hand, belongs to the consonances on account of the plane figures themselves, and also belongs to the solid figures. Therefore it is not kinship but only an affinity which links these pairs of consonances and the more imperfect harmonic divisions with the dodecahedron, which is primary, and the icosahedron, which is secondary.

(17) *That is, that there are two treasure houses of geometry.*] They are two theorems of infinite usefulness, and so of the greatest value; but yet there is a great difference between the two. For the former — that the squares of the sides of a right triangle are equal to the square of the hypotenuse — that, I say, can rightly be compared to a mass of gold; the second, on proportional division, can be called a jewel. For in itself it is indeed splendid, but without the previous theorem it has no force; but it then takes knowledge further, when the previous one which has carried us so far deserts us; that is, on the derivation and discovery of the side of the decagon and related quantities.

(18) *However...the individual harmonies.*] It is not surprising that the fitting of the harmonies to the solids is not obvious; for what is not in the bosom of Nature cannot be drawn out. The two things, though delineated by this number and this quantity, are disparate. It is true that in the *Harmonice*, Book V, Chapter 9, I associate the harmonies with the solids, but that does not constitute a reason for one arising from the other, but a reason for their use in the displaying of the universe. There are indeed in Chapter 2 many arguments for the association even from the formal considerations, referring to both the solids and the harmonies. However, these arguments are always common to many harmonies, and particular harmonies are not assigned to particular solids by them. External arguments of a different kind are therefore added, or arguments drawn from a comparison of the proportions of the figures with the harmonic proportions. Eventually not the harmonies mentioned, but many with greater intervals than those are associated with the solids. However, even this association is not direct; but the harmonies are attributed to the motions of those planets of which the orbits in pairs have been allotted single regular solids. Thus the harmonies are removed to the neighborhood of the five solids, to territories marked out between their boundaries, and are not accepted under the same roofs.

(19) *Only one thing is obvious, that to the pyramid should be attributed the...fifth.*] Or rather not even that is absolutely true. It is true that none of those which are smaller than an octave is more akin to the pyramid on account of the triangle, which provides the base for the pyramid, and its origin for the fifth. However, the place for the fifth cannot be in the interval where the pyramid is placed; but the way in which the harmonies fit the figures must be computed from other indications, on which point see Book V of the *Harmonice*, Chapter 2. Indeed not even the fifth is the closest offspring of the triangle alone, but the octave over the fifth takes precedence over it: see Book IV of the *Harmonice*, Chapter 6, page 154. Indeed I inserted into this very part of the text unknowingly the truest cause of this statement, that is, its being one-third of a circle.

(20) *That the secondaries accept those which describe lines.*] In other words, that with the secondary solids should be associated those concords which are represented by dividing a string in such a way that if the string, marked out by the division, is made into a circle, the straight line which connects the marks does not become the side of some perfect figure, but either remains as one solitary line, or becomes the side of one of the overflowing figures which in Books I and II of the *Harmonice* I have chosen to proclaim as stars from the resemblance. It is indeed a splendid fiction of a cause, and a splendid distribution of the harmonies among the five solids in accordance with it, if you look at the correspondence of the number; but in itself neither does this have the appearance of a cause, nor does the sixth above the octave have any relation to the icosahedron.

143

erat comminiscendum,quo stella Octangularis associaretur Diametro,sub eodem, quasi genere,reclamante Naturâ.Rectè igitur factum,quod non acquieui huic distributioni.

(22) An malumus Octaedro quartam.] *Hoc plane sum secutus in lib. V. Harmonic. sed in instituto diuerso.* Hic enim quærebam ortum Harmoniarum singularum:at lib.V.Harmonicorum;delectus inter iam ortas est institutus , quæ Harmonia , quibus Planetis , qua mediante figura solida,consociaretur.Cubo igitur etsi non rectè hic adscribitur ipse ortus consonantiæ Diapason ; rectè tamen dicto lib.V.Harmonicorum,associatur ipsum Diapason;non causa ortus, sed causa cohabitationis inter Planetas cosdem; rectè associatur Octaedro, quod Cubi coniunx est, Disdiapason , cui in harmonica sectione adhæret Diatessaron.Vide lib.V.cap.IX.Prop.VIII.& XII.

(23) Relinquere Icosaedro priorem imperfectam.] *Hic iterum fortuito (quippe in speculatione non propria)in verum incidi quadamtenus.Nam Prop.XV.& XXVII.dicti capitis IX.Dodecaedro quidem,Diapente obtigit,Icosaedro vero,vtraq, Sextarum, Tertiis locum nullum esse,probatur Prop.VI.*

(24) Veniamus modo ad Aspectus.] *De hac materia est meus liber IV. Harmonicorum.*

(25) Quomodo tres perfectæ Harmoniæ cum tribus.] *Parum aliquid in hac comparatione emendandum,vide lib.IV.Harm.cap.VI.fol.154.*

(26) Quemque debilissimum esse ferunt.] *Nequaquam vero debilem experientia testatur,sed fortiorem sæpe ipso Trino;causam ex meis principijs do lib.IV.Harm.*

(27) Qualem quidem Ptolemæus non dedit.] *Puta in Tetrabiblo de Astrologia scripto.At in Harmonicis,quæ tunc nondum videram,causam hanc tangit, sed male, vt ex meis notis ad Ptolemæum patebit:Omnino enim,& vnum,& quinq,signa, aspectus constituunt efficaces, quos appello,Semisextum,& Quincuncem.*

(28) Nullam talem in vocibus agnoscit Natura concordiam.] *Hoc ad literam falsum est.Nam inter chordas 1.& 12.est Trisdiapason epi diapente ; sic inter chordas 5.& 12.est Tertia minor supra Diapason.Aliud igitur habebam in animo,cum hæc verba scriberem: scilicet,nullam esse sectionem tripliciter Harmonicam,quæ respondeat hisce diuisionibus circuli : quia etsi 1. 12. item 5. 12. consonent : at residua 11.& 7.abhorrent ab vtrisque terminis. At non esse eandem rationem Aspectuum , quæ est Consonantiarum , doceo per totum librum IV. Harmonicorum ,præcipue cap. VI.*

(29) Causa vtrinque,&c.ex quinque corporibus.] *Minime ex his , at bene,ex figuris planis,quarum non ignobilissima,Dodecagonus.*

(30) Cum igitur omnes.] *Hoc initio facto,cæpi augere numerum aspectuum : etsi male adscini Sesquadrum,seu gradus 135.male omisi semisextum,seu gr.30. Vide sæpe allegatum cap.VI. lib.IV.Harmon.*

(31) Causæ quidem quas probabiliter.] *Frustra: Nam confirmat experientia Quintilem,& Biquintilem;De Sesquadro vero,cur ille minus sit efficax,quam reliqui omnes,causæ lib.IV. Harm.cap.V.traduntur longe diuersæ.Ista vero,hic recensita quinq, causæ,sunt nobis iterum refutandæ,ne teneant Quintilem & Biquintilem.*

Nam quod causam primam attinet ; sicut cum Trino sextilis implet circulum,cum quadrato quadratus alius,sic etiam cum quintili Tridecilis, cum Biquintili decilis; cum sesquadro sequadrus implent semicirculum,nec repudiat hos Musica. Non est igitur efficacia ab hac adæquatione semicirculi.

Secunda causa ad rem est: at illa non penitus repudiant Quintilem , sed solummodo imperfectiorem facit Trino& sextili;quantum quidem ipsa pollet,cum sola non sit.Irrationale autem sic nucupo cum vulgo,quod in Harmonicis mihi dicitur,Ineffabile.

Tertia causa coincidit cum prima;omnis enim in semicirculo angulus rectus est. Et si aliter informetur hæc causa,quod bini semper aspectus efficiant summam duorum rectorum, nunc semicirculus iterum est eorum mensura.

Quarta causa futilis est, Si enim Tertia mollis ideo est quodammodo perfecta, quia vtitur eadê diuisione cum perfectis,scil.Duodenaria;sane & diuisio vicenaria constituitur adiumento quaternariæ,& sexagenaria ternaria. Si Tertia dura non quadrat ad duodenarium , maiori termino 5.sane neque tertia mollis quadrat ad Vicenarium , maiori termino 6. Rursum si tertia mollis ideo habetur pro perfecta,quia est dimidium ipsius Diapente,magis tertia dura habebitur pro perfecta, quia & ipsa est dimidium ipsius Diapente superans tantum,quantum tertia mollis deficit à dimidio. Itaq, cauendum

dum

(21) *Makes only lines.*] Implying that the "stars" are not also figures? Forsooth there should have been some fictitious association made between the octagonal star and the diameter, as if included in the same class, though Nature would object. I was therefore right not to agree with this distribution.

(22) *Or do we prefer (to allot) the fourth to the octahedron.*] I have plainly followed this course in Book V of the *Harmonice*, but with a different intention. For here I was seeking for the origin of individual harmonies; but in Book V of the *Harmonice* my intention was to select, among those already originated, which harmony was the partner of which planets, and with which solid figure in between. Therefore, although the attribution here of the origin of the consonance of the octave to the cube is not correct, yet as is said in Book V of the *Harmonice*, the octave is correctly associated with the cube, not as the cause of its origin, but as the cause of its dwelling with it between the same planets. The association with the octahedron, which is the spouse of the cube, of the double octave, to which the fourth is linked in the harmonic division, is correct. See Book V, Chapter 9, Propositions 8 and 12.

(23) *To leave for the icosahedron the former of the imperfect pairs.*] Here again I accidentally (that is, in an investigation which was irrelevant) stumbled on the truth to a certain extent. For by Propositions 15 and 27 of the aforesaid Chapter 9 the fifth indeed has fallen to the dodecahedron, and both sixths to the icosahedron; but it is proved that there is no place for the thirds by Proposition 6.

(24) *Let us now come to the aspects.*] Book IV of my *Harmonice* is about this matter.

(25) *How the three perfect harmonies (can be most beautifully related) to the three.*] There is a small point to be corrected in this comparison: see Book IV of the *Harmonice*, Chapter 6, page 154.

(26) *And which they say is the weakest.*] Experience does not by any means testify that it is weak, but that it is often stronger than the trine itself. I give the reason from my own principles in Book IV of the *Harmonice*.

(27) *Which Ptolemy at least did not give.*] Understand this to refer to what he wrote in the *Tetrabiblos* about astrology. But in his *Harmony*, which I had then not yet seen, he touches on this reason, but wrongly, as will be evident from my notes on Ptolemy. For in all respects both one sign and five signs constitute potent aspects, which I call the semisextile and the quinduodecile.

(28) *Nature recognizes no such harmony among musical notes.*] Taken literally this is untrue. For between strings in the ratio 1:12 is a triple octave and a fifth; and similarly between strings in the ratio 5:12 is the minor third above the octave. Thus I had something else in mind when I wrote these words, that is, that there is no ratio of division which is harmonic in three ways, which would correspond with these divisions of the circle, since although 1:12 is concordant, yet the remainders 11 and 7 are repugnant to both terms.[17] However, I explain in the whole of Book IV of the *Harmonice*, especially Chapter 6, that the reasoning is not the same for the aspects as it is for the consonances.

(29) *The cause in both cases, etc...on the five solids.*] Hardly on them but decidedly on the plane figures, of which the dodecagon is not the most ignoble.

(30) *Since, then, all.*] Having started in this way, I began to increase the number of aspects. Although I was wrong to include the trioctile, or 135°, I was wrong to omit the duodecile, or 30°. See Chapter 6 of Book IV of the *Harmonice*, to which I have often referred.

(31) *The reasons which (one may) with probability.*] In vain. For experience confirms the case of the quintile and biquintile; whereas quite different reasons are reported in Book IV of the *Harmonice*, Chapter 5, why the trioctile is less potent than all the rest.[18] Indeed we must once again refute the five reasons listed here, so that they do not include the quintile and biquintile.

For as far as the first reason is concerned, just as a sextile with a trine makes up a circle, along with a quartile and another quartile, so also a tridecile with a quintile, a decile with a biquintile, and an octile with a trioctile make up a semicircle; and Music does not repudiate them.[19] Consequently potency does not come from this property of equalling a semicircle.

The second reason is relevant; yet it does not entirely repudiate the quintile, but merely makes it more imperfect than the trine and the sextile, as far as the effect of this reason itself is concerned, though it is not alone. However I here use the common word "irrational" for what in the *Harmonice* I call inexpressible.

The third reason coincides with the first. For every angle in a semicircle is a right angle. And if this reason is rephrased as "a pair of aspects always make up the sum of two right angles," in that case a semicircle is again the measure of them.

The fourth reason is worthless; for if the soft third is to a certain extent perfect on account of using the same division as the perfect harmonies, that is, into twelfths, certainly division into twentieths is also established with the help of quarters, and into sixtieths with the help of thirds. If the hard third does not fit its greater term, 5, in respect of divisibility into twelfths, certainly neither does the soft third fit its greater term 6 in respect of divisibility into twentieths. Further, if the reason why the soft third is taken as perfect is that it is half the fifth, the hard third will more readily be taken as perfect, because it is itself half the fifth, as much in excess as the soft third falls short of the half. Thus we must here beware of the

dum hic à collusione istæ accidentaria, quod etiam sextilis sit præcise dimidiatus Trinus, & Sextilis
Tertiæ molli respondeat. Nam docui cap. VI. lib. IV. Harmonicorum, Sextili respondere, non Tertiam
mollem, sed diapente epidisdiapason: ipsam vero Tertiam mollem communem esse sobol. in tam quinquanguli, quam sexanguli, quia his numeris 5. 6. comprehenditur. Est quæ causa ditiss̄sima, quæ Tri-
num in duos perfectos sextiles diuidit, ab illa causa, quæ Diapente in duas Tertias, maiorem & minorem diuidit. Id quidem vel ex hoc apparet, quod partes sunt illic æquales, hic inæquales. Nihil igitur
detrahitur nobilitati Tertiæ duræ, nihil accedit Tertiæ molli, quod Sextilis est dimidium de Trino;
Quintilis non item; & posset non minoris hoc æstimari, quod Quintilis sit dimidium de biquintili,
&c. Equidem non minima pars est solertiæ, ab huiusmodi concursibus accidentariis cauere, qui, vt
quondam Siren sicula nauigantes cantu, sic ipsi philosophantes voluptate apparentis pulchritudinis,
aptique responsus (siquidem hic adhæres̄cant admiratione capti, vbi causa nulla est alterius in altero)
detinent, vt ad scopum præsinitum scientiæ peruenire non possint.

    Quinta causa est effectus secundæ, & efficit, vt Quintilis imperfectior aspectus, Tertia durá
imperfectior (potius alterius generis) consonantia sit: non efficit, vt ille aspectus plane nullius efficacia,
hæc consonantia nullius sit suauitatis. Nam hoc iam dudum de omnibus quinque obiectionibus erat
dicendum; quod si valerent, in Musica æque valerent, ac in negotio aspectuum: nec ratio vlla redditur,
cur hæ causæ valeant illic, non valeant hic.

    (32) Quare nec Octangularis.] De Octangula stella res est alia. Cur enim illa, cum
sesquadro eliminetur, seu magis postponatur ex aspectibus, non item è Musica eliminetur Sexta minor ex Octangulo nota: eius rei causas ego explicui lib. IV. Harmon. cap. VI. Scilicet etiam circa hunc
æqua sunt, tam in Musica, quam inter aspectus, quoad proportiones ipsas 3. & 5. ad 8. sunt enim vtrinque viles: at propter concursum in vna sectione trium proportionum 3. 5. & 5. 8. & 3. 8. cuius ratio
inter aspectus habetur nulla; nobilior est hæc Octogonica secta in Musica.

    (33) Nec Duodecangularis radius.] Imo vero & hic operatur, teste experientia, &
contrariam Octangulari experitur fortunam, in Musica; nullam enim sectionem peculiarem consti-
tuit. Vide sæpe allegatum cap. VI. lib. IV. Harmon. Vides igitur causam illam quintam esse de nihilo;
quasi, qui non implent planitiem, ij non possint fieri aspectus. Nam etsi singularum specierum non implent, at implent iunctarum.

    (34) Atque hic fere separo.] Separatio aliqua necessaria fuit, sed illa ob causas longe
alias, quam quæ hic loco quinto commemoratur.

    (35) Certe enim, quæ ex angulis fit, genuina radijs.] Optime: valet enim hoc ipsum
etiam in vera causa. Vide Harmon. lib. IV.

    (36) Non vero propter figuram.] Hoc nimium est, & contrarium præmisso. Si pro-
pter angulum, vtique etiam propter figuram; Nam & figura per angulos constituitur, & angulorum
delectus per figuras fit. Sed vide scrupulum de figura centrali & de circumferentiali, excussum lib. IV.
Harm. cap. V.

    (37) Possunt tamen.] Hic paragraphus complectitur totam fere dispositionem Harmo-
nicorum meorum. Nam commune illud Geometricum, tanquam causam archetypicam, præmisi lib.
I. & II. quid vero illud causetur in Musica, explicaui lib. III. quid in aspectibus lib. IV.

    (38) Ptolemæi Musica.] Frustra has causas, ex Ptolemæi Musicis expectatas à me esse,
lector ipse dicet, si quando auctores hi cum meis notis edantur, Deo vitam prorogante. Hæret enim Ptolemæus in numeris, vt causa, sine respectu figurarum, vt numeri numerati: itaque & Harmonias non-
nullas cum veteribus iniuste proscribit, & interualla quædam inter concinna recipit nullo illorum
merito. Vide Harm. mea lib. III. fol. 27.

    (39) Quid ex Euclidis Musicis.] De his præter propositiones à Dasypodio exscriptas
nihil vidi. Neque tamen spes est, in Euclide repertum iri, quæ Ptolemæus, qua Porphyrius, ætate poste-
riores, non habent.

G

CA-

accidental coincidence that the sextile is also precisely half the trine, and the sextile corresponds with the soft third. For I have explained in Chapter 6, Book IV of the *Harmonice*, that it is not the soft third which corresponds with the sextile, but the fifth above the double octave; whereas the soft third is the common offspring of both the pentagon and the hexagon, because it is compounded of the numbers 5 and 6. Also the reason why the trine is divisible into two perfect sextiles is quite different from the reason why the fifth is divisible into two thirds, major and minor. That is indeed apparent from the fact that the parts are equal in the former case, unequal in the latter. Thus there is no detraction from the nobility of the hard third, and none is added to the soft third, because the sextile is half the trine and the quintile is not; and it could not be reckoned less important that the quintile is half the biquintile, etc. Indeed it is not the least important part of being shrewd to beware of accidental associations of this kind, which, as the Sicilian siren once detained seafarers with her singing, detain those engaged in philosophy by the pleasure of their apparent beauty and their neatness of fit (if indeed they are struck with wonder and cling to them, when there is no cause for the one in the other), so that they cannot attain the predetermined goal of knowledge.

The fifth reason is an effect of the second, and its effect is that the quintile is a more imperfect aspect, the hard third a more imperfect consonance (but with the other kind of imperfection). It does not have the effect that the aspect is absolutely impotent, or that the consonance is entirely disagreeable. For it was long ago necessary to say of all the five objections that if they had any force, they had equal force in Music and in the affair of the aspects; and no reason is given why these reasons have force in one case and not in the other.

(32) *Why neither an octile.*] The octagonal star is a different matter. Take note why it is banished along with the trioctile, or rather excluded from among the aspects, and the minor sixth which comes from the octagon is not banished from Music. I have expounded the reasons for that in Book IV of the *Harmonice*, Chapter 6. That is to say, even in this case they are equivalent, both in Music and among the aspects, inasmuch as they represent the proportions 3:8 and 5:8 in themselves. For both ratios are base. However, on account of the occurrence together in one division of the three ratios 3:5, 5:8, and 3:8, of which no account is taken among the aspects, this octagonal division is more noble in Music.

(33) *Nor a duodecile aspect.*] On the contrary in fact, this aspect also operates, on the evidence of experience, and meets with the opposite fortune to that of the octagonal in Music; for it sets up no particular division. See Chapter 6, Book IV of the *Harmonice*,[20] which has often been referred to. You see, then, that the fifth reason given above is void, as it implies that those which do not complete a plane surface cannot constitute aspects. For although they do not complete one if taken as separate kinds, yet they do if joined together.[21]

(34) *It is precisely here that I make a distinction.*] Some distinction was necessary, but for reasons far different from that which is mentioned here as the fifth reason.

(35) *For the argument based on the angles is sound in the case of the aspects.*] A very good point, for it also applies to the true reason. See the *Harmonice*, Book IV.

(36) *And not to a figure.*] This goes too far, and is contrary to what has already been said. If it is due to an angle, it is certainly due to the figure; for the figure is established through the angles, and the choice of angles is on account of the figures. But see my reservation about both the central figure and the surrounding figure, which is hammered out in Book IV of the *Harmonice*, Chapter 5.[22]

(37) *Yet (concords and aspects) may.*] This paragraph embraces almost the whole of my *Harmonice*. For I have stated at the beginning, in Books I and II, their shared geometrical properties, as being the archetypal cause. Furthermore, I have explained in Book III what it gives rise to in Music, and in the case of the aspects in Book IV.

(38) *Ptolemy's Music.*] You wait in vain for these reasons from Ptolemy's *Music* to come from me, the reader himself will say, if ever these authors are published with my notes, should God prolong my life. For Ptolemy clings to the numbers as the cause, without reference to the figures, inasmuch as they are counted numbers. Consequently he unjustly proscribes some harmonies, as do the ancients, and accepts as melodic[23] certain intervals which do not in the least deserve it. See Book III, page 27, of my *Harmonice*.

(39) *What (can be applied to it) from Euclid's Music.*] On this, apart from the propositions set out by Dasypodius,[24] I have seen nothing. Nor is there any hope that in Euclid will be found what neither Ptolemy, nor Porphyry, have who were later in time.

CAPVT XIII.

*De computandis orbibus qui corporibus inscribuntur, & circum-
scribuntur.*

ACTENVS nihil dictum,nisi consentanea quædam signa,& εἰκότα suscepti Theorematis. Transeamus modo
ad ἀποσήματα orbium Astronomiæ & demonstrationes
Geometricas;quæ nisi cōsentiant,proculdubio omnem
præcedentem operam luserimus. Primum omnium videamus , in quanta proportione sint orbes singulis his
quinque corporibus regularibus inscripti ad circumscriptos.

Et radij quidem siue semidiametri circumscriptorum æquant se-
midiagonios corporum. Nam nisi omnes anguli figuræ tetigerint eandē
superficiē,corpus regulare non erit. Bini autem anguli oppositi mutuo,
& centrum figuræ semper sunt in eadem linea siue axi orbis. Excipitur
vnum Tetraedron, quod habet singulos angulos singulis facierum centris oppositos.

Iam recta connectens centra figuræ & basis est radius,siue semidiameter inscripti per vltimam lib.15. Campani in Euclidem. Orbis enim
inscriptus tangere debet omnia centra figuræ ; & figuræ inscriptæ cum
circumscriptis omnes possident idem centrum.

Quod cum ita sit,facile est videre , potentiam radij, quo circulus
basi circumscribitur,auferendam de potentia radij orbis circumscripti,
vt residua sit potentia quæsitæ lineæ,seu radij orbis inscripti. In adiuncto
schemate HOM est axis circumscripti orbis, cuius vt & figuræ inscriptæ
commune centrum in OHGL planum vnum figuræ,quod hic sit basis,I.
centrum basis,HI radius circumscripti basi. Et recta ex cētro orbis O in I centrum minoris circuli demissa perpendicularis erit circulo & lineæ HI. In triangulo igitur HI
O angulus ad I rectus.Ergo HO potentia æquat potentias
HI IO. Et potentia HI ablata ex HO potentia, relin-
quit IO potentiam quæsitam,per 47.primi.

Hinc apparet, vt habeatur IO in omnibus figuris,
quærendam esse prius HI radium basis.Habetur autem &HI radius co-
gnito latere figuræ, cui circulum circumscribit. Hinc rursum , vt radius
basis habeatur,quærendum prius latus cuiuslibet figuræ.

Assumpto igitur radio circumscripti cuiuslibet in quantitate sinus totius 1000.partium (sufficit nostro instituto hæc radij magnitudo)
potētia lateris cubici per 15.prop.lib.13.elem. Euclidis , est pars tertia potentia axis, vt si axis habet 2000. latus cubi habet 1155. Lateris Octaedri
potentia per 14.eiusdem, est dimidium potentiæ axis. Lateris Tetrae-
drici potentia est per 13. eiusdem,sesquialtera pars de potentia axis. Atque hactenus vsui fuit aureum illud theorema Pythagoræ de potentijs
laterum in triangulo rectangulo, prop.47. lib.1. In cæteris duobus cor-
pori-

CHAPTER XIII.
ON CALCULATING THE SPHERES WHICH ARE INSCRIBED IN THE SOLIDS, AND WHICH CIRCUMSCRIBE THEM

So far all that has been said is that certain signs agree with the theorem proposed and make it probable. Let us now pass to the distances between the astronomical spheres and the geometrical derivations: if they do not agree, the whole of the preceding work has undoubtedly been a delusion.[1] First of all, let us see in what proportion the spheres inscribed in each of these five regular solids are to those circumscribed.

Now the radii or semidiameters of the circumscribed spheres are equal to the semidiagonals of the solids. For unless all the vertices of the figure touch the same surface, the solid will not be regular. However, pairs of vertices mutually opposite to each other, and the center of the figure, are always on the same line, or an axis of the sphere. The only exception is the tetrahedron, which has a vertex opposite to the center of each of its faces.

Now the straight line connecting the centers of the figure and of the base is a radius or semidiameter of the inscribed sphere, by the last theorem of Book XV of Euclid in the edition of Campanus.[2] For the inscribed sphere must touch all the centers of the faces of the figure; and the inscribed figures all have the same center as the circumscribed.

That being the case, it is easy to see that the square of the radius of the circle circumscribing the base must be subtracted from the square of the radius of the circumscribed sphere to give as the remainder the square of the required line, the radius of the inscribed sphere. In the adjoining diagram, HOM is an axis of the circumscribed sphere; the common center of that and also of the inscribed figure is at O; HGL is one face of the figure, which in this case is to be the base; I is the center of the base; HI is the radius of the circle circumscribing the base. Now the straight line from the center of the sphere, O, to I, the center of the smaller circle, will be perpendicular to the circle and to the line HI. Then in the triangle HIO the angle at I is a right angle. Therefore the square of HO equals the squares of HI and IO; and subtracting the square of HI from the square of HO leaves the square of IO, which it was required to find, by Euclid, Book I, Theorem 47.

Hence it is evident that to find IO in all the figures, we first need to find the line HI, the radius of the base. But the radius HI is known if the side of the figure about which the circle circumscribes is known. Hence to find the radius of the base we first need to find the edge of any figure.

Then taking the radius of each circumscribed circle in terms of the whole sine[3] as 1000 units (a value of the radius which will give sufficient accuracy for our purpose), the square of the edge of the inscribed cube by Proposition 15 of Book XIII of Euclid's *Elements* is one-third of the square of the axis, so that if the axis is 2000 units, the edge of the cube is 1155. The square of the edge of the octahedron, by Proposition 14 of the same Book, is half the square of the axis. The square of the edge of the tetrahedron is by Proposition 13 of the same Book one and a half times the square of the axis. So far we have been able to use the golden theorem of Pythagoras on the squares of the sides in a right-angled triangle, Proposition 47 of Book I. For the other two solids we need that other treasury of

poribus altero illo Geometriæ thefauro opus eft, de linea fecundum extremam & mediam rationem feɗa,qui eft propofitio 30.fexti. Nam Dodecaedricum latus eft maior portio lateris cubici feɗi, fecundum extremam & mediam rationem per corollar.17.decimitertij. Sic pro Icofaedrico latere inueniendo primum quæritur radius illius circuli,qui quinq; Icofaedri tangit angulos,qui eft A C in circulo A B. Eius potentia eft quinta pars de potentia axis, per coroll.16. tredecimi. Igitur per 5. & 9. eiufdem, radij iftius A C,fecundum extremam & mediam rationem feɗi,maius fegmentum A D eft latus decanguli,quod eidem A B circulo infcribi poteft. Iunɗę igitur potentię A C radij totius,& A D maioris fegmenti huius,faciunt potentiam E F lateris quinquangularis in illo circulo, per 10. decimitertij. Quod cum fit inter duos Icofaedri angulos, erit vtique latus Icofaedri, per 11. & 16. eiufdem.

 Habemus latera omnium figurarum in proportione ad axin orbis circumfcripti. Sequitur vt radios circulorum qui bafibus circumfcribūtur,inueftigemus ex iam notis lateribus : id quod adminiculo finuum facilime affequetur quilibet, qui reputabit, hîc exquifitiffimis numeris non opus effe. Si tamen alicui placet artificiofius laborare ; ei fundamenta rei ex Euclide apponam. Cum igitur tres faltem formæ fint bafium, triangula, quadrangula, quinquangula : in triangularibus quidem, latus G H poteft triplū quæfiti radij H I, per 12. fæpe allegati ; In quadrato latus G H poteft duplum quæfiti radij : in quinquangulo deniq; G H lateris & K H fubtendentis(datarum linearum)iunɗæ potentiæ poffunt quintuplum radij H I quæfiti, per 4. decimi quarti fecundum Campanum. Habemus radios circulorum in bafibus in eadem proportione,qua latera.

 Subtraɗis igitur potentijs radiorum de potentia finus totius, qui eft quantitas femidiametri fiue radij in circumfcripto: reftabunt, vt fupra probatum eft, potētiæ radiorum,quos quærimus,infcriptorum fc.orbium. Commodius tamen& facilius vtēris,vt dixi,finubus.

 Sed hîc neque alia quædam prætereunda compendia, ne nimium operofe laboremus. Primum orbes infcripti Dodecaedro & Icofaedro funt eiufdem amplitudinis,fi figuræ eidem orbi infcribantur. Habent enim bafes vtriufque figurę eundem radium per 2.decimiquarti. Idem iudicium efto de cubo & oɗaedro. Nam axis poteft triplum cubici lateris, & hoc duplum radij in bafi,ergo axis poteft fextuplum radij in bafi : in oɗaedro viciffim, axis poteft duplum lateris, & hoc triplum radij in bafi. Poteft ergo etiam hic axis fextuplum radij. Cum ergo fit ex hypothefi idem radius circūfcriptorum fiue H M (in primo huius capitis fchemate)fitq; idem etiam radius bafium H I,& I O H femper reɗus : Ergo etiam radius infcriptorum,tertium nempe latus O I, idem erit per 26. primi conuerfam. Quare habitis cubi & Icofaedri infcriptis,de Oɗaedro& Dodecaedro nihil opus inquirere.

 Deinde in cubo cum ipfum latus fit altitudo figuræ: dimidium latus di-

geometry, on the line divided in the extreme and mean proportion, which is Proposition 30 of Book VI. For the edge of the dodecahedron is the larger portion of the edge of the cube divided in the mean and extreme ratio, by Corollary 17 of Book XIII. Thus to find the edge of the icosahedron we require to find the radius of the circle which touches five vertices of the icosahedron, which is AC in circle AB. Its square is a fifth of the square of the axis, by Corollary 16 of Book XIII. Then by Propositions 5 and 9 of the same Book, if the radius AC is divided in the mean and extreme ratio, its larger segment AD is the side of the decagon, which can be inscribed in the same circle AB. Then the sum of the squares of AC the radius of the whole, and AD its larger segment, gives the square of EF, the side of the pentagon in the same circle, by Proposition 10 of Book XIII. In that case the distance between two vertices of the icosahedron will naturally be the edge of the icosahedron, by Propositions 11 and 16 of the same Book.

We have found the edges of all the figures in terms of the axis of the circumscribed sphere. The next thing is to investigate the radii of the circles which circumscribe the bases, from the edges which are already known. That will easily be achieved with the aid of sines by anyone who realizes that here there is no need of very precise numbers. However if anyone wishes to work it out more laboriously, I will append the principles of the method from Euclid. Then inasmuch as there are three shapes for the bases — triangles, squares, and pentagons — in the case of the triangles the square of the side GH is three times that of the required radius HI, by Proposition 12 of the Book frequently cited; in the case of the square, the square of the side GH is twice that of the required radius; and lastly in the case of the pentagon, the sum of the squares of the side GH and the chord KH (which are known lengths) is five times that of the required radius HI, by Proposition 4 of Book XIV according to Campanus. Thus we have found the radii of the circles on the bases in terms of the axis like the edges.

Then subtracting the squares of the radii from the square of the whole sine, which is the measure of the semidiameter or radius of the circumscribed sphere, the remainders will be, as has been proved above, the squares of the radii which we require, that is, the radii of the inscribed spheres. However as I have said it will be easier and more convenient for you to use sines.

But here we must not miss certain other ways of saving effort, to avoid excessive labor. First, the spheres inscribed in the dodecahedron and icosahedron are of the same size, if the figures are inscribed in the same sphere. For the bases of both figures have the same radius, by Proposition XIV.2. The same conclusion must apply to the cube and the octahedron. For the square of the axis is three times that of the edge of the cube, and the latter is twice that of the radius of the circle about the base; therefore, the square of the axis is six times the square of the radius of the circle about the base. In the octahedron on the other hand the square of the axis is twice that of the edge, and the latter is three times that of the radius of the circle about the base. Therefore in this case also the square of the axis is six times that of the radius.

Then since by hypothesis the radius of the circumscribed circles, or OH (in the first diagram in this chapter) is the same in each case, and the radius HI of the circle on the bases is the same, and OIH is in all cases a right angle; therefore the radius of the inscribed sphere, that is the third side OI, will be the same by the converse of Proposition 26 of Book I. Consequently, since the spheres inscribed in the cube and icosahedron are known, there is no need to investigate the octahedron and dodecahedron.

Second, since in the case of the cube the edge is itself the height of the figure,

tus dimidia erit altitudo, nempe linea conneſtens centra figuræ & baſis.
Nihil igitur opus inquiſitione radij in baſi.

(1) Tertio Octaedri & pyramidis æqualium laterum eſt eadem altitudo. Quanto maius igitur latus pyramidis, tanto altior etiam ipſa figura, Ipſa Octaedron & pyramis duplo maiorum laterum habent eundem
orbem inſcriptum. Nam pyramis ſi ſecetur medijs lateribus, concidit in
quatuor pyramidas & Octaedron vnum, duplo minorum laterum. Cumque pyramis habeat quatuor facies, nulli earum reſecta pyramis minor
adimit centrum, vtpote quod ſectione longe inferius eſt; manet igitur in
Octaedro ex ſecto orbis inſcriptus, antiqua quatuor centra, & per definı-
tionem regularis corporis etiam noua quatuor ex ſectione accedentia ſimul tangés. Siue igitur pyramidis, ſiue Octaedri vel cubi inſcriptus prius
habeatur, facilime per proportionem laterum habebitur etiam quantitas alterius inſcripti.

His adde quæ Candalla, & quæ alij de corporibus iam demonſtra-
runt, vt quod potentia N M dimetientis in ſphæra, quæ Tetraedro circū-
ſcribitur, ſit potentiæ H I radij in baſi tetraedri $4\frac{1}{2}$ per coroll. 1. prop. 13. lib. 13. Quod ibidem N I altitudo, ſiue perpendicularis corporis ſit bes N M dimetientis, & illius N I
potentia ſit bes potentiæ lateris G H. Quod inſcripti pyramidi radius O I ſit pars quarta ipſius N I perpendicularis, tertia ipſius N O circumſcripti, vel ſexta N M dimetié-
tis, Coroll. 3. prop. 13. lib. 13. iuxta Candall. Breuiter ſic
ſunt inter ſe Potentiæ. OI. 1. IP. 2. HP. 6. HI. 8. NO. 9. NI. 16. NP. 18. NH.
24. NM. 36.

Qualiū ſemidiame-ter orbis circumſcri-pti cuiliber figuræ eſt par. 1000. talis eſt in Ergo:		longitudo lateris			ſemidiame-ter circuli plano cir-cumſcripti		ſemidiameter inſcri-pti orbis.	
	Cubo		1155			$816\frac{1}{2}$		577
	Pyramide		1633			943		333
	Dodecae.		714			607		795
	Icoſaed.		1051			607		795
	Octaed.		1414			$816\frac{1}{2}$		577

707. quadrato
Octaedri inſcri-
pti circuli.
Quod nota.

IN CAPVT DECIMVMTERTIVM
Notæ Auctoris.

(1) TErtio Oct. & Pyr. æqualium lat. eſt eadem alitudo.] *Pyramidis quidem altitudo*
cenſetur à centro baſis, vſq; ad oppoſitum angulum: Octaedri vero altitudo hic illa conſide-ra-
tur, quæ eſt inter duas baſes parallelas. Demonſtratio facilis eſt; Pyramidis enim lateribus biſectis, &
reiectis quatuor pyramidibus minoribus, reſtat Octaedron, laterum ſubduplorum lateribus Pyrami-
dis magnæ, cuius quatuor plana, vnum infra, tria circum, ſunt partes quatuor baſium magnæ Pyra-
midis: habent igitur tria circum eandem inclinationem cum tribus ſurgentibus à baſi Pyramidis ad
faſtigium anguli: quamuis angulos habeant deorſum verſos recta: ergo eadem eſt proportio perpendi-
cularium in tali plano ad perpendicularem corporis, quæ eſt in Tetraedro perpendicularium illius ad
hanc.

C A-

half the edge will be half the height: that is, a line connecting the centers of the figure and base. Therefore there is no need for investigation of the radius of the circle on the base.

(1) Third, the height of the octahedron and a pyramid of equal edges is the same. Therefore the longer the side of the pyramid, the higher is the figure itself. Also, the octahedron and a pyramid with edges twice as long have the same inscribed sphere. For if a pyramid is divided at the midpoints of its edges, it falls into four pyramids and one octahedron, with edges half as long. A pyramid has four faces; and the smaller pyramid formed by the division does not take away the center from any of them, as the center is far below the point of division. Therefore the inscribed sphere remains within the octahedron formed by the division, and touches the four original centers, and by the definition of a regular solid the four new centers resulting from the division as well. Therefore whether the sphere inscribed within the pyramid or the octahedron or the cube is considered first, by the ratio of the edges the size of the other inscribed sphere will also easily be found.

To these points is to be added what Candalla has proved, and what others have now proved, about the solids, for instance, that the square of NM, the diameter of the sphere circumscribing a tetrahedron, is 4½ times the square of HI, the radius of the circle on the base of the tetrahedron, by Corollary 1 of Proposition 13 of Book XIII ; that in the same figure NI, the height or perpendicular of the solid is 2/3 of NM the diameter, and the square of NI is 2/3 of the square of the edge GH; and that the radius OI of the sphere inscribed in the pyramid is 1/4 of NO, the radius of the circumscribed sphere, or 1/6 of NM the diameter, by Corollary 3 of Proposition 13 of Book XIII according to Candalla.[4] In short, the squares of OI, IP, HP, HI, NO, NI, NP, NH, NM are to each other respectively as 1, 2, 6, 8, 9, 16, 18, 24, 36.

Therefore: in units in which the semidiameter of the sphere circumscribed about each figure is 1000 units,

in	length of edge	semidiameter of circle circumscribing a face	semidiameter of inscribed sphere
the Cube	is 1155	is 816½	is 577
Pyramid	1633	943	333
Dodecahedron	714	607	795
Icosahedron	1051	607	795
Octahedron	1414	816½	577

707 in case of circle inscribed in square of octahedron.[5] Note this.

(1) *Third, the height of the octahedron and a pyramid of equal edges is the same.*] The height of the pyramid indeed is reckoned from the center of the base up to the opposite vertex; whereas the height of the octahedron is here considered as the height between two parallel faces. The proof is easy. For if the edges of the pyramid are bisected, and the four smaller pyramids rejected, there remains an octahedron, with edges half the edges of the large pyramid, of which four faces, one below, and three at the sides, are parts of the four faces of the large pyramid. Then the three faces at the sides have the same inclination as the three faces rising from the base of the pyramid to the vertex at the top, though their vertices are turned straight downwards. Therefore the ratio of the perpendicular to a given face and the perpendicular height of the solid is the same in the pyramid as the ratio of the corresponding perpendiculars in the tetrahedron.

CAPVT XIV.

Primarius scopus libelli, & quod hæc quinque corpora sint inter orbes, Astronomica probatio.

GITVR vt ad principale propositũ veniamus : notum est, vias planetarum esse eccentricas : & proinde recepta physicis sententia, quod obtineant orbes tantam crassitiem, quanta ad demonstrandas motuum varietates requiritur. Et hactenus quidem (1) nostris Philosophis assentitur Copernicus. Verum iam porro nõ paruum cernitur opinionum discrimen. Nam censent Physici ab ima cœli lunaris superficie ad decimam sphæram vsque nihil esse cœlestibus orbibus vacuum; sed tangi semper orbem ab orbe, imamque superioris superficiem cum summa inferioris penitus vniri. Sic enim quærenti,quis exempli causa cœli Martii locus sit Physicus, respondent: interiorem Iouis superficiem. Et apud Ptolemæum, atque vsitatam Astronomiæ descriptionem obtinere fortasse possunt hanc causam : propterea, quod orbium proportiones inuestigandi nulla illic occasio, nullum adminiculum. Quemadmodum enim ijs, qui de nouis Indijs scripserunt, nemo facile contradicit, qui illa loca non ipse lustrauit : sic physicorum ratiunculas de contactu orbium Astronomus reijcere non potest, quem obseruationum experientia & hypothesium conditio in cœlum ipsum, interq; orbes nõ euexit. Iam vero ex Copernici hypothesibus,& ex illo terræ motu sequitur, nullam esse orbium vicinorum differentiam, quæ non multis partibus orbis vtriusque eccentricitatem superet. Atque huius rei cape exemplum ex Telluris & Veneris orbibus, ijs nempe, qui minimum ab inuicem absunt. Qualium Telluris à centro mundi distantia mediocris est 60. talium Veneris ab eodem distantia mediocris est 43½ Differentia 16⅚ scrupula. Iam Tellus in perigæo appropinquat Veneri scrupulis 2½ Venus illi obuiam procedit in Apogæo scrupulis itidem 2½ summa,5.scrupulorum. Ergo duodecim residuis scrupulis hæc duo corpora distant etiam cum proxime ab inuicem absunt. Quod si quis hoc intermedium spacium compleri asserat deferentibus nodos, & circulis latitudinum, is cogitet: posse ea officia etiam à longe tenuioribus orbibus, quam qui tantum hiatum impleant, administrari: neque naturam immani mole tantorum orbium onerandam. Quamuis hercle Copernici hypotheses omnes ita comparatæ,ita aptæ sunt, ita inuicem inseruiunt,vt haud facile vllo orbe, qui vltra planetæ viam euagatur, ad motus reddendos indigere videamur. Sed esto, vt in propinquis spacia his impleantur orbibus: quæso illud quale sit,videamus. Cum à perigæa Iouis distantia ad Martis Apogæam, duplo longius numeretur spatium, quam ab ipso Marte ad centrũ Mundi (Iouis enim distantia tripla est ad Martiam) ergone ad pusilli Planetæ vix ad sensum variandas motiunculas,in longum,in latum,totum hoc spatiũ duplo crassius omni Marte,repletur tam portentosis orbibus: Quæ hæc Naturæ luxuries? Quam inepta

Copern.
lib.5.c. 2
22. Et in
in Tabul

G 3   ptã

CHAPTER XIV.
PRIMARY AIM OF THE BOOK, AND ASTRONOMICAL PROOF THAT THE FIVE SOLIDS ARE BETWEEN THE SPHERES

Therefore let us come to the principal purpose. It is known that the paths of the planets are eccentric, and consequently the received opinion among physicists is that the spheres have the thickness which is required for representing the variations in the motions. So far indeed (1) Copernicus agrees with our philosophers; but from now on a considerable difference of opinions is perceptible. For the physicists believe that from the inner surface of the heaven of the Moon right up to the tenth sphere there is no empty space in the celestial spheres, but sphere is always touched by sphere, and the inner surface of the upper is completely united with the higher surface of the lower. Thus if anyone asks, for example, what the position of the heaven of Mars is, physically, they reply: the inner surface of Jupiter's. And from Ptolemy and the customary astronomical description they may be able to draw the following reason: that in them there is no opportunity for investigating the proportions of the spheres, and no assistance. For just as nobody easily refutes those who have written about the new Indies if he has not inspected those regions for himself, so an astronomer cannot reject the physicists' feeble figuring about the contact of the spheres unless the test of observation and the agreement of the hypotheses has transported him out to the actual sky and between the spheres. Now in fact it follows from the hypotheses of Copernicus and from this motion of the Earth that in no case does the difference in size between neighboring spheres fail to exceed many times over the eccentricity of either sphere. Take as an example the spheres of the Earth and Venus, that is, the ones which are at the smallest distance from each other. In units in which the mean distance of the Earth from the center of the universe is 60, the mean distance of Venus from the same point is 43 ⅓. The difference is 16 ⅚ units. Now the Earth at perigee comes 2½ units closer to Venus, and Venus at apogee similarly goes out 2½ units to meet the Earth, a total of 5 units. Therefore these two bodies are the remaining twelve units apart even when they are at their closest distance from each other. But if anyone were to assert that this intervening space is filled by the deferents of the nodes, and the circles of latitude, let him reflect that those duties could be performed by far thinner spheres than those which would fill such a large gap, and that Nature should not be burdened with the vast bulk of such large spheres. Yet I swear the hypotheses of Copernicus are so neatly adjusted, fit so closely and support each other so well, that we do not readily seem to need any sphere which wanders outside the planet's path to account for its motions. But suppose that the spaces between neighboring planets are filled with these spheres: I should like us to see how it would be. Since the space between the distance of Jupiter at perigee and the distance of Mars at apogee amounts to twice as far as that from Mars itself to the center of the universe (for the distance of Jupiter is three times that of Mars), then for the scarcely detectable variations of the tiny motions in longitude and in latitude of a feeble planet, is this space which is twice as thick as the whole of Mars's filled with such portentous spheres? What extravagance of Nature is this, so out of place, so pointless, so little like herself?

Copern. Bk. V, Ch. 21, 22; & below in Plate.

pta?Quam inutilis?Quam minime ipsi vsitata? Atque ex hoc videre est, in Copernico nullum orbem ab alio tangi,sed ingentia relinqui systematum interualla vtique plena cœlesti aura, sed ad neutrum tamen propinquorum systematum pertinentia. (Hac tabula ab oculos propono tibi *Huc perti-* orbium & interstitiorum magnitudines iuxta veras proportiones; vti eæ *net Tabu-* numeris à Copernico expressæ sunt.) Eorum autem spaciorum cũ ini- *la quarta.* tio professus sim causas ex 5. corporibus reddere, cur tanta singula inter binos planetas relicta sint à Creatore Opt.Maximo, nempe quod singulæ figuræ singula interualla efficiant: videamus modo, quam id feliciter tentatum sit, causamque hanc coram Astronomia Iudice, & interprete Copernico disceptemus. Orbibus ipsis tantam relinquo crassitiem,quátam requirit ascensus descensusque planetæ; quæ tamen vtrum sufficiat, infra,cap.22.videbis. Quodsi figuræ interiectæ sunt,vt dixi: oportet imã superioris orbis superficiem æquari circumscripto figuræ,summam inferioris inscripto;figuras autem censeri eo ordine, quem supra rationibus confirmaui. Quare

Lib. 5. Copern.

$$\text{Si ima}\begin{Bmatrix}\hbar\\ \text{⚹}\\ \sigma\\ \text{terræ}\\ \text{♀}\end{Bmatrix}\begin{Bmatrix}\text{est 1000.}\\ \text{debebat es-}\\ \text{se summa}\end{Bmatrix}\begin{Bmatrix}\text{Iouis} & 577\\ \text{Martis} & 333\\ \text{Telluris} & 795\\ \text{Veneris} & 795\\ \text{Mercurij} & 577\\ & \text{vel 707}\end{Bmatrix}\begin{matrix}\text{pernicum}\\ \text{At est secundũ Co-}\end{matrix}\begin{Bmatrix}635\\ 333\\ 757\\ 794\\ 723\end{Bmatrix}\begin{matrix}\text{Cap.9.}\\ \text{Cap.14.}\\ \text{Cap.19.}\\ \text{Cap.21.\& 22.}\\ \text{Cap.27.}\end{matrix}$$

Quod si crassitiei orbis terreni accenseatur systema lunare: ergo si ima superficies orbis terreni, etiam Lunæ cœlum comprehendens, est 1000.summa Veneris est in Copernico 847. Et terreni orbis cum Luna summus margo est 801.si ♂ ima habet 1000. Hic velim te identidem respicere ad tabellam capitis secundi,nempe ad huius interpositionis qualemcunque imaginem.

En numeros (2) parallelos propinquos inuicem, & Martis quidẽ atque Veneris eosdem.Telluris vero & (3) Mercurij non admodum diuersos,solius Iouis immodice discrepantes, sed quod in tanta distantia nemo miretur.Et in Marte quidem atque Venere, vicinis orbi Telluris, vides quantam efficiat diuersitatem orbiculus Lunæ accensitus crassitiei orbis terreni: (4) qui tamen orbiculus vix 3′.scrupula æquat, qualium orbis terræ habet 60.

Vnde colligere potes,quam facile animaduersum fuisset, quantaq; numerorum extitisset inæqualitas:si hæc contra cœli naturam tentarentur,hoc est, si Deus ipse in Creatione non ad has proportiones respexisset.Certe enim fortuitum hoc esse non potest, vt tam propinquæ sint interuallis hisce proportiones corporum;cum propter alia, tum maxime, quia idem ordo est interuallorum, quem supra rationibus optimis, corporibus ascripsi,vide cap.3.Nam etsi 635.à 577.discrepat:nulli tamen propinquior est,atque
huic ipsi.

IN

Chapter XIV

From this it may be seen that in Copernicus no sphere is touched by another, but huge gaps are left between the various systems, certainly full of the air of heaven, but concerning neither of the neighboring systems. In this plate I set before your eyes the sizes of the spheres and the intervening spaces according to their true proportions, just as they were elicited numerically by Copernicus. However since I claimed at the start to draw from the five solids the reasons why such large spaces were left between each pair of planets by the best and greatest of Creators, namely that particular figures produce particular intervals, let us now see how successfully this has been undertaken, and argue the case with astronomy as judge and Copernicus to plead for us. For the actual spheres I leave as great a thickness as the upward and downward movement of the planets requires. Whether that is enough you will see below in Chapter 22. But if the figures are interposed, as I have said, the inner surface of the sphere above ought to be equal to the circumscribed sphere of the figure, the outer surface of the sphere below to the inscribed sphere, the figures being ranged in the order which I have established above by argument.[1] Therefore

Here belongs Plate IV.

						Bk. V of Copernicus	
If lowest point of	Saturn Jupiter Mars Earth Venus	is 1000 highest point should be	of Jupiter Mars Earth Venus Mercury	577 333 795 795 577 or 707	But according to Copernicus it is	635 333 757 794 723	Ch. 9 Ch. 14 Ch. 19 Ch. 21 & 22 Ch. 27

But if the lunar system is allocated to the thickness of the Earth's sphere, then if the inner surface of the Earth's sphere, also including the Moon, is 1000 units, the outer surface of the sphere of Venus is 847 units in Copernicus. Also the outer boundary of the sphere of the Earth[2] along with the Moon is 801 units if the lowest point of Mars is 1000. Here I should like you to look repeatedly at the plate in the second chapter, that is, at the representation such as it is of these interpolations.

(2) Notice that corresponding numbers are close to each other, and indeed in the cases of Mars and Venus, the same. Indeed in the cases of the Earth and (3) Mercury they are not very different: only in the case of Jupiter is there an undue discrepancy, which however at such a great distance should surprise nobody. Also for Mars and Venus, which are next to the Earth's sphere, you see how great a difference is made by the allocation of the little sphere of the Moon to the thickness of the Earth's sphere, (4) although that sphere scarcely amounts to three units where the Earth's sphere is 60.

Hence you can realize how easily it would have been noticed, and how greatly unequal the numbers would have been, if this undertaking had been contrary to Nature, that is, if God himself at the Creation had not looked to these proportions.[3] For certainly it cannot be accidental that the proportions of the solids are so close to these intervals, for various reasons but particularly because the order of the intervals is the same as that which I ascribed to the solids above for excellent reasons; see Chapter 3. For although 635 differs from 577, nevertheless there is no number to which it is closer than to that.

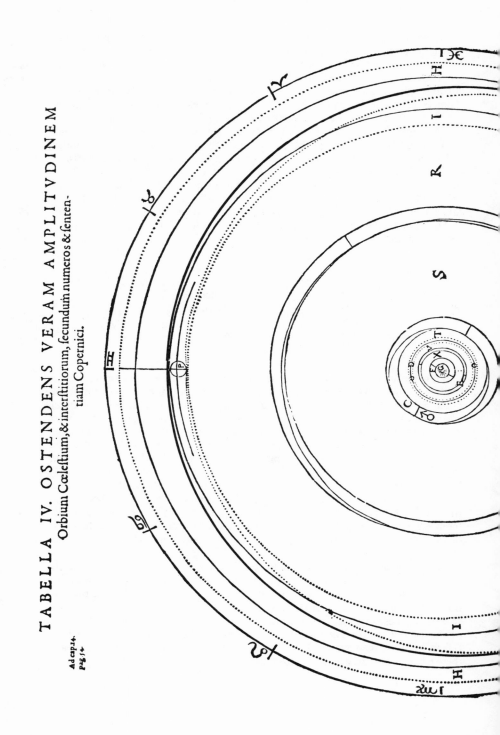

TABELLA IV. OSTENDENS VERAM AMPLITVDINEM
Orbium Cœlestium, & interstitiorum, secundum numeros & sententiam Copernici.

Ad cap. 4.
pag 54

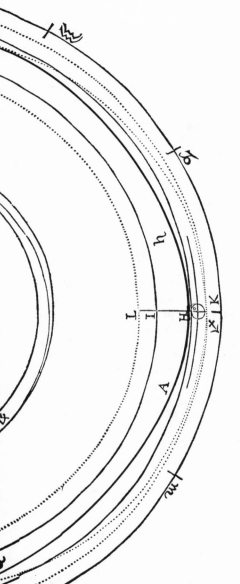

Extremus circulus Zodiacum refert in Orbe stellato, descriptus ex centro Mundi vel Orbis magni, vel etiam ex globo Terreno, quis totus Orbis Magnus ad eum insensibilis sit.

A Saturni systema, concentricum ex G centro Orbis magni.
B Systema Iouis.
C Martis.
D Circulus, siue via centri globi terreni concentrica ex centro G, cum sphaerula Lunari duobus locis appicta. Due caecae lineae circulares orbis terrae cum inserta Luna crassitiem denotant.
E Duo circelli delineantes crassitiem systematis Veneris, intra quam omnis eius motuum varietas perficitur.
F Spatium inter duos circellos, in quo omnis motuum stellae Mercurij varietas perficitur.
G Centrum omnium, & prope ipsum corpus Solare.

Circulus per O & P transiens (cuius hic tantum duo arcus comparent) est concentricus epicyclus Saturni est.

Linea curua per Q, atq; per perigaeum epicycli in O apogaeo eccentrici positi, & per apogaeum eiusdem in P perigaeo eccentrici, est via planetae eccentrica. Circulus quidem non est, sed tamen à circularis lineas sensibiliter non differt.

HI Crassities duobus circulis concentricis inclusa, quam via Saturni eccentrica sibi vindicat.

Linea curua, vel quasi circulus per M, & per apogaeum epicycli in O, atque per perigaeum eiusdem in P transiens, eccentrica est, quam Ptolemaeus Aequantem vocat.

KL Crassities duobus caecis circulis concentricis intercepta, quam totus epicyclus, & aequans ille requirunt.

Planeta vero ultra H I nunquam ascendit, nec infra I descendit.

Similibus particularibus orbibus caeterae sphaerae etiam distinctae intelligantur, qui tamen, ne multitudo linearum negotium potius obscuraret, quam declararet, hic omittuntur. Ideo in Ioue & Marte viae totum eccentrica, duoque tam continentes circuli concentrici, in caeteris soli concentrici, scripti sufficiunt.

T Dodecaedri.   R Locus Cubi.   S Tetraedri.
   Spatia intermedia. V Icosaedri.   Z Est spatium
inter Saturnum & fixa, insinuè simile.   X Octaedri.

See p. 229.

IN CAPVT DECIMVMQVARTVM
Notæ Auctoris.

(1) N Oſtris Philoſophis aſſentitur Copernicus.] *Intelligc de ſp.itio Orbium Geometri-*
co : de materia enim, hoc eſt, de corpulentia adamantina ne Ptolemæus quidem adeo craſſe
philoſophatur.

(2) En Numeros parallelos.[*E regione ſitos,vt* 577.635.ſic 333.333.

(3) Mercurij non admodum diuerſas.] *Si in ☿ non ſum.is* 577. *radium inſcripti*
Octaedro , ſed 707. *radium inſcripti quadrato Octaedri : tunc iſte non multum diſcrepat à*
703.

(4) Qui tamen orbiculus.] *Hic proportio Orbium Solis & Lunæ aſſumitur ea quæ* 20.
ad 1.*quantam tradit Aſtronomia antiqua circiter. At doceo lib.* 4. *Epitomes quod illa ſit ſere triplo*
maior;etſi in Ephemeridibus modeſtia quadam vſus,vſurpaui illam ſeſquiplo maiorem,ſcil.eam quæ
30.*ad* 1.*interim dum plane concluderem.*

CAPVT XV.

Correctio diſtantiarum & diuerſitas proſthaphæreſeon.

N E vero tibi, Lector amice, occaſionem vllam præbeam totum hoc negotium propter leuiculam diſcordi am reij-
ciendi, monendus hîc es, quod te probe meminiſſe velim;
Copernici intentum non in Coſmographia verſari , ſed
in Aſtronomia; hoc eſt, vtrum nonnihil in veram orbium
proportionem peccet, parum ipſi curæ eſt: modo nume-
ros ex obſeruationibus eos conſtituat,qui ſint ad demonſtrandos motus,
Planetarumque loca computanda , quantum fieri potuit, maxime apti.
At ſi quis aptiores dare conetur, & hos Copernici numeros ita corrigat,
vt nihil interea aut parum in proſthaphæreſeſi turbet ; id illi per Copernicum facile licebit.

Vt igitur ſummam denique huic negotio manum imponam , atq;
vt appareat, quid quantumque penes ſingulos Planetas in parallaxibus
orbis terreni mutetur;nouum ſtruam mundum;& cum prius inueſtiga-
ta fuerit ab artificibus cuiuslibet ἐκκεντρότητ῍Θ. ad orbis ſemidiametrum
proportio : ideo ſi quid in longiſſima vel proxima orbis à centro mundi
diſtantia mutabitur per interpoſitionē corporum;id inἐκκεντρότητι anim-
aduertendum erit proportionaliter.Initium erit à maxima terræ diſtantia ſurſum,minima deorſum,centrum verſum.

Ante omnia autem retexendi numeri Copernici, atque peculiariter accommodandi ſunt ad præſens inſtitutum. Nam etſi ille ſine dubio
centrum totius vniuerſi in corpore ſolari conſtituit; tamen vt calculum
iuuet compendio , & ne nimium à Ptolemæo recedendo , diligentem
eius lectorem turbet : (1) diſtantias omnium Planetarum maximas
atque minimas, vt & loca earum in Zodiaco (quæ Apogæorum & Perigæorum nomen retinuerunt)computauit non à centro Solis,ſed à centro orbis

Chapter XV

(1) *Copernicus agrees with our Philosophers.*] Understand this to refer to the geometrical space of the spheres: for on their material, that is, on their solid corporeality, not even Ptolemy offers such a crass piece of philosophizing.

(2) *Notice that corresponding numbers.*] Located according to the region to which they refer. For instance 577 against 635, 333 against 333.

(3) *(Of) Mercury they are not very different.*] If in the case of Mercury you take not 577, the radius of the sphere inscribed in the octahedron, but 707, the radius of the circle inscribed in the square of the octahedron, then the discrepancy from 723 is not great.

(4) *Although that sphere.*] Here the ratio of the Spheres of the Sun and of the Moon is assumed to be as 20:1, about as reported by ancient astronomy. But I explain in Book IV of the *Epitome*[4] that it is about three times larger; though in the *Ephemerides* I leaned to the modest side and took it as one and a half times larger, that is as 30:1, for the time being until I came to a definite conclusion.

CHAPTER XV.
CORRECTION OF THE DISTANCES AND VARIATION OF THE EQUATIONS

In case, friendly reader, I should offer you any occasion for rejecting the whole of this enterprise because of a trifling dispute, I must here mention to you something which I should like you to remember carefully: Copernicus's purpose was not to deal with cosmography, but with astronomy. That is, he is not much concerned whether there is a mistake relating to the true proportion of the spheres, but only with establishing from the observations the values which are best suited for deriving the motions of the planets and computing their positions, as far as possible. But if anyone should try to give better suited values, and rectify Copernicus's in such a way as to upset in the process nothing, or very little, in the system of equations, that will readily be permitted as far as Copernicus is concerned.

Therefore, to put the finishing touch to this enterprise and to make clear the alterations in the parallaxes of the Earth's orbit produced by the particular planets, and their amounts, I shall construct a new universe; and as the proportion of each eccentricity to the semidiameter of the orbit has previously been investigated by the authorities, if by the interpolation of the solids any alteration is produced in the greatest or smallest distance of the orbit from the center of the universe, it will be noticed proportionately in the eccentricity. The starting point will be from the greatest distance of the Earth upwards, from the smallest distance of the Earth downwards, towards the center.

First of all Copernicus's values must be reworked, and particularly adapted for the present undertaking. For although he undoubtedly established the center of the whole universe in the solar body, yet to help his calculation by shortening it, and to avoid upsetting the diligent reader by too great a departure from Ptolemy, he computed the maximum and minimum (1) distances of all the planets, and also their positions on the zodiac (for which he retained the name of apogees and perigees) not from the center of the Sun, but from the center of the Earth's orbit,

tro orbis Magni,quaſi illud eſſet Vniuerſitatis centrum;cum tamen illud
à Sole tanto ſemper interuallo diſtet,quanta eſt quouis tempore Telluris
(vel Solis)maxima ἐκκεντρότης. Quos numeros ſi retinerem in preſenti negotio;illud incommodum ſequeretur, quod aut error committeretur in
inſcriptione,dum terræ orbis pro corpore cenſeretur, qui ſuperficies ſaltem eſſet;vt videre eſt in præced.Tabella IV.aut orbi terreno nullam, vt
cæteris relinquerem craſſitiem. Eſſent igitur Dodecaedricorum planorum centra & Icoſaedrici anguli in eadem ſuperficie ſphærica ; atque ita
totus mundus arctius conſideret,fieretque longe anguſtior, quam experientia motuum & obſeruationes patiuntur. Atq;hunc ſcrupulum cum
ego Michaeli Mæſtlino,præceptori meo Clariſſimo aperirem,explora-
turus,an probare vellet modo poſitum hoc Theorema: is inſperato mei
iuuandi ſtudio hunc laborem in ſe ſuſcepit, & non tantum ex Prutenicis
Tabulis ipſas Planetarum diſtâtias de nouo computauit, ſed etiam præſentem Tabulam mihi confecit; atq; ſic me tum alijs non paucis occu-

Huc perti-
net Tabula
quinta.

pationibus detentum magno & difficili atq; moleſto labore ſubleuauit.
Quam tabulam ipſo permittente Auctore tecum, Lector, communico:
tibique ſic eam commendo, vt quæ non tantum in præſenti negotio tibi
profutura, ſed etiam intricatiſſimum nodum ad oculum ſolutura, atque
adeo te in ipſa Prutenicarum atque Copernici adyta,quaſi manu,ductu-
ra ſit. Etenim ex ea iucundum eſt diſcere, quomodo Auges Planetarum
diuerſæ,in diuerſa Zodiaci loca cadant;quod in Venere plus integri triē-
tis diuerſitatem,parit.Nam eius Apogeum eſt in ♉ & ♊ἀφήλιον in ♄ &♒.
Videre etiam eſt,longe alias eſſe lineas diſtantiarum à Sole, quam à centro terreni orbis.Quæ diuerſitas in ♄ maxima eſt:propterea quod integra Telluris ἐκκεντρότης eius diſtantiæ accedit. In Ioue autem parum mutatur,quia is,non vt Saturnus è regione Solis ſit altiſſimus, ſed in ♎, vbi
fere æqualiter abeſt ab vtroque centro Solis & Orbis magni. Atque inde etiam ad oculum patet demonſtratio eius,quod Copernicus lib.5.Re-
uol.cap.4.16.& 22. ſub finem , de mutabili Eccentricitate Martis & Veneris ad mutationem terrenæ,breuiſſimis verbis innuit ; Rheticus vero
in ſua Narratione copioſius perſequitur. Aliud etiā eſt, cuius nos iſthæc
tabula admonet, quod quia commodius alio loco dici poteſt, nunc dif-
feram. Nunc ad rem. Pandam autem quadruplicem ordinem numero-
rum.In primo erunt Planetarum abſceſſus à centro magni Orbis;ſicut ij
abſceſſus & numeri ex Copernico & Prutenicis ſimpliciter & ſine mutatione eliciuntur.In ſecundo erunt abſceſſus orbium à centro Solis,qui
proueniunt ex Copernico poſt illam reſolutionem numerorum, de qua
modo vidiſti tabulam. In tertio & quarto venient rurſum abſceſſus planetarum à ☉, prout illi per interpoſitionem corporum mutati ſunt. Et
tertius quidem ordo erit ex ſtructura mundi ea, quæ pro fundamento
habebit orbis terreni craſſitiem ſimplicem , non accenſito ſyſtemate Lunari. Quartus denique prodet craſſitiem orbis terreni tantam,quæ ſupra
& infra ſemidiametrum orbis Lunaris contegere poſſit.

♄ Altiſſ.

as if it were the center of the whole universe, even though that is always separated from the Sun by a distance which depends on the maximum eccentricity of the Earth (or of the Sun) at any given time.[1] If I retained those values in the present enterprise, the inconvenient consequence would be that either an error would be committed in the interpolation, the Earth's sphere being counted as a body when it was really a surface, as may be seen in Plate IV above, or I should leave no thickness for the Earth's sphere, such as was left in the other cases. Then the faces of the dodecahedron and the vertices of the icosahedron would be on the same spherical surface; and thus the whole universe would be more restricted in size, and would become far narrower, than our knowledge of the motions and observations allow. When I revealed this difficulty to Michael Maestlin,[2] my famous teacher, to find out whether he wished to verify this theorem which has been just proposed, he took that labor on himself, with an enthusiasm for helping me which was greater than I had hoped, and not only computed the actual distances of the planets afresh from the Prutenic Tables but also executed the present plate for me; and thus he relieved me, at a time when I was busy with quite a number of other commitments, of this great and difficult and troublesome labor.[3] This plate, reader, I communicate to you with the author's permission; and I commend it to you as something which will not only be of great help to you in the present enterprise, but will also disentangle a most intricate knot at sight, and thus lead you as if by the hand into the very tabernacles of the Prutenic Tables and of Copernicus. For it is pleasing to learn from it how the various apsides of the planets fall in different places on the zodiac, which in the case of Venus brings about a difference of more than a whole thirty degrees. For its apogee is in Taurus and Gemini, its aphelion in Capricorn and Aquarius. It may also be seen that the lines of the distances from the Sun are far removed from those from the center of the Earth's orbit. This difference is greatest in the case of Saturn, because the complete eccentricity of the Earth is added to its distance. In the case of Jupiter, however, it is little changed, since unlike Saturn it reaches its highest point not opposite the Sun but in Libra, where it is almost equally distant from both the center of the Sun and from the Great Orbit of the Earth. Moreover the derivation is obvious at sight of what Copernicus hints at in very brief words in Book V of the *De Revolutionibus*, in Chapters 4, 16, and 22 at the end, on the variable eccentricity of Mars and Venus compared with the variation in the Earth's eccentricity, but Rheticus pursues at greater length in his *Narratio*. There is also another point of which this plate reminds us, which I shall postpone for the present because it can more conveniently be made in another place. Now to business. I shall set out a fourfold table of values.[4] In the first column there will be the distances of the planets from the center of the Earth's Great Orbit, just as these distances and values are extracted from Copernicus and the Prutenic Tables, simply and without alteration. In the second column will be the distances of the orbits from the center of the Sun, which emerge from Copernicus after the revision of the values illustrated by the plate which you have just seen. In the third and fourth columns again will come the distances of the planets from the Sun, as they have been altered by the interpolation of the solids. Thus the third column will be derived from the arrangement of the universe which will have as its basis the thickness of the Earth's sphere on its own, without the addition of the lunar system. Finally the fourth column will give the Earth's sphere a thickness sufficient to cover the semidiameter of the lunar sphere above and below.

Plate V belongs here.

TABELLA V. OSTENDENS POSITVS CENTRORVM ECCENTRICARVM
fphærarum Mundi, fecundum fententiam Copernici, & numeros Tabularum Prutenicarum.

Adpap. 15.
pag.56.

Ad tempora Copernici, circa Annum
Chriſti 1525.

Ad tempora Ptolemæi, circa Annum
Chriſti 140.

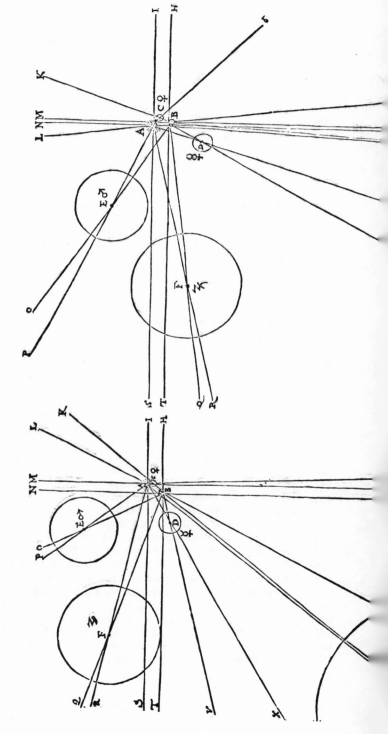

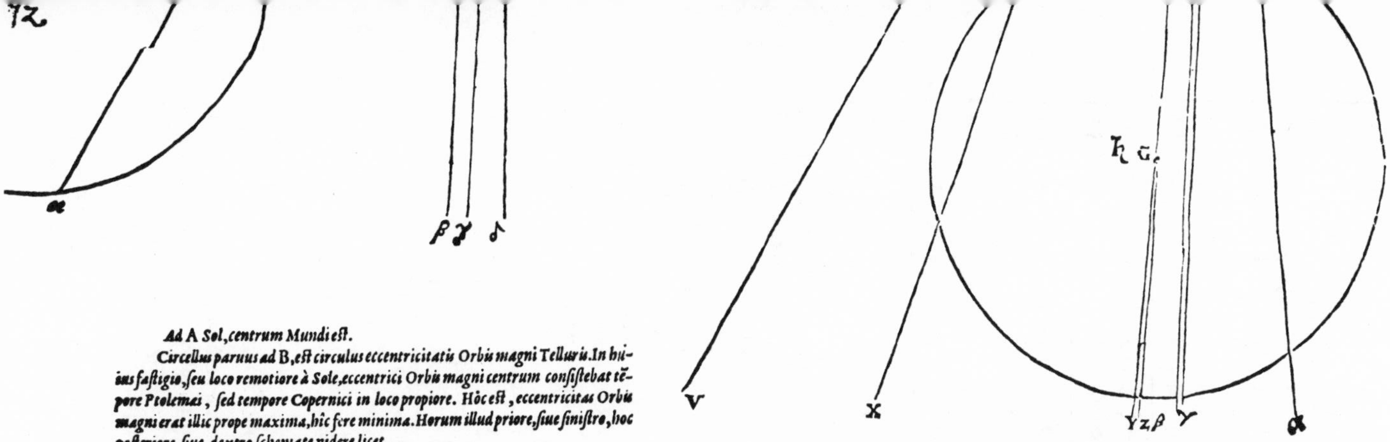

Ad A *Sol, centrum Mundi eſt.*

Circellus paruus ad B, *eſt circulus eccentricitatis Orbis magni Telluris. In huius faſtigio, ſeu loco remotiore à Sole, eccentrici Orbis magni centrum conſiſtebat tẽpore Ptolemai, ſed tempore Copernici in loco propiore. Hôc eſt, eccentricitas Orbis magni erat illic prope maxima, hîc fere minima. Horum illud priore, ſiue ſiniſtro, hoc poſteriore, ſiue dextro ſchemate videre licet.*

A B *priore ſchemate eſt* 4170. *qualium ſemidiameter Orbis magni eſt* 100000. *Hinc maxima Terræ à Sole remotio eſt* 104170. *& minima* 95830. *Sed in altero ſchemate illa eccentricitas prope minima, eſt* 32195.

A C *eſt circulus paruus eccentricitatis* ♀. *Huius ſemidiameter (qualium orbis magni ſemidiameter, eſt* 100000) *eſt* 1040. *&* BC (*dextra figuræ*) *eccẽtricitas centri parui circuli à centro orbis magni* B, *eſt* 3120. *Sed* A C, *eiuſdem eccentricitas à Sole* A, *eſt* 1262. *Hinc maxima Veneris à* ☉ *diſtantia* 74232. *& minima* 69628.

D *centrum eſt circelli eccentricitatis* ☿. *Huius ſemidiameter eſt earundem; quæ ſupra, partium* 2114½. *eiuſq; eccentricitas à centro orbis magni* D B 7345½ *ſed* D A, *eccentricitas eius à Sole* 10270. *Vnde maxima Mercurij diſtantia à* ☉ *inuenitur* 48114½. *& minima* 23345½.

E *centrum eſt parui circuli eccentricitatis* ♂. *Huius ſemidiameter eſt* 7602½. *&* BE *eccentricitas ab orbis magni centro* 22807½. *Sed* A E *eccentricitas à Sole* 20342. *Vnde diſtantia* ♂ *à* ☉ *maxima* 164780. *minima* 139300.

F *centrum eſt parui circuli eccentricitatis* ♃. *Huius ſemid. eſt* 12000. *&* BF *eccentric. à* B 36000. *Sed* A F *à* ☉ 36656. *Iouis maxima diſtantia à* ☉ 549256. *minima* 499944.

G *centrum eſt parui circuli eccentricitatis* ♄. *Huius ſemid. eſt* 26075. BG *eſt* 78225. *&* A G *eccentricitas à* ☉ 82290. *Saturni maxima remotio, à Sole eſt* 998740. *& minima* 834160.

Recta H B T *eſt linea æquinoctialis reſpectu Terræ. Sed* I A S, *reſpectu Solis. Sic recta* N B β *eſt linea ſolſtitialis reſpectu Terræ, &* M A γ *reſpectu Solis.*

ἀπόγειον

	tempore Ptolemai,		*Copernici.*	
♄ BGY	23	♏	27 42	♐
♃ BFQ	11	♍	6 21	♎
♂ BEO	25 30	♋	27	♌
♀ BCK	25	♉	15 44	♊
☿ BDV	10	♎	28 30	♏
☉ BAL	68	♋	6 40 69	

ἀφήλιον

	tempore Ptolemai		*Copernici.*	
♄ AGZ	23 40	♏	28 3	♐
♃ AFR	17 31	♍	11 30	♎
♂ AEP	4 27	♌	4 21	♍
♀ ACδ	4 39	♑	19 48	♒
☿ ADX	29 42	♎	13 40	♐
Terr. ABα	6 8	♐	6 49	♑

See pp. 230-31.

		o ' "	o ' "	o ' "	o ' "
♄	Altiſſ.	9 42 0	9 59 15	10 35 56	11 18 16
	Humil.	8 39 0	8 20 30	8 51 8	9 26 26
♃	Altiſſ.	5 27 29	5 29 33	5 6 39	5 27 2
	Humil.	4 58 49	4 59 58	4 39 8	4 57 38
♂	Altiſſ.	1 39 56	1 39 52	1 33 2	1 39 13
	Humil.	1 22 26	1 23 35	1 18 39	1 23 52
terra.	Altiſſ.	1 0 0	1 2 30	1 2 30	1 6 6
	Humil.	1 0 0	0 57 30	0 57 30	0 53 54
♀	Altiſſ.	0 45 40	0 44 29	0 45 41	0 42 50
	Humil.	0 40 40	0 41 47	0 42 55	0 40 14
☿	Altiſſ.	0 29 24	0 29 19	0 30 21	0 28 27
	Humil.	0 18 2	0 14 0	0 14 0	0 13 7
☉	Altiſſ.	0 2 30	0 0 0	0 0 0	0 0 0
	Humil.	0 1 56	0 0 0	0 0 0	0 0 0

Hæ diſtantiæ. Iam porro ſubiungam laterculum arcuum, qui ſinu-
bus debentur ijs, quos efficiunt Veneris quidem & Mercurii altiſſimi abſceſſus, ſi media terræ diſtantia ſit ſinus totus: Telluris vero media diſtā-
tia, ſi ſuperiorum abſceſſus longiſſimi ſint ſinus totus; quorum arcuū illi
quidē elongationibus maximis Veneris & Mercurij à Sole, hi vero proſthaphæreſibus ἀπογείοις Saturni Iouis & Martis proximi erunt. In primo
ordine ſunt arcus, qui proueniunt ex corporibus excluſa Luna, in ſecundo arcus, qui proueniunt ex diſtantijs à Sole Copernicanis, in tertio denique, arcus qui ex corporibus, adiuncta Telluri Luna ſequuntur; Et in-
terponentur vtrinq; differentiæ.

		o '	o '	o '	o '	o '
♄		5 25	—0 20	5 45	— 0 41	5 4
♃		10 17	—0 12	10 29	— 0 6	10 23
♂		40 9	✛ 2 47	37 22	✛ 0 20	37 52
♀		49 36	✛ 1 45	47 51	— 2 18	45 33
☿		30 23	✛ 1 4	29 19	— 1 1	28 18

In Caput XV. Notæ Auctoris.

(1) **Diſtantias omnium Planetarum.**] *Quid peccetur per hanc veluti luxationem Syſte-
matis Planetarij, & quomodo peccatū hoc redarguatur obſeruationib. Brahcanis in Marte, diligenter explicaui in Comment. de motibus illius Planetæ, idq̃, ex profeſſo, parte prima, quæ eſt de
æquipollentia hypotheſium. Et quia ad declinandos hos errores, neceſſe fuit fundamentū veluti mundi
in ipſum ſolu centrum reponere: hinc adeo factum, vt loca Zodiaci quibus planetæ fiunt altiſſimi &
humilimi, non iam amplius Apogæorum & Perigæorum nomen retinere poſſent, vt quidem in Copernico retinuerunt abuſiue: ſed proprie & ſignificanter indigetarentur à me Aphelia & Perihelia.*

H   CA-

		° ′ ″	° ′ ″	° ′ ″	° ′ ″
Saturn	Highest	9 42 0	9 59 15	10 35 56	11 18 16
	Lowest	8 39 0	8 20 30	8 51 8	9 26 26
Jupiter	Highest	5 27 29	5 29 33	5 6 39	5 27 2
	Lowest	4 58 49	4 59 58	4 39 8	4 57 38
Mars	Highest	1 39 56	1 39 52	1 33 2	1 39 13
	Lowest	1 22 26	1 23 35	1 18 39	1 23 52
Earth	Highest	1 0 0	1 2 30	1 2 30	1 6 6
	Lowest	1 0 0	0 57 30	0 57 30	0 53 54
Venus	Highest	0 45 40	0 44 29	0 45 41	0 42 50
	Lowest	0 40 40	0 41 47	0 42 55	0 40 14
Mercury	Highest	0 29 24	0 29 19	0 30 21	0 28 27
	Lowest	0 18 2	0 14 0	0 14 0	0 13 7
Sun	Highest	0 2 30			
			0 0 0	0 0 0	0 0 0
	Lowest	0 1 56			

Those are the distances. I shall now also append a list of the angles subtended, which correspond with the sines defined by the greatest distances of Venus and Mars, taking the mean distance of the Earth as the whole sine or with the mean distance of the Earth, if the furthest distances of the superior planets are taken as the whole sines. Of these angles the former will be very close to the greatest elongations of Venus and Mercury from the Sun, the latter to the equations for Saturn, Jupiter, and Mars in apogee. In the first column are the angles derived from the solids with the omission of the Moon; in the second the angles derived from the Copernican distances from the Sun; in the third, finally, the angles which follow from the solids, with the Moon added to the Earth; and in between them are the differences in either direction.[5]

	° ′	° ′	° ′	° ′	° ′
Saturn	5 25	−0 20	5 45	−0 41	5 4
Jupiter	10 17	−0 12	10 29	−0 6	10 23
Mars	40 9	+2 47	37 22	+0 20	37 52
Venus	49 36	+1 45	47 51	−2 18	45 33
Mercury	30 23	+1 4	29 19	−1 1	28 18

AUTHOR'S NOTES ON CHAPTER FIFTEEN

(1) *Distances of all the planets.*] What is at fault in this, so to speak, dislocation of the planetary system, and the way in which this fault is at odds with the observations of Brahe on Mars, I have carefully explained in my *Commentary* on the motions of that planet, and specifically in the first part, which is on the equivalence of the hypotheses.[6] And because in order to avoid those errors it was necessary to restore the foundation, so to speak, of the universe to the actual center of the Sun, the result was that the positions in the zodiac at which the planets are highest and lowest could no longer retain the name of apogees and perigees, though indeed in Copernicus they did retain it, improperly. But they were appropriately and significantly named by me aphelia and perihelia.

CAPVT XVI.

De Luna peculiare monitum, & de materia corporum & orbium.

ON ergo exiguum fcrupulum Lunę Orbis, vtut exiguus fit, mouet. Quare porro de Luna tempus eſt, vt aliquid dicam. Et incipio quidem fine ambage, tibi Lector; fincere meam mentem exponere; fecuturum nempe me in hac cauſa, quocunque propinquitas numerorum præit. Vt ſi interpoſitio Lunæ numeros & arcus Copernici verius reddit: dicam accenfendum illud ſyſtema craſſitiei orbis magni. Sin autem eiecta Luna melius nobis cum Copernico conuenire poteſt: etiam ego dicam, orbem magnum non tam craſſum eſſe circumcirca, vt cœlum lunare tegat; ſed eminere interdum furfum, interdum deorſum, integrum Lunæ hemiſphærium ſupra vel infra margines orbis magni, interdum & plerumque quidem minus hæmiſphærio extare; omnino prout ipfum corpus telluris, quod eſt Orbis Lunæ centrum vel afcenderit, vel defcenderit per orbis fui ſpiſſitudinem. (1) Nec hercle fcio, quorfum magis inclinent Coſmographicæ vel etiam Metaphyſicæ rationes. (2) Concinnum quidem negotium eſſe videtur; vt non ſit in cœlo orbis aliquis, qui talem gerat nodum, velut annulus gemmam, cuius eminentia obſit, quo minus abſolutiſſima conſtet orbi rotunditas. Ac viciſſim in cenſenda figura orbis quid attinet Lunæ rationem habere, cum illa non proprie ad orbem terræ veluti cæterorum Planetarum euagationes in altum, in profundum (quæ phyſice commodiſſime per epicyclia demonſtrantur) velut, inquam, hæc epicyclia ad fuum quodque orbem pertineat? Tellus enim eſt cui Orbis ille tertius à Sole debetur, ipſa eius remigio inter cæteros Planetas Solem circumit, ipſa per ſe, perque ſua epicyclia nullo ad hoc Lunæ vſa miniſterio ſuas perficit varietates, vt docent Copernici placita: Luna vero hanc circa tellurem exiguam domunculam quaſi precario aut conductam obtinet, Luna ſequitur vel trahitur potius, quocunque Tellus quacunque varietate graditur. Finge Tellurem quieſcentem, nunquam Luna viam circa Solem inueniet, nedum circumueniet. Diſcurſitat enim hinc inde anguſtis incluſa fpacijs circa terram, lucis humorumque Telluri miniſtra, veluti Atrienſis aliquis circa herum, aut veluti qui in naui obambulant, neque tamen ſeſe fatigando proficiunt in itinere, niſi magna vis aquarum incertos quorfum eant, & vel quietos promoueat. Atque vt ſpatium Luna ex orbe terreno, motumque ſortita eſt, ſic & * multas conditiones globi terreni adeptam, puta, continentes, maria, montes, aerem, vel his aliqua quocunq; modo correſpôdentia, multis côiecturis Mæſtlinus probat, nec nullas ego habeo; vt vel ob hoc ſolum veriſimilior ſit Copernicus, qui eandem loci motuſq; communionem duobus hiſce corporib. largitur. Ac certe φιλαίθρωπος Creator vltimo veſtiuiſſe videtur Tellurem hoc orbe Lunari; quia ſimilé ei ſitû attribuere voluit, ſitui Solis; vt ſi & ipſa orbis alicuius centrum eſſet (vt Sol eſt centrû omniû) inſtar Solis cuiuſ-

CHAPTER XVI.
A PARTICULAR COMMENT ON THE MOON, AND ON THE MATERIAL OF THE SOLIDS AND SPHERES

It is therefore by no means a small doubt which the Moon's orbit raises, however small it is itself. It is time, then, for me to say something about the Moon. And I begin indeed without prevarication, reader, by frankly revealing my intention to you, which is to follow in this debate wheresoever the closeness of the numbers leads. Thus if the insertion of the Moon makes the numbers and angles of Copernicus more accurate, I shall say that that system should be added to the thickness of the Earth's sphere. But if the rejection of the Moon can give us better agreement with Copernicus, I shall also say that the Earth's sphere is not so thick all round as to cover the lunar heaven, but that the complete hemisphere of the Moon sometimes juts out upwards, sometimes downwards, above or below the boundaries of the Earth's sphere, and sometimes, usually indeed, projects by less than its hemisphere, in general in proportion as the body of the Earth, which is the center of the Moon's orbit, either ascends or descends through the breadth of its sphere. (1) Nor, great heavens, do I know in which direction the cosmographical or even the metaphysical arguments tend more. (2) Indeed it seems to be a question of tidiness, about whether there is not in the heaven some sphere which carries such a lump, like the jewel on a ring, that its protuberance prevents the sphere being perfectly round. On the other hand, in reckoning the shape of the sphere, how is it relevant to take account of the Moon, although it does not properly belong to the sphere of the Earth, as the wanderings of the other planets in height and in depth (which physically are very conveniently explained by epicycles), as, I say, these epicycles belong each to its own sphere? For it is the Earth to which this third sphere from the Sun is allocated, the Earth by its impulsion goes round the Sun among the other planets, the Earth on its own and by its own epicycles with no assistance from the Moon for this purpose performs its variations, as Copernicus's theories tell us, but the Moon holds its tiny home round the Earth as if as a favor or on lease, the Moon follows, or rather is dragged, wherever the Earth goes in any of its variations. Imagine the Earth at rest; the Moon will never find its way round the Sun, much less make its way round. For it keeps flitting to and fro, shut into its narrow spaces round the Earth, and serves the Earth with light and moisture like some steward about his master, or like people who wander about in a ship, who although they tire themselves out make no progress in their journey unless the great power of the waters carries them on, uncertain where they are going to and even if they rest. And just as space and motion have been assigned to the Moon from the terrestrial sphere, similarly Maestlin proves by many inferences, of which I have not a few, that it has also got * many of the features of the terrestrial globe, such as continents, seas, mountains, and air, or what somehow corresponds to them; so that on this account alone Copernicus is more convincing, as he endows these two bodies with a common position and motion. And certainly the Creator, loving Man, seems finally to have clothed the Earth with this lunar sphere, because he wished to allot to it a position similar to the Sun's, so that if it too was the center of a sphere (as the Sun is the center of all things), it could be considered as like a

cuiufdam haberi poffet,ob quod ipfa totius vniuerfi commune centrum
communiter quafi habita fuit.

Eft omnino,vt denuo ludam Allegoria, homo quidam quafi Deus
in mundo,& eius domicilium Tellus;ficut Dei,fi vllum corporeum;cer-
te Sol illa lux inaccefla. Vt igitur homo Deo, fic Tellus Soli refpondere
debuit. Argumento eft huius rei (3) eadem fere proportio globi Telluris ad orbem Lunæ, quæ globi Solaris ad mediam Mercurij digreffionem à Sole.

Neque vero metuendum eft, ne lunares orbes à vicinis corporum
proportionibus compreffi elidantur, fi non fint in orbe ipfo abfconditi
atque inclufi. Nam abfurdum & monftrofum eft, corpora hæc materia
quadam veftita, quæ alieno corpori tranfitum non præbeant, in cœlum
collocare. Certe multi non verentur dubitare, an omnino fint in cœlo
eiufmodi Adamantini orbes ; an diuina quadam virtute, (4) moderante curfus intellectu proportionum Geometricarum,ftellæ per cam-
pos & auram æthercam liberæ iftis orbium compedibus tranfportentur.
Nullum equidem pondus dubios & titubantes motori greffus efficiet,
quo aliquando à circulo fuo exorbitet.

(5) Nullum enim punctum,nullum centrum graue eft. Centrum
vero omnia eiufdem cum corpore naturæ fequuntur. Nec pondus ex eo
acquirit centrum,quod cætera ad fe allicit,aut ab illis appetitur: (6) nõ
magis atque Magnes,dum actu ferrum trahit,ingrauefcit. Vel hæc tellus,quam omnino cum Copernico vehi ftatuimus, quibus vectibus,qui-
bus catenis,quo Adamante cœlefti in orbem fuum inferta eft? Eo nempe quem omnes circumcirca in fuperficie Telluris homines haurimus
(fermétatum & commixtum vaporibus)aerem ; quem manu,quem corpore penetramus,neq; tamen difcludimus,aut femouemus cum fit influxuum (7) cœleftiú in media corpora vehiculú.Hoc n. cœlum eft,in quo
viuimus,mouemur & fumus nos & omnia mundana corpora. Quamuis
quid opus tot verbis?Nam etfi orbiculus Lunæ fupra Telluris orbé emi-
neat: quid eft de Dodecaedro vel Icofaedro, quod illum tranfitu prohibeat? Vidifti fupra cap. XI. quo loco Zodiaci planum hæc duo corpora
fecat,nullum angulum,nullum faciei centrum occurrere, fed exiftere ex
fectione decangulum vtrinque,cuius quæ ex centro ad latus perpendicularis cadit,longe maior eft in Dodecaedro, radio infcripti, longe bre-
uior in Icofaedro radio circumfcripti:& adeo longa quidem, vt non cœ-
lulum illud Lunæ tantum, fed longe maius aliquid fupra orbem extans,
per mediam illam viam interque illa decangula tranfire poffet. Sed hæc
omnia quamuis fuo loco relinquantur,nihilo peius fe res habet.

Vides enim per interpofitionem Lunæ præterquam in
Venere quam proxime accedi ad proditos,
per finus Copernici,numeros
arcuum.

Sun; and on that account it was in effect commonly considered the common center of the whole universe.[1]

In general, to indulge in allegory once again, a man is like God in the universe, and Man's dwelling place is the Earth, just as God's, if he has any material dwelling, is certainly the inaccessible light of the Sun. Then as Man to God, so the Earth should have corresponded with the Sun. That is evidenced by the fact that (3) the proportion of the Earth's globe to the sphere of the Moon is almost the same as that of the Sun's globe to the mean distance of Mercury from the Sun.

Nor should we be afraid that the lunar spheres may be compressed and squeezed out by the close proportions of the solids, if they are not hidden away and shut into the actual sphere of the Earth. For it is absurd and monstrous to locate these solids in the heaven if they are clothed in such a material that they do not allow an alien body to pass.[2] Certainly many have no fear of doubting whether the spheres in the heaven are entirely of that adamantine sort, or whether it is by some divine power, (4) the understanding of the geometrical proportions governing their courses, that the stars are transported through the ethereal fields and air free of the restraints of the spheres. No weight, indeed, will make the progress of the mover doubtful and tottering, so that it deviates at some time from its circle. (5) For no point or center has weight. Everything, in fact, which is of the same nature as a body seeks its center. Nor does a center acquire weight from the fact that it draws other things to it or is sought out by them, (6) any more than a magnet, when it attracts iron by its action, becomes heavy. Or by what bars, what chains, what heavenly adamant, has this Earth, which with Copernicus we have completely established to be moving, been brought into its sphere? With that air, no doubt, which (fermented and mixed with vapors) all we men drink in all round the surface of the Earth, which we penetrate with our hand, with our body, but do not part or separate, though (7) it conveys the heavenly influences right into our bodies. For this is the heaven in which we and all worldly bodies live, move and have our being. Yet what need is there for so many words? For even though the little sphere of the Moon projects above the sphere of the Earth, what is there about the dodecahedron or icosahedron to prevent it from passing through? You have seen above in Chapter XI that where the plane of the zodiac cuts these two bodies, there is not a vertex, and there is not a center of a face, but on either side of the division a decagon is formed, in which a line from the center perpendicular to the side is much longer than the radius of the inscribed sphere in the case of the dodecahedron, but much shorter than the radius of the circumscribed sphere in the case of the icosahedron; and so long indeed, that not only the Moon's little heaven, but something far larger which stood out above the sphere could easily pass through that intervening path and between the two decagons. But although all this is left for its proper place, the argument does not suffer. For you see that by the interpolation of the Moon, except in the case of Venus, we come as close as possible to the values published for the angles, in accordance with Copernicus's sines.

IN CAPVT DECIMVMSEXTVM
Notæ Auctoris.

(1) NEc hercle scio,quorsum magis inclinent rationes.] *At iam in lucem prolatis contemplationibus Harmonicis,decisa est hæc controuersia,lib.V.Harmon. Primum enim corporibus ipsis quinque adempta sunt proportiones Orbium ex parte: vltima sc. & absolutißima Orbiũ proportio communis est facta & corporibus & Harmoniis Prop.XLVIII.& XLIX. cap.IX. Quo nomine nihil ex solis corporibus in hanc vel illam partem de Luna disputari potest. Deinde si maxime ex Solis quinque corporibus formarentur proportiones orbium;huius tamen formationis modus alius, vt in quo inscriptio orbium Physica gradus perfectionis proportionum Geometricarum æmularetur stabilitus est Prop.XLVI.XLVII. Tertio constat ex omnibus illius libri axiomatibus & propositionib. vltimam limitationem proportionis diastematum fieri necessariam,propter motus Planetarum; vt sc.inter extremos motus esse possent harmoniæ certæ.Si hoc;nulla igitur potest haberi ratio Lunæ,terram circumcursitantis, vt quæ nihil conferat ad incitandum vel retardandum vllius Planetæ motum, nec curriculum suum circa Solem exercet,nec ex Sole regularis apparet eius motus. Nam ex Sole inspectus Lunæ motus videretur saltuatim incedere.Sic igitur de orbe Telluris est disputandum,ac si Lunæ cœlum nullam ei crassitiem adderet.*

(2) Concinnum quidem, vt non sit talis orbis cum Nodo.] *Hæc gemino sensu possunt accipi;primus,textui conueniens,est hic: vt sit quidem Orbis cum nodo, sed includatur Orbita Planetæ,tantæ spißitudini,vt nodus hic, seu Lunæ cœlum,lateat totum intus, nihil impediens exma intimæque superficiei rotunditatem absolutam. Alter sensus horum verborum; posset arripi iste: quod in genere absurdum sit Lunam circumire Terram,dum hæc interim circa Solem incedit.Vt igitur hanc etiam obiectionem diluam: dico,quod hoc tunc concinnum videri potuerit,cum nondum detecti essent Iouiales Planetæ, & cætera in cœlo noua. At ex quo illa scimus, concinnum nequaquam amplius videri debet,non esse,quod omnino est,Nodus sc.quadruplex circa Iouem, si pro Nodo corporeo spatia curriculorum intelligas, sic circa Iouem ordinatorum, vt circa Terram Lunæ curriculum ordinatum est. Nam de corporea Orbium soliditate supra satis cautum, & cauetur etiam in textu sequenti.*

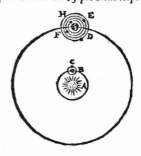

* Multas conditiones globi terreni adeptam.] *Consensus in hoc multorum per omnes ætates philosophorum, qui supra vulgus sapere sunt ausi. Diogenes Laertius Anaxagoræ tribuit;libro meo,cui Titulus, Ad Vitellionem Paralipomena, capite de Luce siderum,allegaui Plutarchum de facie Lunæ. Citatur & Aristoteles ab Auerröe.Verum hoc dogma postremus Galilæus Telescopio Belgico confirmatißimum reddidit.Videetiam dissertationem meam cum nuncio siderio Galilei.*

(3) Eadem fere proportio globi Telluris ad Orbem Lunæ.] *Certa quidem est proportio ista,sc.quæ 1.ad 59.circiter: at proportio corporis Solis ad orbem Mercurij est paulo alia; sc. non medius orbis Mercurij, sed intimus & angustißimus est assumendus; cui in Tabella capitis XV. tribuuntur gr.14. cum Solis semidiameter ex eadem Tellure inspectus, contineat minuta 15.quare fere est proportio quæ 1.ad 56.*

(4) Moderante cursus,intellectu proportionum.] *Ita quidem tunc censebam; at postea in Comment. de Marte, ne hoc quidem intellectu in motore opus esse demonstraui. Nam etsi proportiones certæ sunt præscripta motibus omnibus, idque ab Intelligentia ipsa suprema & vnica, hoc est,à Deo creatore: illa tamen proportiones motuum inde à creatione hucusque conseruantur inuariabiles,non per intellectum aliquem Motori concreatum, sed per duas res alias;prima est, æquabilißima & perennis rotatio corporis solaris,cum specie sui immateriata,in totum mundum emanante, quæ specie vicem motoris præstat;altera causa, sunt libramenta & magneticæ directiones corporum ipsorum mobilium immutabilia & perennia. Vt sic aque non magis sit opus creaturis istis intellectu ad tuendas motuum proportiones,atq̃ libræ lancibus & ponderibus mente est opus ad prodedam proportionem ponderum.Etsi sunt alia argumenta quibus probatur,inesse in corporibus Planetarum,saltem Telluris & Solis, intellectum aliquem,non quidem ratiocinatiuum vt in homine; attamen in-*

stinctum

Chapter XVI

AUTHOR'S NOTES ON CHAPTER SIXTEEN

(1) *Nor, great heavens, do I know in which direction the...arguments tend more.*] But now that my meditations on harmony have been brought out into the light of day, this controversy has been settled, in Book V of the *Harmonice*. For, first, the proportions of the spheres have been partly removed from the five solids. That is to say, in its final and most finished state the proportion of the spheres belongs to both the solids and the harmonies in common, by Propositions 48 and 49 of Chapter 9. On this showing, no argument for one side or the other about the Moon can be drawn from the solids alone. Second, even if the proportions of the spheres were formed chiefly from the five solids alone, yet it has been determined by Propositions 46 and 47 that the mode in which they were formed was different, and such that the physical inscribing of the spheres reflected the degrees of perfection of the geometrical proportions. Third, it is established from all the axioms and propositions of that book that there must necessarily be a final limit to the ratios of the intervals on account of the motions of the planets, so that there could be definite harmonies between the extreme motions. If that is so, then no proportion can be found for the Moon's flitting round the Earth, as it contributes nothing to hastening or retarding the motion of any planet, and does not perform its circuit about the Sun, nor have a motion which appears regular from the Sun. For the motion of the Moon viewed from the Sun would seem to proceed in jumps. Therefore we must argue on the sphere of the Earth as if the heaven of the Moon added no thickness to it.

(2) *Indeed (it seems to be a question of) tidiness, about whether there is not such a sphere with a lump.*] These words can be taken in two senses. The first, agreeing with the text, is this: that there is indeed a sphere with a lump, but it is contained within the orbit of the planet, the thickness of which is so great that this lump, or the heaven of the Moon, is completely concealed within it, and does not hinder at all the absolute roundness of the outer and inner surface. The other sense which could be forced into these words is the following: that it is categorically absurd for the Moon to go round the Earth, while the latter at the same time proceeds round the Sun. Then to clear away this objection as well: I say that this could have appeared tidy when the satellites of Jupiter, and the other new objects in the sky, had not yet been detected. But ever since we came to know of them, it should no longer seem at all tidy that what decidedly exists should not exist, that is to say the quadruple lump round Jupiter, if by corporeal lump you understand the spaces occupied by the courses which have been appointed round Jupiter in the same way as the course of the Moon has been appointed round the Earth. For on the corporeal solidity enough caution has been shown above, and caution will be shown in the following text.

* *It has (also) got many of the features of the terrestrial globe.*] The consensus of many philosophers on this point throughout the ages, who have dared to be wise above the common herd. Diogenes Laertius[3] attributes it to Anaxagoras; and in my book entitled *Paralipomena Ad Vitellionem*, in the chapter "On the Light of the Stars," I have referred to Plutarch's *On the Face of the Moon*. Aristotle is also cited by Averroes. However Galileo has at last throughly confirmed this belief with the Belgian telescope. See also my *Conversation with Galileo's sidereal messenger.*[4]

(3) *The proportion of the Earth's globe to the sphere of the Moon is almost the same.*] Indeed this proportion is certain, that is about as 1 to 59; but the proportion of the body of the Sun to the sphere of Mercury is slightly different, that is, not the mean measurement of the sphere of Mercury, but its inside and smallest measurement must be assumed. To that in the table in Chapter 15 are attributed 14°, whereas the radius of the Sun, as seen from the same Earth, contains 15′; so that the ratio is about 1 to 56.

(4) *The understanding of the (geometrical) proportions governing their courses.*] So indeed I then supposed; but later in my *Commentaries on Mars*[5] I showed that not even this understanding is needed in the mover. For although definite proportions have been prescribed for all the motions, and that by the supreme and unique Understanding himself, in other words, by God the Creator, yet those proportions between the motions have been preserved unchanged from the Creation right up to the present not by some understanding created jointly with the Mover, but by two other things. The first is the completely uniform perennial rotation of the solar body, along with its immaterial emanation, which is diffused to the whole universe, an emanation which takes the place of a mover. The other cause is the weights and magnetic directing forces of the moving bodies themselves, which are immutable and perennial properties. Thus there is no more need for these created things to have understanding to observe the proportions of their motions than there is for the scales and weights of a balance to have intellect to declare the proportions of weights. Nevertheless there are other arguments by which it is proved that there exists in the bodies of planets, at least of the Earth and Sun, some understanding, not indeed rational as in Man, but

stindum vt in planta, quo conseruatur species floris, & numerus foliorum. De hoc vide Epilogos libro-
rum IV. & V. Harmonices nostrae.

(5) Nullum enim punctum graue est] *Ita conceptum est hoc argumentum, vt au-*
dire velim physicos, quid contra dicere possint. Nam ab his 25. *annis nemo quod sciam extitit, qui illud*
excuteret. At me candor solus mouet, vt ipse excutiam. Vides igitur Lector, quid voluerim, Centrum
solum esse quod primo circa Solem agatur in gyrum: Id vero vel solo nutu fieri posse, cum graue non sit,
vt cuius pars nulla. Hanc propositionem non potest mihi eripere physicus, qui contendit, quod hic sequi-
tur, omnia centrum sequi. Et quia vulgata doctrina physica tenet hoc de centro mundi, quod omnia
grauia id centrum quærant, ideo existimaui ego, posse grauia eâdem opera centrum sui corporis quæ-
rere. Verum in Epitomes Astronomiæ lib. 1. *demonstraui, falsum esse hoc physicorum axioma, quod*
grauia quærant vllum centrum vt tale, falsissimum quod centrum totius mundi; verum, sed per acci-
dens, quod centrum Telluris appetant, non quam id punctum est, sed quia corpus Telluris appetunt;
quod cum sit rotundum, ex eo fieri vt appetentia ista feratur versus medium, & sic versus centrum;
adeo quidem, vt si terra figuram haberet distortam sensibiliter; Grauia non versus vnum vndiq́; pun-
ctum tensura fuerint. Hoc igitur fundamento corruente, structura etiam euertitur huic nimia. Sci-
licet corpora Planetarum in motu, seu translatione sui circa Solem, non sunt consideranda vt puncta
mathematica, sed plane vt corpora materiata, & cum quodam quasi pondere (vt in libro de stella noua
scripsi) hoc est, in quantum sunt prædita facultate renitendi motui extrinsecus illato, pro mole corpo-
ris, & densitate materiæ. Nam quia omnis materia ad quietem inclinat in loco illo in quo est (nisi cor-
pus vicinum vi magnetica illam & se alliciat) hinc adeo fit vt virtus Solis motoria pugnet cum hac
inertia materiæ, sicut in lance pugnant duo pondera, ex que virtrarumque virium proportione tandem
enascatur celeritas vel tarditas Planetæ. Vide introductionem in Comment. Martis, & ipsa Com-
mentaria passim; præcipue vero librum IV. Epitomes Astronomiæ.

Neque tamen ex eo sequitur, quod hic per falsam ratiocinationem amolitum ibam, dubios &
titubantes motoris gressus effici, si laborat in pondere, vincitque in pugna. Nam certa & constans est
proportio virium inter se vtrarumque, & victoria partibilis, pro virium modulo; vt neque Planeta
in eodem hæreat loco; neque rotationis Solaris celeritatem assequatur.

(6) Non magis atque magnes, dum actu ferrum habuit, ingrauescit.] *Manifes-*
tis experimentis hoc falsum deprehenditur. Pondera seorsim ferrum, seorsim & Magnetem; collige
pondera in vnam summam. Suspendatur deinde ferrum à Magnete vi illa inuisibili; Magnes vero ne-
ctatur à lance, aut inijciatur, quia vis permeat lancem, si non sit ferrea: videbis, Magnetem, dum actu
tenet attractum ferrum, æque ponderaturum vtrisque, prius ab inuicem separatis.

(7) Influxuum cœlestium in media corpora vehiculum.] *Non equidem, quod*
influxus cœlestes indigeant aliqua materia, qua ad nos deuehantur; falsum enim est illud Aristotelis,
aëre opus esse, ad sensionem corporis Solaris transportandam vsque ad oculum; vt in Opticis demon-
straui: quin potius, quo minus occurrit materia, in itinere medio, hoc minus impeditur lux in traie-
ctione sua. Hoc igitur sibi volunt ista verba: sicut corpora non impediant, quo minus influxus cœlestes
in intima penetrent: sic etiam Motorias facultates non indigere corporibus aliquibus intermediis,
quibus veluti catenis aut vectibus mouenda Planetarum corpora prehendant. Ludere placuit in voce
aëris paulo audacius. Quid Orbis vel cœlum? Quid nisi aër? Et quid aër? Quid nisi species immateria-
ta corporis, quod motum Planetis infert, in gyratione versantis? Atqui seposito Lusu, concedamus, aë-
rem nostrum esse corpus materiatum, permeabile à facultatibus magneticis, motoriis, calefacto-
riis, illuminatoriis, & similibus: vt sit vapor non toto genere diuersum ab aëre,
sed saltem gradibus crassitiei distinctus à circumfusis
aëris campis.

instinctive as in a plant, by which the type of flower and the number of leaves are perpetuated. On this point see the Epilogues to Books IV and V of our *Harmonice*.

(5) *For no point...has weight.*] This argument has been conceived in such a way that I should like to hear from the physicists what they can say against it. For in the course of the ensuing 25 years no one as far as I know has come forward to examine it. However I am moved by honesty alone to examine it. Therefore, reader, you see what I mean, which was that it is the center alone which is propelled in a circle round the Sun; and that indeed could be brought about by a single nudge, as it has no weight; as something which has no parts. This proposition cannot be snatched from me by a physicist who argues what follows here, that everything conforms with its center. Also because the commonly received doctrine in physics holds with respect to the center of the universe, that everything which has weight seeks that center, for that reason I supposed that things which have weight by the same token seek the center of their own body. However in the *Epitome of Astronomy*, Book I, I showed that the axiom of the physicists stating that things which have weight seek any center as such is false, and that they seek the center of the whole universe is more false. It is true, but by accident, that they strive towards the center of the Earth, yet not insofar as it is a point, but because they strive towards the body of the Earth; since it is round, as a result of that it comes about that the direction of this striving is towards the middle, and thus towards the center. In fact, it follows that if the Earth had a shape which was sensibly distorted, things which have weight would not tend towards the same point from every direction. Consequently with the collapse of this foundation, the edifice, which was excessive for it, is also demolished. Clearly the bodies of the planets in motion, or in the process of being carried round the Sun, are not to be considered as mathematical points, but definitely as material bodies, and with something in the nature of weight (as I have written in my book *On the New Star*), that is, to the extent to which they possess the ability to resist a motion applied externally, in proportion to the bulk of the body and the density of its matter. For because all matter tends to remain at rest in the place where it is (unless a neighboring body attracts it to itself by magnetic force), it comes about as a result that the motive power of the Sun contends with this inertia of matter,[6] as two weights contend on a balance, and from the relative strength of the two forces is at length produced the quickness or slowness of the planets. See the introduction to the *Commentaries on Mars*, and the Commentaries themselves throughout, but especially Book IV of the *Epitome of Astronomy*.[7]

However, it does not follow from that, as I was setting out to refute here by false reasoning, that it makes the progress of the mover doubtful and faltering, if it struggles in the balance and wins the contest of weights. For the proportion of the two forces to each other is definite and constant, and the victory can be shared in accordance with the rating of the forces, so that the planet neither sticks in the same place nor matches the speed of the Sun's rotation.

(6) *Any more than a magnet, when it attracts iron by its action, becomes heavy.*] This is detected as false by clear experiments. Weigh separately a piece of iron and a magnet, and add the weights together. Then let the iron be suspended from the magnet by this invisible force, and let the magnet be fastened to the scale of a balance, or thrown into it, since its force would pass through the scale, if it were not of iron. You will see that the magnet, while it attracts and holds the iron by its action, will weigh the same as the two did previously, when they were separate from each other.

(7) *It conveys the heavenly influences right into our bodies.*] Not, that is, because the heavenly influences need some matter to carry them to us; for Aristotle is wrong when he states that air is needed to transport the sensation of the solar body to the eye, as I have shown in the *Optics*.[8] Rather, the less matter the light meets in the course of its journey, the less it is impeded in its passage. The meaning of these words, then, is as follows: just as bodies do not impede the heavenly influences from penetrating into our inner parts, so also powers capable of producing motion do not require intermediate bodies, by which to take hold of the bodies of planets which are to be moved as if by chains or bars. I chose to make rather too bold a play with the word "air." What is a sphere or a heaven? What but air? And what is air? What but an immaterial emanation of the body, which imparts motion to the planets, as it turns in its gyration? But, laying aside the play on words, let us concede, that our "air" is a material body, through which magnetic, moving, heat-producing, light-producing, and similar powers can pass, so that it is a vapor not totally different in kind from air, but rather distinguished by degrees of thickness from the expanses of air which surround it.

C A P V T XVII.

Aliud de Mercurio monitum.

I L L V D magis mirabere, cum promiserim, velle me corporibus ipsis inscribere Planetas, cur Mercurium non Octaedro inscripserim; sed passus sim eum in circulo aliquo vltra orbem inscriptilem ad quadrati Octaedrici amplitudinem expatiari. Nam supra cap. 13. & 14. pro 577. numero orbis inscripti vsurpaui 707. numerum circuli inscripti quadrato. Causam dicam. Primum, quia eius à Sole digressio longior minime pati potuit tam angustos carceres: deinde quia & Octaedron inter corpora, & motus Mercurij inter Planetas peculiare quid, & commune inuicem habent. Nam in solo Octaedro super angulum erecto vsu venit, vt quadratum directis lateribus viam aliquam môstret ampliori circulo, quam est orbis inscriptus, per medium transeundi. Id quod in nullo alio corpore quomodocunque voluto vsu venit. Semper enim transuersa per medium & impedita incedent latera.

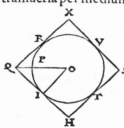

In hoc schemate quatuor lineæ extremæ sunt quatuor perpendiculares totidem planorum in Octaedro. R I T V sunt eorum planorum centra, determinantia amplitudinem orbis inscripti, de quo hîc vides Circulum maximum. Qui orbis si intelligatur volui super punctis ad X H, duos angulos figuræ, reperiet in P Quadrante à polis circumcirca amplitudinem aliquam maiorem, quam est O I, vel O P semidiameter orbis, nempe O Q. Differentia eius est P Q. Et tanta est latitudo circuli, qui vltra orbem excurrens, instar Horizontis alicuius in sphæra armillari, per medium Octaedri transire potest. Q enim & S sunt media puncta duorum laterum, proinde & proxima orbi.

Quomodo si animatus quidam planeta per medium Octaedrum currere iuberetur, & angulos duos pro polis, amplitudinem inscripti pro curriculo obseruare; non hercle mirum, si inuitatus illa amplitudine, vbi nullæ illi metæ obstarent per totum ambitum, exorbitaret aliquando, vt Phaethon ille, tantisper, dum repelleretur ab occurrenti latere. Quod per iocum dixi, id serio aiunt Artifices euenire Mercurio. Cum enim cęteri omnes in singulis reuolutionibus describant eiusdem amplitudinis circulos (quantum enim ab vna parte discedunt, tantũ ex altera viæ parte accedunt ad Solem) (1) solus Mercurius ab Artificibus obtinuit, vt aliquando maiorem, aliquando minorem circulum describere diceretur: idque priuilegium merum haberet. Dicunt enim illum accedere & recedere à Cẽtro sui orbis O per lineam rectam Y Z, vbi semidiameter O Y longe minorem Circulum describit, quam O Z. Nam cęteras inæqualitates omnes cum alijs æqualiter sor-

titus

CHAPTER XVII.
ANOTHER COMMENT ON MERCURY

You will wonder all the more, since I have promised that I intend to inscribe the planets within the actual solids, why I have not inscribed Mercury within the octahedron, but have allowed it to diverge to the full breadth of the square in the octahedron in a circle outside the inscribed sphere. For above, in Chapters 13 and 14, I have taken instead of 577, the value for the inscribed sphere, 707, the value for the circle inscribed in the square. I will explain the reason. First, it is because its further deviation from the Sun could in no way permit such narrow confines; and secondly because the octahedron among the solids and the motion of Mercury among the planets have a feature which is peculiar and shared by both.[1] For only in the case of an octahedron standing on a vertex does it happen that the square built from the perpendicular edges shows a path by which to go round in a circle wider than the inscribed sphere. That occurs in the case of no other solid however it is turned. For edges always cross the path and block it.[2]

In this diagram the four outer lines are four perpendiculars of the same number of faces on the octahedron. RITV are the centers of the faces, which determine the size of the inscribed sphere of which you here see a great circle. If that sphere is understood to be turned on the points at X and H, two vertices of the figure, there will be traced out at P, a quadrant from the poles, a width all round which is greater than OI or OP the semidiameter of the sphere, that is, OQ. The difference is PQ. And that is the size of the circle which can run round outside the sphere, like a horizon on an armillary sphere, and go round inside the octahedron. For Q and S are the midpoints of two edges, and consequently the nearest points to the sphere.

Thus if some sentient planet were ordered to follow a path inside the octahedron, and to take two vertices as the poles of its orbit, and the breadth of the inscribed sphere as the limit of its track, heavens! it would not be surprising if attracted by such a breadth, where no boundary markers obstructed it round the entire circuit, it should sometimes leave its orbit, as Phaethon did, up to the point where it was driven back by finding an edge in its way. Although I have said this jokingly, the practitioners say seriously that this happens to Mercury. For whereas all the rest in their individual revolutions describe circles of the same size (for if they move outwards on one side, on the other side of their path they move in towards the Sun by the same amount), (1) Mercury alone has persuaded the practitioners that he should be said to describe a circle which is sometimes larger, sometimes smaller, and that he should have that privilege without penalty. For they say that he moves towards and away from the center of his orbit O along the straight line YZ, where the semidiameter OY describes a far smaller circle than OZ.[3] For he has been allotted his share of the other irregularities equally with the rest, and he has not exchanged any of them for this departure from his orbit.

titus est; nullamque cum hac exorbitatione commutauit. (2) Et cum
cæterorum eccentrotetes omnes,si non proportionaliter,sic tamen de-
crescant; vt minoris semper minor sit eccentricitas: solus Mercurius im-
manem habet,nempe decuplum Veneris,cum ipsi vt inferiori minus etiam deberetur. Quare etsi illam inæqualitatem priuatam nondum cum
hac circuli ab orbe differentia conciliauerim, nec ea fortasse conciliari
possit,vt prodita est ab Artificibus,ad amussim: Nihilominus ego nõ dubito,quin creator ad figuræ huius prescriptum in motibus Mercurio tri-
buẽdis respexerit. Quo diuinior magis magisq; mihi & Astronomia &
Copernici placita,& hæc ipsa 5. corpora videntur.

(3) Quærant alij,qui voluerint,cæterarum etiam eccẽtricitatum causas ex suis quasque corporibus. Cum enim neq; hæ exorbitationes à Deo
temere & sine causa tantæ singulis Planetis indultæ sint: non desperanda
est neq; harumcausarum inuestigatio.

Porro vt varietas Mercurij ad Octaedron accommodetur, sic agi
posset. Sumeretur proportio eccentr. ☿ ad distantiam mediam à ⊙ pro
certa,vt quia in Copernico distantia (sicut vides in tab.V.cap.15.) longis-
sima est 488. breuissima 231. media igitur erit 360. & crassities tota 257.
Hęc iam crassities cotrigeretur proportionaliter,vt quia circulus Octaedri pro 488.numero Copernici largitur non plus 474.ergo crassities erit
in hac proportione 250.& media correcta distantia 349. Iam vide,quid
orbis in Octaedro admittat,scil.387. Differentia igitur inter 387.altissi-
mam orbis,& 349.mediam est 38. & duplum 76. crassities orbis ad modũ
cæterorum,maior quidem adhuc quam Veneris, sed tamen non ita im-
manis.Reliqua differentia inter altissimam orbis 387.& altissimam circuli 474.quæ est 87.debetur peculiari exorbitationi Mercurij Hoc ὀπηχεί-
ρεμα, an abijciendum, an conciliandum cum ἐπιϰέτῳ forma motuum in
☿,an noua motuum ratio constituenda,considerent Artifices.Nec enim
ita bene explorati sunt errores huius sideris, vt eius orbis correctione
non egeat.

In Caput XVII. Notæ Auctoris.

(1) *S* Olus Mercurius obtinuit.] *Quale sit illud, quod Artifices peculiariter adscribunt Mer-*
curio,rectius petes ex Ptolemæo ipso,exque Purbachÿ & Mastlini Theoricis: denique quomodo
Copernicus illud duplici via (quia sibi ipse non satisfecit) in formam suarum hypothesium transtule-
rit,seipsum tamen considerit,plus aliquid præstans (per suos motus triangulationis alicuius æmulos)
quam ex Ptolemæo sibi proposuerat exprimendum: id totum, nec adeo necessarium est hoc loco expli-
cari,cum sit de opinionibus hominum,non de veritate rerum;& si quid vtiliter dici potest, rectius al-
lorsum reijcitur. In recensim,hoc est,quod Mercurius facit enormem Eccentricitatem circuli sui à Sole
quem circulum Ptolemæus Epicyclum, ego eccentricum dico, quodque in illo etiam eccentrico moue-
tur inæqualiter,ad proportionem eccentricitatis. Ex hisprincipiis, & ex eccentricitate Telluris, quo-
modo conflata sit phantasia illa duplicis in Mercurio perigæi,& sic motus quasi triangularis:id expli-
cabitur in demonstratione motuum Mercurij;nec plane prætereo summam rei in Epit. Astr.lib.6. Suf-
ficit hoc loco,illud monere, non esse huius singularitatis Mercurialis causam aliquam Archetypicam
ex Octaedro;eoque falsam huius capitis Hypothesin: iucundissimam tamen recordationem huius E-
pichirematis, vt appareat, quibus ignorantiæ gradibus ad Astronomiæ scientiam & constitutionem
ascenderim.

(2) And whereas the eccentricities of the others all decrease, if not in proportion to their size, at any rate in such a way that the smaller orbit always has a smaller eccentricity, Mercury alone has a huge one, in fact ten times that of Venus, although as it is the lower planet it ought also to have had a smaller eccentricity. For this reason although I have not yet linked this special irregularity with the difference between the orbit and the sphere, and perhaps it is impossible to link them, as has been claimed by the practitioners, precisely, nevertheless I have no doubt that the Creator oberved the pattern of this figure in allocating motions to Mercury. All the more, and ever yet more divine do astronomy, and the theories of Copernicus, and these very five solids, seem to me.

(3) Let those who wish seek the reasons for the other eccentricities also in the appropriate solid in each case. For since such large departures from their orbits have not been conceded to the individual planets by God at random and without a reason, neither should we despair of finding out even those reasons.

Further, adjustment of the variation of Mercury to the octahedron could be achieved at this point in the following way. The ratio of the eccentricity of Mercury to its mean distance from the Sun would be taken as certain, that is, since in Copernicus the longest distance (as you see in Plate V,[4] Chapter 15) is 488, the shortest 231, the mean distance will therefore be 360, and the total thickness 257. Now this thickness would be corrected by proportion, that is, since the circle of the octahedron instead of Copernicus's value, 488, concedes no more than 474, therefore the thickness will be 250 in that ratio, and the corrected mean distance 349. Now consider what the sphere in the octahedon permits, that is to say, 387. Then the difference between 387, the highest level of the sphere, and 349, the mean, is 38, and twice that is 76, which is the thickness of the sphere reckoned in the same way as the others, still larger indeed than that of Venus, but yet not so huge. The remaining difference between the highest level of the sphere, 387, and the highest level of the circle, 474, which is 87, is due to the peculiar departure of Mercury from its orbit. Whether this attempt should be rejected, or reconciled with the pattern of motions assumed in the hypothesis for Mercury, or whether a new system should be established for the motions, let the practitioners examine. For the deviations of this star are not so well investigated that its orbit does not need correction.

AUTHOR'S NOTES ON CHAPTER SEVENTEEN

(1) *Mercury alone has persuaded.*] The nature of what the practitioners ascribe as a peculiarity to Mercury you can more properly find in Ptolemy himself, and in the *Theories* of Peurbach and Maestlin, and lastly in the way in which Copernicus incorporated it into the pattern of his hypotheses by two methods (because he was not satisfied about it himself) yet confounded himself, providing something more (by these motions which emulate a triangular arrangement) than he had planned to draw from Ptolemy. It is not so essential for this whole matter to be explained at this point, as it concerns the opinions of men, not the truth of things; and if anything can usefully be said, it is more properly dismissed to somewhere else. For the fact of the matter is that there is an enormous eccentricity in Mercury's circle about the Sun, a circle which Ptolemy calls the epicycle, but I call the eccentric, and that on that circle it also moves non-uniformly in proportion to the eccentricity. From these principles and from the eccentricity of the Earth, the way in which the fantasy of a double perigee of Mercury and thus of its motion's being, so to speak, triangular, was concocted, will be explained in the derivation of the motions of Mercury; and I certainly do not omit to mention the summary of the matter in the *Epitome of Astronomy*, Book VI.[5] It is sufficient at this point to make the comment that there is not an archetypal cause of this singular property of Mercury which arises from the octahedron; and hence the hypothesis of this chapter is false; yet it is very pleasant to recall to mind this argument for it, so that it is made apparent by what steps of ignorance I have ascended to the knowledge and establishment of astronomy.

(2) Et cum cæterorum eccentrotetes omnes.] *Neque hoc vndiquaque sic ha be Saturni quidem vera Eccentricitas maior est Iouiali : Iouis vero multo minor , Martiali inferiori.*

(3) Quærant alij.] *Nemo extitit,qui quæreret. Quærite & inuenietis. Quæsiui,& ecce inueni,lib.V.Harmonicorum, causas præstantißimas. Adeo bonum & fidum hoc omen fuit :* Non desperare : *adeo pollens & prægnans axioma hic vsurpatum :* Nihil à Deo temere constitutum.

C A P V T   XVIII.

De discordia προθαφαιρέσεων *ex corporibus à Copernicanis in genere,& de Astronomiæ subtilitate.*

V P R A cap. XIV. & XV cum alicuius prope falsitatis teneri viderer indicio distantiarum, quas Copernicus diuersas ab his figuralibns prodidit : prouocaui ad προθαφαιρέσεις ἀπογέιχς: neq; condemnationem deprecatus sum, si meæ à Copernicanis aliquantum recederent. Atqui postquam sub finem XV. capitis arcus similes προθαφαιρέσεων ex elongationibus à Sole, veluti testes coram hoc iudicio stiti : visi sunt illi contra me deponere.Nullus enim Planetarum fuit,qui tributum à Copernico arcum retineret. Saturno ademi 4´1.Ioui,´6. Marti apposui 3´0,Veneri vero immane quantum dempsi 2.gr.1´8.& Mercurio 6´1. Existimabunt igitur qui exactius omnia examinare volunt, quia non ad vnguem consentiat calculus corporum cum placitis Copernici, cumque eius numeris,omnem operam à me lusam esse. Quod nisi contra excepero, meapte sententia causam perdidero. Et Physicis quidem siue Cosmographis , qualem hoc libello personam ego sustineo, nullam de hac differentia rationem debeo.Nam etsi illi suorum placitorum argumenta mutuantur ab Astronomis; ea tamen non ita subtiliter, vt Astronomi, ad calculos reuocant ; nec adeo sunt perspicaces aut morosi, vt hac leuicula differentia moueantur. Quare causam meam coram Cosmographis obtinui.

Astronomorum vero vulgus etsi iure metuo; tamen cum iudicio Artifices præesse par sit, non despero,neque contra illud, victoriam. Ac primum ipsos bene de calculo sperare iubeo.Nam etsi interdum grandiuscula est differentia , meminerint tamen numeros excerptos ex locis totius circuli euidentißimis,atque ex concursu omnium inæqualitatum. Nec enim per totum circulum tanta est discordia locorum ex corporibus,& ex Copernico Planetis assignatorum, nec æqualis etiam in omnibus reuolutionibus.Atque ego sic existimo,etsi certißimæ essent Prutenicæ, atque verißime per hanc corporum interpositionem errores isti committerentur ; non posse tamen iure abiici tam concinnum επιχέιρημα, propterea quod error ille in minimis esset. (1) Atqui non tantum incertum est,vtrorum vitio differentia hæc existat;sed contra magna suspicio & multa argumenta, calculum ipsum & Prutenicas tabulas in culpa versari;

(2) *And whereas the eccentricities of the others all (decrease).*] Nor is this so in all cases: indeed the true eccentricity of Saturn is greater than that of Jupiter, whereas that of Jupiter is much smaller than that of Mars, which is below it.

(3) *Let those (who wish) seek.*] No one has come forward to seek them. Seek and ye shall find. I have sought, and you can see that I have found in Book V of the *Harmonice* outstanding causes. So good and reliable was this omen, "Never despair"; so powerful and pregnant was the axiom here adopted: "Nothing has been established by God at random."

CHAPTER XVIII.
ON THE DISAGREEMENT BETWEEN THE EQUATIONS DERIVED FROM THE SOLIDS AND THOSE OF COPERNICUS IN GENERAL, AND ON THE PRECISION OF ASTRONOMY

Above in Chapters 14 and 15 since I seemed to be almost convicted of some error on the evidence of the distances which Copernicus published and which are different from those which I have derived from the figures, I appealed to the equations in the apogee and I did not defend myself against any charge if my values departed to some extent from those of Copernicus. But after I set up, at the end of Chapter 15, angles corresponding to the equations calculated from the elongations from the Sun, like witnesses in the face of this charge, they seemed to give evidence against me. For there was none of the planets which retained the angle attributed to it by Copernicus. I took 41' from Saturn, 6' from Jupiter; I added 30' for Mars; and from Venus I removed the huge amount of 2° 18', and from Mercury 61'. Consequently those who wish to scrutinize everything too exactly will think that because the reckoning from the solids does not agree to the last detail with the theories of Copernicus, and with his values, the whole of my work has been made ridiculous. On the other hand, unless I remove this charge, in my own opinion I shall have lost the case. To the physicists, indeed, or cosmographers, according to the role which I maintain in this little book, I owe no explanation of this difference. For although they borrow their arguments for their theories from the astronomers, yet they do not check them by calculation as precisely as the astronomers, and they are not so acute or so critical as to be influenced by this trivial difference. Consequently I have won my case among the cosmographers.

Now I rightly fear the mass of astronomers; yet since it is right that the practitioners should take precedence in giving judgment, I do not despair of victory, even in the face of that. First, I tell them that they should be hopeful about their calculations. For although the difference is sometimes rather large, nevertheless let them remember that the values are taken from the most noticeable positions in the whole orbit, and where all the irregularities are combined. However round the whole orbit the disagreement between the positions designated according to the solids and according to the planets of Copernicus is not so great, and also is not the same in every revolution. Thus it is my opinion that even if the Prutenic Tables were completely reliable, and these errors were quite genuinely brought about by this interpolation of the solids, yet it would not be right to reject an argument which fits so neatly, because that error was insignificant. (1) Yet it is not only uncertain which of the two is responsible for this difference, but on the contrary there is a great suspicion and many arguments that the actual calculation

versaris;adeo vt magna coniectura contra me fuisset, si cum numeris Copernici penitus consensissem.

Eorum autem argumentorum hoc primum esto,quod Prutenicus calculus non raro in colligendis Planetarum locis fallitur. Multa quidem restaurauit nobis Copernicus in collapsa motuum scientia : multoque nostra, quam patrum memoria;purior est Astronomia. Veruntamen si rem ipsam penitus inspiciamus,fateri vtique cogemur, nos ab illa beata & optabili perfectione haud multo propius abesse, quam ab hodierna vetus abest Astronomia.Longa via est,& variæ ambages ad hanc veritatem. Monstrarunt illam nobis veteres,ingressi sunt maiores nostri, nos illos anteuertimus, & gradu propiori consistimus, sed metam nondum attigimus.Non ego hæc in Astronomiæ contemptum dico: Est aliqua prodire tenus, si non datur vltra ; sed ideo, ne quis temere grauius quid in hanc discordiam statuat,& dum me petit,& hæc quinque corpora ; in ipsa fundamenta Astronomiæ insultet. Ad omnium Artificum obseruationes prouoco : ex quibus videre est, quanta sæpe sit inter verum locum, & inter eum, quem calculus indicat,differentia, quæ interdum (2) in quibusdam ad secundum integrorum graduum longitudinem excrescit. Quod cum ita sit, expedit mihi nonnihil à Copernici numeris discedere ; & iam porro diligentium obseruatorum iudicio relinquitur,vtri arcus cum cœlo propius conueniant,mei, an Coperni-cani.

Alterum argumentum,quo differentiæ huius culpam in ipsas Prutenicas transfero, præbent mihi suspectæ Planetarum Eccentricitates; quod eo tendit, vt quamuis nec mei arcus omnino perfecti & certi sint (sicuti fateri cogor)tamen vitium ex contagione Eccentricitatum contraxerint.Si corpora super mediæ planetarum distātiæ superficies sphæricas struerentur,vt eadem superficies circumscripti corporis centra, & inscripti angulos tangeret;tum nihil mihi rei esset cum orbium crassitie, quam requirunt viæ Planetarum Eccentricæ.

(3) Cum autem illud fieri non potuerit,& nondum similiter causa Eccentricitatum,vt & differentiarum,explorata sit;oportuit me orbium spissitudines à Copernico,tanquam certas mutuari ; quas tamen non certissimas esse in confesso est. Quamuis enim omnis cœlestium motuum historia lubrico est aditu,per diuturnas, & difficiles obseruationes ; præcipue tamen hoc in constituendis Eccentricitatibus & locis Apogæorum appareat. Solaris (vel terrestris) Eccentricitas omnium rectissime habere debebat; Nam & vicinissima stellarum est Tellus nobis incolis, (4) & paucioribus quam cæteræ motibus vehitur. In mundo vero per interiecta corpora struendo,supra cap.XV. vidimus,quantum afferat momentum ad omnes sphæras artandas aut laxandas solius ἀεγνίσκε lunaris appositio,vel exemptio, qui valde exigua portiucula terrestris orbis crassitiem excedit. (5) Hic igitur orbis, quē certissime dimensum habere oportebat, & posse verisimile erat; hic, inquam,vide,in quanta versetur difficultate apud Copernicum qui ipse lib.3. Reuol.cap. 20.queritur, (6) quod *per minima quædam & vix apprehensibilia magna ratiocinari cogimur,quod interdum sub vno diuersitatis scrup.5.vel 6.gr.prætereant, & modicus error in immensum sese propaget.* Quanto peius igitur habe-

I   bunt

and the Prutenic Tables are at fault; so that there would be a large query against me if I agreed completely with Copernicus's values.

Now of those arguments let this be the first, that the Prutenic calculation is not seldom wrong in determining the positions of planets. There are indeed many things which Copernicus repaired for us in our ruined knowledge of the motions, and our astronomy is much purer than our fathers remember. However, if we thoroughly examine the facts of the matter, we are certainly obliged to confess that we are not much nearer to that blessed and desirable state of perfection than the ancient astronomy was to the modern. The way to the truth of the matter is long and has many windings. The ancients have shown it to us; our predecessors have started on it; we go on ahead of them, and stand on a closer level, but we have not yet reached the goal. I do not say this to show contempt for Astronomy —

"You can get somewhere, if you can't get further"[1] —

but to prevent anyone rashly putting a more serious construction on this disagreement, and while aiming at me, and the 5 solids, scoffing at the very foundations of astronomy. I appeal to the observations of all the practitioners. From them it can be seen how great a difference there often is between the true position and that which calculation indicates, amounting sometimes (2) in certain cases to as much as two complete degrees in longitude. In that case, it is helpful for me to depart to some extent from Copernicus's values; and from this point on, it is left to the decision of careful observations which angles agree more closely with the heaven, mine or those of Copernicus.

The second argument by which I transfer the blame for the difference to the actual Prutenic Tables is provided for me by the suspect eccentricities[2] of the planets. This tends to show that although my angles are not completely correct and certain (as I am obliged to confess) yet they have contracted the fault by infection from the eccentricities. If the solids were constructed over the spherical surfaces of the mean distance of the planets, so that the same surface touched the centers of the faces of the circumscribed solid and the vertices of the inscribed solid, then I should not be troubled by the thickness of the spheres which the eccentric paths of the planets require.

(3) However, since that has not been possible, and neither the cause of the eccentricities, nor that of their differences, has yet been investigated, I had to borrow the thicknesses of the spheres from Copernicus as if they were certain; yet it is admitted that they are not entirely certain. For though the approach to the whole history of the heavenly motions is slippery, and requires lengthy and difficult observations, yet it is particularly apparent that in establishing the eccentricities and the positions of the apogees, the solar (or terrestrial) eccentricity should be the most accurately known of all. For the Earth is both the nearest of the stars to us who inhabit it, and travels (4) with fewer motions than the others. Now in constructing the universe by the interpolation of the solids, we have seen above in Chapter 15 what a great effect on the narrowing or widening of all the spheres the addition or removal of only the tiny heaven of the Moon has, which goes outside the thickness of the terrestrial sphere by a very tiny little fraction. (5) Then, in the case of this sphere, of which we ought to have, and probably could have, an absolutely certain measurement, consider, I say, what great difficulty is found over it in Copernicus, who himself in Book III, Chapter 20 of the *De Revolutionibus* complains (6) that "we are forced to work out from very small and almost imperceptible quantities large quantities (the errors in) which sometimes exceed 5 or 6 degrees for a difference of a single minute, and a small error is immensely magnified."[3] How much worse off, then, will be the thicknesses of the spheres

bunt fpiſſitudines orbium & remotiorum à nobis,& qui pluribus motuū
varietatibus ſunt obnoxij.Quod ſi aut orbium illa πιχη certiſſime explorata,aut cauſæ ſaltem probabiles patefactæ fuerint,cur tanta ſingulis at-
tributa ſint à Conditore: (7) tum ego ſpondeo me producturum ex his
corporibus arcus per omnia motibus conſonos.Sic enim exiſtimo,quic-
quid hanc proportionem cœlorum inuentam adhuc impediat,quo minus ad exactam motuum cognitionem veniatur: (8) id omne in eccentricitatum vitia conferendum;quibus ſublatis, (9) magno adiumento
Artificibus futura puto ſolida hæc quinque, ad correctionem motuum
quam paſſim meditantur non pauci.

Vt hoc illis ſpondeam de eccentricitatibus,mouit me & hoc,quod
(10) vbique de minori particula,quam eſt πιχης orbis integrum contro-
uertitur.Eripe namque omnibus ſex orbibus ſua πιχη nota,aut dupla ſingulis attribue;videbis mundum & προσαφαιρέσεις omnes in immenſum illic conſidere & augeri,hic diſtrahi & deminui. Vt ita veritas inter nihil
& duplum conſiſtat, neque metuendum ſit, ne nimiam habeat Artifex
licentiam eccentricitates mutādi ; ſi quis illas his figuris aptare conetur.
Atque ſic hæc altera ratio eſt, quæ me de diſcordia inter meos & Copernici numeros excuſare poteſt.

Tertiam mihi præbent ipſi numeri Prutenicarum etiamnum craſ-
ſi,nec ita expreſſi,vt non poſſit aliquādo bona cum venia vel ſemiſſe gradus ab iis diſcedi. Rheinholdus quidem in Prutenicis omnia diligentiſ-
ſime diſpoſuit. Sed nolim aliquis hac ſpecie ſcrupuloſitatis ineſcatus,
craſſiuſculos numeros in Aſtronomia faſtidiat;rem exactius cenſeat. Illa ſummi viri minuta & ſcrupuloſa cura aut eſt propter certitudinē calculi, aut non neceſſaria in partibus numerorum, ipſos vero totos numeros,quos tam ſcrupuloſe diduxit, è Copernico excerpſit, ſicuti illos re-
perit.

Ac ipſe quidem Copernicus quam humanus ſit in recipiendis qua-
libuſcunque numeris qui quadamtenus ex voto obueniunt,& ad inſtitutum faciunt:id experietur diligens Copernici lector. Numeros qui per
diuerſas operationes vi demonſtrationis penitus conuenire debebant,
non repudiat,quamuis diſcrepent aliquot ſcrupulis. Obſeruationes in
VValtero , in Ptolemæo & alibi ſic legit, vt ijs eo commodioribus vtatur ad extruendum calculum, vnde in tempore horas, in arcubus qua-
drantes graduum & amplius interdum negligere vel mutare nulla illi
religio. Alicubi, vt in mutata eccentricitate Martis & Veneris, ſinus
etiam diſcrepantes à veritate acceptat, tantum ideo,quia parum per ad
eos, quos optat, digitum intendunt. Multa quæ ex ipſius confeſsione
emendanda fuiſſent, integra & ſincera ex Ptolemæo depromit, muta-
tis cæteris ſimilibus ; atque ijs poſtea fundamenta nouæ Aſtronomiæ
extruit. Quorum omnium mihi plurima documenta dedit Mæſtlinus:
quæ breuitatis cauſa mitto aſcribere. Atque adeo in reprehenſionem in-
currere iure videretur ; niſi conſulto feciſſet, eo quod præſtaret,im-
perfectam quodammodo habere Aſtronomiam, quam penitus nullam.
Nam eiuſmodi quidem difficultates occurrent, dum ſidera currēt : quas
ſuperare, & non impeditum ad conſtitutionem ſcientiæ cum minimo
damno aſpirare, vt auſus eſt Copernicus,id viri fortis eſt; ignaui ſubter-
fugere,

which are both farther away from us, and subject to more variations in their motions. But either if these thicknesses of the spheres were investigated with complete certainty, or at least if the probable reasons why such large thicknesses were allocated by the Creator to each of them were revealed, (7) then I pledge that I should produce from the solids angles which would agree with the motions in all cases. For it is my opinion that after the discovery of this proportion in the heavens (8) everything which still prevents us from attaining exact knowledge of the motions is to be attributed to errors in the eccentricities; and if those were removed, (9) I think that the five solids would be of great assistance to the practitioners for the correction of the motions, which not a few of them in various places are contemplating.

Another thing which impelled me to make that pledge to them on the eccentricities is that (10) everywhere the controversy is about a minor part, which is less than the complete thickness of the sphere. For take away their known thicknesses from all the six spheres, or grant double the thickness to each of them: you will see that there is an immense contraction of the universe and an increase in all the equations in the former case, or swelling of the universe and diminution of the equations in the latter. Thus the truth stands between nothing and double, and we need not fear that the practitioner will have too much license to alter the eccentricities, if he should try to make them fit these figures. And so this is a second line of argument which can excuse me for the disagreement between my values and those of Copernicus.

A third is afforded to me by the actual values of the Prutenic Tables, which are even now rough and not so refined that one cannot sometimes pardonably depart from them by even half a degree. Rheinhold indeed set everything out with great care in the Prutenic Tables. But I should not like anyone to be enticed by this kind of pedantry into disdaining rather rough values in astronomy. Let him consider the point in more detail. The minute and pedantic precision of that great man is either appropriate on account of the accuracy of the calculation, or inappropriate in the fractional parts: but the actual whole numbers which he so pedantically divided he took from Copernicus just as he found them.

Indeed the human failings of Copernicus himself in accepting any sort of figures which suit him up to a point and help his case will be found out by the careful reader of Copernicus. He does not repudiate values which though derived in different ways ought according to a theoretical proof to have agreed exactly, even though they differ by a few minutes. He selects observations in Walter,[4] in Ptolemy and elsewhere in such a way as to make all the more convenient use of them in building up his calculation, so that he has no scruple sometimes in neglecting or altering hours in time, quarters of degrees in angles, or more. In some places, as in altering the eccentricity of Mars and Venus, he even accepts sines which are at variance with the truth, simply because they point a little towards those which he wants. Many things which ought on his own admission to have been corrected he takes complete and as they are from Ptolemy, though he has altered others like them; and on these he later builds the foundations of the new astronomy. Of all this Maestlin has given me a great many examples, which I omit from this account in the interests of brevity. And consequently he would rightly seem to incur criticism, if he had not done it deliberately, because it was better to have an astronomy with some imperfections than none at all. For difficulties of that kind will occur while the stars run their courses. To overcome them, and unhampered to aspire to the establishment of knowledge with the least possible detriment, as Copernicus dared to do, is the part of a brave man. It is for a lazy

fugere,timidi defperare,& omnem hanc curam abijecte Quemadmodum & ipfe Copernicus hæc modo recenfita σφάλματα deie neque difsimulat,neque cum pudore fatetur. Exemplo Ptolemæi & vetorũ fe munit,difficultate obferuandi excufat, atque vbique alijs exemplo præit,in præclarorum inuentorum confirmat.one minutulos hofce defectus cõtemnendi;quod nifi factum antea fuiffet;nunquam Ptolemæus illam μεγάλην σύνζαξιν, Copernicus τῶν ἀνελιτθϫϭῶν libros, Rheinholdus Prutenicas nobis edidiffet.

Neq; nullam excufationem mihi quarto loco fuppeditat illa Mæftlini tabula in cap. XV. inferta. Copernico,cum eccentricitates Planetarum à Ptolemæo mutuaretur, nihilminus , quam de hac diuina cœlorum proportione fuboluit ; vt non iniuria vehementer quis miretur, ipfum tam prope ad eam accelfiffe; neque fore putauit , vt neceffitas aliquando cogeret inquirere diftantias à Sole,& ἀφηλίων loca.Quid mirum igitur,fi in hac ad viuum refectione , & ἀναλύσα mundi multa deprehendantur rudia,cum artifex ad minima non refpexerit? Quafi in parua pictura , quæ vix integram faciem ad fenfum exprimit, fi quis oculi aut pupillæ veram proportionemquærat,eum falli neceffe eft. Neglexit enim hanc pictor ob exilitatem,contentus fi, quæ funt euidentiora, quodammodo reprefentaret. Sic ad hanc ἀνάλυσιν quamuis optima ratione accefferim , cogente me vi demonftrationis, & conditione rei propofitæ: nolim tamen, vt quis fibi perfuadeat,abfolute certiffimos numeros fe inde retuliffe. Fieri namque poteft, vt hæc ipfa refectio erroris vlterioris caufa fuerit. Ecce non leuia indicia. Cauffam, cur mutentur Eccentricitates Martis & Veneris, Copernicus in mutationem terrenæ confert. Non igitur mutatur vera eorum à Sole Eccentricitas; Demonftrationem ad oculum habes in tabula. Quod fi ita eft, oportebat Eccentricitates à terra,quæ Ptolemæi feculo, & quæ noftro fuerunt, eodem deducere, atque ex vtrifque eandem à Sole Eccentricitatem concludere. Atqui calculum confule, videbis hoc non, vt par erat,fieri. Difcrepantes enim inuicem prouenient etiam ἀφήλιοι Eccentricitates. Idem de locis ἀφηλίων dictum efto, quia hæc mutuo connexa funt : Atque hoc vnum eft.

Deinde facile colligitur ex afpectu tabulæ, cum inæqualiter procedant, & ἀφήλια & ἀπόγεια, magnam inde fucceffu feculorum extituram ἐϰϰενϯϱϭϯήϯων diuerfitatem. Hodie Saturni & Telluris abfides prope coniunctæ funt,quare integra Telluris Eccentricitate minor eft Saturni à centro orbis terreftris,quam à Sole,diftantia. Vbi quadrâte diftiterint, æqualis erit vtraq; & à ☉ & à Terra, crefcet nempe Copernico fua Eccêtricitas Saturnia vfque dum opponentur inuicem Saturni &Telluris abfides. Quem ad euentum etfi mundus non durabit;tamen fi perfecta effet Aftronomia , tales debebat hypothefes vfurpare, quæ quafi æterno mundo fufficerent. Atqui nihil horum monet neq; Copernicus,neque Rheinholdus. Nõ igitur perfectiffimi funt eorum numeri,neq; integras planetarũ fpheras nobis explicant,quibus illos feros motus accidere poffe intelligamus.Hæc & huiufmodi fimilia cũ me nõnihil conturbarent, atq; ego hererê inops cõfilij,quafi qui difiectas automatis rotulas in ordinem

I 2

nem

man to shirk it, for a coward to give up hope and to reject all this trouble. Hence even Copernicus himself neither tries to hide his own failings, which have just been recounted, nor shows shame in admitting them. He arms himself with the example of Ptolemy and the ancients; he excuses himself by the difficulty of observing; and everywhere he sets a precedent for others of scorning these petty little short-comings in the process of establishing splendid discoveries. If that had not been done previously, Ptolemy would never have brought forth for us his *Almagest*, Copernicus his books *On the Revolutions*, Rheinhold his Prutenic Tables.

Also a not inconsiderable excuse, in the fourth place, is provided for me by Maestlin's table inserted in Chapter 15. So it is for Copernicus, since he borrowed the eccentricities of the planets from Ptolemy, even though he had an inkling of the divine proportion of the heavens; so that there may rightly be great wonder at his having approached so near to it; and he did not suppose that he would ever be com-pelled by necessity to inquire into the distances from the Sun, and the positions of the aphelia. Then why is it surprising, if in this pruning to the live wood, and analysis of the universe, many awkwardnesses are detected, since the craftsman did not consider details? Thus in a miniature, which scarcely enables us to see the whole face, anyone who looks for the true proportion of eye or pupil must necessarily be disappointed. For the painter neglected that on account of its minuteness, content if somehow he portrayed the more obvious points. So although I have approached this analysis by the finest logical process, governed by the force of the demonstra-tion and the requirements of the premises, yet I should not want anyone to persuade himself that absolutely certain values have been inferred by this means. For it can happen that this very pruning of error is a cause of further error. Consider the following weighty evidence. The cause of the changes in the eccentricities of Mars and Venus is ascribed by Copernicus to the change in the Earth's. Therefore there is no change in their true eccentricity from the Sun. You have a visual demonstration in the plate. If that is the case, their eccentricities from the Earth, as they were in Ptolemy's time and in our own, ought to have led to the same conclusion, and the same eccentricity from the Sun ought to have resulted from both. But refer to the calculation: you will see that that does not come about, as it should have done. For there will also turn out to be discrepancies in the eccentricities with respect to the Sun. The same must be said of the positions of the aphelia, because these are mutually interdependent; and this is one and the same thing.

Further, one easily gathers from a glance at the plate that when both the aphelia and the apogees advance irregularly, with the lapse of time the effect of that will be a great variation in the eccentricities. Today the apsides of Saturn and the Earth are nearly in conjunction, so that the distance of Saturn from the center of the Earth's sphere is less than that from the Sun by the whole of the Earth's eccentricity. When they are a quarter of a circle apart, the distance from the Sun and the Earth will be equal and indeed according to Copernicus Saturn's own eccentricity will increase until the apsides of Saturn and the Earth are in opposition to each other. Although the universe will not endure until that comes about, yet if astronomy were correct, it ought to adopt hypotheses which would be satisfactory if the universe were eternal. But neither Copernicus nor Rheinhold tells us any of these things. Consequently their values are not absolutely correct, and they do not set out for us the complete spheres of the planets, to show us that those motions can occur in the end.

As these and similar points disturbed me considerably, and I was hesitating in doubt of how to proceed, like a man who does not know how to put together

nem redigere nescit;Mæstlinus me consolatus,imo dehortatus est ab his
subtilitatibus : Non posse nos,aiebat, omnes naturæ thesauros exhauri-
re, non mouendum esse malum bene conditum, & tolerandam potius,
atque sustentandam leuaminibus quibusdam hanc veluti rupturam humani corporis, quam vt tam exquisita anatome conijciatur æger in præ-
sentissimum vitæ periculum. Proferebat mihi exemplum Rhetici, curá-
que eius ad vnguem meæ similiter curiosam, & increpantem pro se Copernicum. Epistola est Rhetici Ephemeridi 1551. præfixa, quæ quia non
passim est obuia,& totum hoc caput multis locis mirifice iuuat, præcipua
inde pro colophone huic capiti subiungam. Sic igitur Rheticus ad lectorem inter cætera. *Suas autem* (Copernicus) *exquisitiones mediocres ,nō nimias*
esse voluit. Itaq, consulto,non inertia aut tædio defatigationis, eas comminutiones
vitauit,quas nonnulli etiam affectarunt, & sunt qui exigant , qualis est Purbachij
in Eclipsium tabulis subtilitas. Videas autem quosdam in his omnem curam pone-
re , vt plane scrupulose loca siderum scrutentur , qui dum secundanis, & tertianis,
quartanis,quintanis minutiis inhiant,integras interim partes prætereunt , neque
respiciunt ,& in momentis τῶν Φαινομένων *sæpe horis , non etiam nunquam diebus*
totis aberrant. Hoc nimirum est,quod in fabulis Æsopicis fit ab eo, qui iussus bo-
uem amissam reducere,dum auiculis quibusdam captandis studet, neque his potitur,
& boue etiam ipso priuatur. Recordor cum & ipse inueni. curiositate impellebar,&
quasi in penetralia siderum peruenire cupiebam. Itaque de hac exquisitione interdū
etiam rixabar cum optimo & maximo viro Copernico. Sed ille,cum quidem animi
mei honesta cupiditate delectaretur , molli brachio obiurgare me , & hortari solebat,
vt manum etiam de tabula tollere discerem : Ego,inquit, si ad sextantes, quæ sunt
scrupula decem,veritatem adducere potero,non minus exultabo animis , quam ra-
tione normæ reperta Pythagoram accepimus. Mirante me,& annitendum esse ad
certiora dicente: huc quidem cum difficultate etiam peruentum iri demonstrabat,
cum aliis,tum tribus potissimum de causis. Harum primam esse aiebat, quod anim-
aduerteret , plerasque obseruationes veterum synceras non esse, sed accommodatas
ad eam doctrinam motuū,quam sibiipsi vnusquisque peculiariter constituisset. Itaq,
opus esse attentione & industria singulari,vt quibus aut nihil, aut parum admodum
opinio obseruatoris addidisset,detraxissetve,ea à corruptis secernerentur. Secūdam
causam esse dicebat , siderum inerrantium loca à veteribus non vlterius , quam ad
sextantes partium exquisita:Et secundum hæc tamen præcipue errantium positus ca-
pi oportere;pauca excipiebat,in quibus declinatio sideris ab æquinoctiali annotata rē
adiuuaret,quod de hac locus ipse sideris certius constitui iam posset. Tertiam causam
hanc memorabat : Non habere nos tales auctores, quales Ptolemæus habuisset post
Babylonios & Chaldæos,illa lumina artis,Hipparchum,Timocharem, Menelaum,
& cæteros , quorum & nos obseruationibus , ac præceptis niti ac confidere possimus.
Se quidem malle in iis acquiescere,quorum veritatem profiteri posset,quam in ambi-
guorum dubia subtilitate ostentare ingeny acrimoniam. Haud quidem longius certe,
vel etiam propius omnino abfuturas suas indicationes sextáte,aut quadrante partis
vnius à vero,cuius defectus,tantum abesse vt se pæniteat,vt magnopere lætetur,huc
vsq, longo tempore,ingenti labore,maxima contentione,studio & industria singu-
lari,procedere potuisse. Mercurium quidem, quasi secundum prouerbium Græcorū,
relinquebat in medio communem ; quod de illo neque suo studio obseruatum esse di-
ceret,neq, ab aliis se accepisse,quo magnopere adiuuari,aut quod omnino probare pos-
set. Me quidem multa monens , subijciens, præcipiens, in primis hortabatur, vt
stellarum

again the dismantled wheels of a machine, Maestlin consoled me, or rather dissuaded me from such minute precision. "We cannot," he said, "exhaust all the treasuries of Nature; the deeply seated flaw cannot be removed; and we must rather tolerate, and endure by some palliatives, this, so to speak, injury to the human body, than throw the sick man by so radical an operation into immediate danger of his life." He pointed out to me the example of Rheticus, and his attention to every last detail, as laborious as mine, and in itself a criticism of Copernicus. There is a letter of Rheticus prefixed to his Ephemeris for the year 1551; and as it is not to be found everywhere, and gives wonderful support to the whole of this chapter in many places, I shall append its main points as a tailpiece to this chapter.[5] This, then, among other things, is what Rheticus says to the reader.

"However he (Copernicus) wanted his investigation to be moderately thorough but not excessive. So it was on purpose, and not from laziness or distaste for sustained effort, that he avoided the minuteness of detail which several have even striven for, and some demand, such as Peurbach's precision in his tables of eclipses. However you can see certain people giving all their attention to examining the positions of the stars with absolute pedantry, and while they peer at seconds, and third, fourth, or fifth minutes,[6] meanwhile overlooking complete degrees, and not considering them; and in the precise times of phenomena they are often out by hours, and sometimes even by days. This is exactly what the man in Aesop's fables did when he was told to bring back a lost ox, but in his eagerness to catch some little birds he both failed to take them and missed the ox itself. I remember when I myself was also driven by a youthful inquisitiveness and desire to penetrate, as it were, the inner fastnesses of the stars. Consequently on the score of precision I sometimes even quarrelled with that great and good man Copernicus. But he, since indeed he was delighted by my mind's creditable desire, used to upbraid me with a gentle hand, and encourage me to learn to lift up my eyes from the page. 'If,' said he, 'I can get as close to the truth as sixths of a degree, which are ten minutes, I shall be no less glad at heart than we have been told Pythagoras was when he discovered the right-angle theorem.' I was surprised, and said that we should strive for greater accuracy; but he showed that even that point would only be reached with difficulty, for three reasons in particular, among others. Of these he said that the first was that he noticed many observations of the ancients were not genuine but adjusted to fit the particular theory of the motions which each had decided for himself. Particular care and attention were therefore needed to separate those in which the opinion of the observer had added or subtracted nothing, or very little, from the corrupt ones. He said that the second reason was that the positions of the fixed stars had not been investigated by the ancients with greater accuracy than sixths of degrees, though it was chiefly by reference to them that the positions of the planets had to be determined. He made an exception for a few cases in which noting the declination of the star from the equator had helped matters, because the actual position of the star could now be determined more accurately from it. The third reason which he told me was the following: we do not have authorities such as Ptolemy had after the Babylonians and Chaldeans, those ornaments of the science, Hipparchus, Timochares, Menelaus, and the rest, on whose observations and injunctions we also could depend and rely. He himself preferred silent acceptance of points for the truth of which he could vouch to a display of the quickness of his wits over questionable precision in uncertainties. Certainly his own indications would be no more, or even less, than a sixth or a quarter of a degree from the truth; and he was so far from regretting that deficiency that he was extremely pleased to have been able to make so much progress after a long time, vast toil, a very great struggle, assiduity, and particular application. Mercury indeed, as if according to the Greek proverb, he left open to all comers, as he said that no observation had been made of it either by his own assiduity, nor had any been obtained from others, which was of great assistance, or which he could completely corroborate. Indeed in giving me many admonitions, injunctions, and instructions, first and foremost he

ſtellarum incrrantium obſeruationi operam darem, illarum potiſſimum, quæ in ſi-gnifero apparent, quod cum his errantium congreſſus notari poſſent, &c. Hactenus ex epiſtola Rhetici ea, quæ ad rem fuêre. Quid tu iam, amice Lector, de Copernico ſentis? Si de hoc negotio fuiſſet monitus, atque deprehendiſſet, quam prope abſit ab eo cum ſuis rationibus, quid putas non tentaturus fuiſſet, quem laborem non ſumpſiſſet, vt corpora cum ſuis orbibus conciliaret? Atque hoc ſi daretur, qui conſenſus, quæ perfectio non ſperanda eſſet. Quain re quid alij, quid ipſe Mæſtlinus aliquando, fauente Deo, præſtiturus ſit, tempus docebit. Interea nolim, quis temere contra me pronunciet; & æquo animo hanc litis dilationem ferat.

IN CAPVT DECIMVMOCTAVVM
Notæ Auctoris.

(1) ATqui non tantum incertum eſt, vtrorum vitio.] *Etſi verum eſt, Prutenicas pecca-re, cum alias, tum etiam in Proſthaphæreſibus Orbis annui; potiſſima tamen cauſa, nõ huius tantum rei, quod interualla Orbium non exacte quadrant ad proportiones quinque corporum Geo-metrices; ſed etiam alius maioris rei, quod ſcilicet Planetarij orbes habent tantas ſinguli, tamque dif-ferentes Eccentricitates, vtrinſque inquam rei cauſa eſt in archetypo exornationis motuum, ſecun-dum rationes Harmonicas: vbi cum non poſſent exactæ proportiones figurales ſtare iuxta proportiones Harmonicas; neceſſe fuit illis, vt magis ad rationes materiæ declinantibus, derogari parum aliquid, vt proportiones Harmonicæ iuxta locum haberent, illæ quidem in ſpaciis mundi, iſta vero inter motus per ſpacia. Vide hunc Ornatum ornatiſſimum, lib.V.Harmon. cap. IX. Prop.à XLVI. in XLIX. ad longum.*

(2) In quibuſdam ad 2.integrorum gr. *Imo in Marte tres in Venere quinque gradus in tranſuerſum, in Mercurio 10. vel 11. gradus (ſi etiam de iis locis, vbi Planeta hic videri nequit, ex hypotheſi Theoriæ Mercurij à me conſtitutâ licet aliquid affirmare) certis Orbium locis, in errore ſunt, apud Prutenicas.*

(3) Cum autem illud fieri non potuerit.] *Centra planorum figuræ circumſcripta, & anguli figuræ inſcriptæ, non potuerunt eſſe coniuncti in hoc archetypo mundi. Cauſa dicta eſt in ſupe-rioribus. Nimium enim conſiderem Orbes: fierent maiores Proſthaphæreſes Orbis magni apud ſingu-los, quantos non obſeruamus. Ergo fuit reſpiciendum ad diſtantias Planetarum à Sole non mediocres, ſed apheliam duorum interioris, & periheliam exterioris; id eſt, ad Eccentricitates planetarum, quæ diſtantias, aphelium & Periheliam, formant. Atqui ſic ad incerta reſpiciebam: nondum enim erat co-gnita Eccentricitatum cauſa, cur tanta eſſet penes ſingulos Planetas Eccentricitas; cur tanta differen-tia; cur Saturnus, Iupiter, mediocres haberent, Mars, Mercurius maximas, Tellus, Venus, minimas. Ignorata cauſa, quantitatem ignorari neceſſe erat à priori, remittebar ad nudas obſeruationes.*

(4) Et paucioribus quam cæteræ motibus.] *Ita quidem tenet Ptolemæus, & ex illo Copernicus. Sol enim (ſeu Terra) non tantum Epicyclo caret, ſed etiam Æquante, vt illi putabant. At ſecundum rei veritatem, in motu illo tranſlationis circa Solem ſimilis eſt Terra vnicuique reli-quorum Planetarum in omnibus; vt demonſtratum eſt à me in Comment. Martis, parte tertia: & E-pit. Aſtr. lib. 7.*

(5) Hic igitur Orbis, quem certiſſime.] *Hic Orbis Ptolemæo Solis, Copernico Terræ Prutenicis Annuus dictus.*

(6) Quod per minima quædam.] *Hæc Copernici querela potiſſimum attingit loca Apogæorum (quæ loca nihil attinent hoc negotium proportionis Orbium) non eadem eſt de Eccentri-citatibus. Itaque non petis, ſed melius habent ipſæ Orbium ſpiſſitudines.*

(7) Tum ego ſpondeo me producturum.] *Audaciam ecce ſponſionis, ſuffultam dif-ficultate conditionis hic propoſitæ. Vide tamen & ſælicitatem: explorata ſunt à me quantitates Eccen-tricitatum ex Obſeruationibus Brahei, patefacta in Harmonicis cauſa Eccentricitatum ſingularum: & ecce productos, non quidem ex ſolis 5. figuris, ſed potiſſimum ex cauſis Eccentricitatum (Harmoniis) arcus per omnia motibus conſonos.*

I 3   (8) Id

exhorted me to attend to the observation of the fixed stars, especially those which are to be seen in the zodiac, since the conjunctions of those with the planets could be noted, etc."

This is the end of the part of Rheticus's letter which was relevant. What is your opinion about Copernicus now, dear reader? If he had been told about this undertaking, and had understood how close it is to his own thinking, what do you suppose he would not have attempted, what toil would he not have undertaken, to reconcile the solids with his spheres? And if that were achieved, what agreement, what accuracy could not be hoped for? In this matter what others, what Maestlin himelf some day, with God's favor, will produce for us, time will show. Meanwhile I should not wish anyone to pronounce against me hastily, and this postponement of judgment should be accepted without dismay.

AUTHOR'S NOTES ON CHAPTER EIGHTEEN

(1) *Yet it is not only uncertain which of the two is responsible.*] Though it is true that there are mistakes in the Prutenic Tables in various places including the equations of the annual orbit, yet the chief cause not only of the fact that the intervals between the orbits do not exactly square with the geometrical proportions of the five solids, but also of a more important fact, which is that the individual planetary orbits have such large and such different eccentricities, the cause I say of both facts is in the archetype of the display of the motions according to the harmonic ratios. There, since the exact proportions of the figures could not stand alongside the harmonic proportions, it was necessary for the former, as leaning more towards the arguments from the material side, to be moderated somewhat, so that the harmonic proportions might find a place beside them, the former indeed in the spaces of the universe, the latter however among the motions through the spaces. See this display most elegantly displayed in Book V of the *Harmonice*, Chapter 9, Propositions 46 to 49 throughout.

(2) *In certain cases to as much as two complete degrees.*] Rather in the case of Mars three, in the case of Venus five degrees in longitude, in the case of Mercury 10 or 11 degrees (if I may make a statement, from the hypothesis established by me for the theory of Mercury, even about those positions in which the planet cannot be seen here) is the amount of the error in the Prutenic Tables at certain positions on the orbits.

(3) *However since that has not been possible.*] The centers of the faces of the circumscribed figure and the vertices of the inscribed figure could not have been linked in this archetype of the universe. The reason has been stated in what precedes. For too much consideration would be given to the orbits: the equations for the Great Orbit would become too great in particular instances, of a size which we cannot observe. It was therefore necessary to have regard not to the mean distances of the planets from the Sun, but to the distance at aphelion of the inner of the two, and at perihelion of the outer: that is, to the eccentricities of the planets, which regulate the distances, at aphelion and at perihelion. But I was consequently having regard to what was uncertain; for it was not yet known what the cause of the eccentricities was; why the eccentricity was so great in the case of particular planets; why the difference was so great; why Saturn and Jupiter had intermediate eccentricities, Mars and Mercury the greatest, and the Earth and Venus the smallest. As the cause was unknown, it was inevitable that I did not know the amount *a priori*, and I was driven back to the bare observations.

(4) *With fewer motions than the other.*] This indeed is what Ptolemy holds, and following him Copernicus. For the Sun (or the Earth) not only has no epicycle, but also no equant, as they thought. But according to the truth of the matter, in its motion of translation round the Sun the Earth is similar in all respects to each of the remaining planets, as has been shown by me in my *Commentaries on Mars*, Part 3, and my *Epitome of Astronomy*, Book VII.

(5) *Then, in the case of this sphere,. . .an absolutely certain.*] This is the orbit called in Ptolemy the Sun's, in Copernicus the Earth's, and in the Prutenic Tables annual.

(6) *That ". . .from very small."*] This complaint of Copernicus chiefly applies to the positions of the apogees (which do not affect this business of the proportion of the spheres at all). It does not apply in the same way to the eccentricities. Therefore the thicknesses of the spheres are not in a worse state, but better.

(7) *Then I pledge that I should produce.*] You can see the audacity of this pledge, beset with the difficulty of the condition here proposed. However notice also how fortunate it was. The amounts of the eccentricities have been investigated by me from the observations of Brahe; the causes of the eccentricities have been made clear in the *Harmonice*; and you can see that arcs which agree with the motions in all respects have been inferred, not indeed from the five figures alone, but chiefly from the causes of the eccentricities (the harmonies.)

(8) Id omne in Eccentricitatum vitia.] *Laudabis opinor etiam puerulum trimulũ, præſumentem animo pugnam cum gigantibus.Non enim omnes Aſtronomiæ naui,imo minima illorũ pars,ſunt ex vitioſis Eccentricitatibus ſingulorum.De Solis vel Terræ Eccentricitate poſt dicetur.*

(9) Magno adiumento futura ſolida hæc quinque ad correctionem motuũ.] *Nullo equidem,ne minimo quidem ; quia non formant Orbes, nec præſcribunt metas Eccentricitatũ. Sed vbi prius inuentæ fuerint Eccentricitates , vt τὸ ὂ η , ex Obſeruationibus Braheì:iam denique locum habet inquiſitio cauſarum , ſeu τῦ διόλι ex his quinque figuris, & iunctis proportionibus Harmonicis.*

(10) Vbique de minori particula,quam eſt πῆχος Orbis,controuertitur.] *Cum enim Harmoniarum ſit aliqua copia , electæ fuerunt pro ſingulis bigis Planetarum vicinorum, quæ quantitate quamproxime reſponderent proportionibus harum quinque figurarum.*

CAPVT XIX.

De ſingulorum in ſpeiie Planetarum reſidua diſcordia.

HÆc igitur in genere fuere, quæ cauſam meam releuare poſſunt. Nunc in ſpecie videamus, ecquid excuſari amplius poſſit. Initium à Saturno ſumamus. Atque eius quidem ἀπωσίμαλι magna facta eſt acceſſio;ſed quę tamen differentiam proſthaphæreſeos cauſata eſt non maiorem 41.ſcrupulis. Nam ſicut ingens eius diſtantia facilimam errori cauſam præbet in obſeruatione ; ſic error in diſtantia quamuis luculentus exiguam & opinione minorem efficit in προϑαφαιρέσι diuerſitatem. Et tamen neque huius ſideris motus certiſſime diméſos eſſe Aſtronomos,vel ſola præterita hyeme cernere erat.Nam die ⁷⧸₂ Nouemb.anno 1594. Saturnus viſus eſt exacte inter Ceruicem & cor Leonis, vbi eſſe debebat ſecundum calculum die ²⁷⧸₁₁ Octob.præterita. Differentia long. 37.ſcrup.plus minus. Quodſi hanc quantitatem non excedat eius à Copernico diſcordia προϑαφαιρέσεως , correcta modo diſtantia ; exiſtiment Aſtronomi ſibi abunde ſatisfactum.

In Ioue nihil iure deſiderari poteſt. Nam exiguam habet differentiam;atque minorem ſextante gradus.

Quod autem etiam in Marte ſemiſſis gradus abundat,nihil mirum, nec me mouet; mouet id potius,maiorem non eſſe diuerſitatem. Teſtatur enim in præfatione Ephemeridis ad annum 1577. Mæſtlinus;ſideris huius errores à calculo intra duorum graduum anguſtias cogi non poſſe.

Iam ad inferiores ♀ & ☿ quod attinet, etſi præ ſuperioribus nonnihil commoditatis habere videntur;propterea,quod ex elongatione maxima facilius eſt, quam ex ἀκρονυχία obſeruatione, ipſorum orbes dimetiri,ipſa tamen obſeruandi via mihi ſuſpecta eſt. Quamuis rectius Aſtronomis hoc æſtimandum relinquo ; nempe vtrum non in his planetis (1) vaporum denſitate & phyſica parallaxi, quam nec Sol nec Luna effugit,interdum fallantur. Certe Mæſtlinus in Diſputatione de Eclipſibus , theſi 58.de Venere affirmat , quod non rato viſa fuerit eius à Sole prope horizontem diſtantia notabiliter minor vera. Quanto magis id de Mer-

(8) *Everything...to errors in the eccentricities.*] You would praise, I think, even a little boy of three years old who had the spirit to take on a battle with giants. For not all the blemishes of astronomy, indeed only the smallest part of them, are due to erroneous eccentricities of particular planets. The eccentricity of the Sun or the Earth will be spoken of later.

(9) *The five solids would be of great assistance for the correction of the motions.*] Of no assistance, in fact, not even the smallest, because they do not regulate the spheres, nor prescribe the limits of the eccentricities. But now that the eccentricities have already been found, as knowledge "that," from the observations of Brahe, at last there is room for a search for causes, or knowledge "why," from these five figures and the linked harmonic proportions.

(10) *Everywhere the controversy is about a minor part which is less than the thickness of the sphere.*] For since there is an abundance of harmonies, for the individual couples of neighboring planets have been chosen those which would correspond as nearly as possible quantitatively with the proportions of these five figures.

CHAPTER XIX.
ON THE REMAINING DISAGREEMENT IN THE CASE OF PARTICULAR INDIVIDUAL PLANETS

Those, then, were the general arguments which may save my case. Now let us see what further defense can be made in individual cases. Let us start with Saturn. Now a great increase has been made in its distance; but this has been made the reason for a difference in the equation not greater than 41 minutes. For just as its vast distance provides a very easy cause of error in observation, so an error in the distance even if it is considerable produces a tiny, and less than expected variation in the equation. Yet the fact that astronomers have not very accurately measured the motions of this star was easily to be perceived even in the passage of a single winter. For on the 2nd/12th November in the year 1594 Saturn appeared exactly between the neck and the heart of Leo, where it should have been according to calculation on the 21st/31st of the previous October. The difference in longitude is 37 minutes more or less. But if that amount were not exceeded by the discrepancy between Copernicus and its equation, the correction now having been made in the distance, the astronomers would think they had given thorough satisfaction.

In the case of Jupiter nothing can rightly be desired, for it has a tiny difference, less than a sixth of a degree.

That there is also half a degree too much in the case of Mars, however, is not at all surprising, and does not influence me. I am influenced rather by the fact that the variation is not greater. For Maestlin bears witness in the preface to his Ephemeris for the year 1577 that the irregularities of this star cannot be confined within the limits of two degrees.[1]

Now as far as the inferior planets Mercury and Venus are concerned, although compared with the superior planets they seem to have considerable convenience, because it is easier to measure their orbits from the maximum elongation than from an observation at opposition, yet the actual way of observing seems to me suspect. However I leave one point to the astronomers to evaluate: that is, whether in the case of these planets they are not sometimes led astray (1) by the density of the atmosphere and the physical parallax, which are not escaped by the Sun or the Moon either. Certainly Maestlin in his *Disputation on Eclipses*, in thesis 58, asserts of Venus that its distance from the Sun near the horizon has not infrequently seemed to be noticeably less than the truth.[2] That can be said all the

de Mercurio dici poterit, qui fere femper fub folis radijs eft; & quamuis interdum emergat: nunquam tamen, nifi prope horizontem per interiectam exhalationum copiam noftro fe vifui præfentat. Et quamuis Veneri opitulentur fixæ, fimul & prope apparentes: Mercurius tamen frequentius in culpa manet, qui ipfe raro cernitur, & rarius fixæ prope ipfum. Cumque hæc hodie accidant; credibile eft & veteribus quantifcunque Artificibus accidere potuiffe. Nam quod Lectorem de eo non monent, id ipfum fufpicionem de horum Planetarum dimenfionibus vitiofis auget. Hoc enim indicio eft; nec animaduerfum ipfis nec correctum effe, fi quid ex eo vitij extitit. Quare in lectione veterum imprimis fpectandum effe puto, vtrum fingularum obferuationum, quæ allegantur, inftrumenta & modi huic errori obnoxij effe potuerint.

Deinde non iniuria metuo, vt multa adhuc in ratione hypothefium his duobus Planetis relicta incerta fint. Copernicus (vt colligitur ex modo pofita Rhetici, & infra ex Mæftlini epiftola) plus Ptolemæi placita, quam obferuationum neceffitatem fequutus eft in emendandis theorijs. Qua in re quo minus reprehendi poffet, Rheticus in fua narratione effecit; vbi monet, religiofiffime veterum veftigijs inhærendum, nec facile quid mutandum, donec obferuationum extrema neceffitas vrgeat. Quod igitur adeo exquifitæ obferuationes haberi non poffent, ea fortaffe fatis magna caufa fuit Artifici prudentiffimo, præter accommodationem ad fua placita nihil vlterius in Planetas hofce tentandi.

Quod igitur in Venere magnam vides arcuum diuerfitatem eius rei culpam inter cætera, quæ in genere præmifi (quæ te probe meminiffe velim) etiam in hæc modo allegata offendicula confer; & magnitudinem difcordiæ æquanimitate tua, fi bene fingula perpendifti, facile fuperabis. Qua in re magno tibi folatio erit, quod numerus Copernicanus medius eft inter arcus ex interpofita, & ex omiffa Luna prodeuntes. Nam fi orbem magnum fyftemate Lunæ farcias: Icofaedron Venerem longius à terra dimouet, atque Copernicus prodidit; fin exempta Luna tenuiorem efficias orbem magnum: figura Venerem nimium prope admittit, maioremque, quam eft in Copernico, effe patitur. Quare aliquid minus Luna rem iuuare poterit, fi tenendus Copernicus eft.

De Mercurio vero tantum iam dictum eft, dicique amplius poteft, vt exiftimem te, Lector æque, fi aliquid amplius etiam deeffet, concocturum, atque excufaturum. (2) Neque mihi digna videtur eius motus diuerfitas, de qua magnam litem moueam. Quamuis melius fe gerit, quam Venus; facit enim vnius tantum gradus differentiam, quod mirum eft; adeo nunquam non fallaci eft ingenio. Certe vnus hic eft, qui Aftrologorum famam maxime proftituit, & meteororum rationem omnem turbat.

(3) Et in ventis quidem prædicendis (quos certiffime concitat, quotiefcunque locis eft idoneis) fæpe adeo conftanti numero dierum aberrat; vt parum abfit, quin tum eius in Ephemeride vitiofe proditum circulum corrigere poffim; Itaque fi quem Aftronomum cernerem nimium

more of Mercury, which is almost always close to the rays of the Sun; and although it sometimes emerges, yet it never presents itself to our sight except near the horizon with a quantity of vapors interposed. And although Venus is succored by the fixed stars, which appear at the same time and close to it, yet Mercury more frequently remains at fault, as it is rarely to be seen itself, and the fixed stars are more rarely to be seen near it. And since that happens today, we may believe that it could have happened to the ancient practitioners, however great. For the very fact that they do not comment on it increases the suspicion that the measurements of these planets may be faulty. For a sign of that is their failure to mention or correct any fault which resulted from it. Consequently in reading the ancients I think the first thing to look at is whether in particular observations the instruments which are mentioned and their methods could have been liable to this error.

Secondly it is not unfair of me to fear that in the case of these two planets much has still been left uncertain in the reasoning of the hypotheses. Copernicus (as may be gathered from the letter of Rheticus just quoted, and that of Maestlin below) followed the beliefs of Ptolemy more than the requirements of the observations in correcting the theory of the inferior planets. On this point Rheticus managed to defend him from criticism in his *Narratio*, where he remarks that we should adhere scrupulously to the path marked out by the ancients, and alter nothing lightly, until driven to it by the unavoidable requirements of the observations. Thus the fact that such refined observations were impossible was perhaps a great enough reason for that most careful practitioner to attempt nothing more on these planets beyond fitting them to his beliefs.

Thus for the great variation of angles which you see in the case of Venus, attribute the blame, among the other things which I have already stated (which I should like you to remember thoroughly) to those minor stumbling blocks just mentioned; and you will easily rise above a discrepancy of that size without disturbing your calm of mind, if you have considered the individual cases well.[3] In this connection it will be a great consolation to you that the Copernican value is half way between the angles resulting from the interposition of the Moon and the omission of the Moon. For if you stuff the Great Orbit with the system of the Moon, the icosahedron moves Venus further from the Earth than Copernicus reported; but if you leave out the Moon and make the Great Orbit thinner, the figure lets Venus too near, and allows the orbit to be greater than it is in Copernicus. Consequently something smaller than the Moon will help matters, if Copernicus is to be retained.[4]

About Mercury indeed so much has already been said—and more can be said—that I think, friendly reader, if anything further were still missing, you would put up with it, and excuse it. (2) Nor do I think that the variation in its motion is worth stirring up a great dispute about. However, it conducts itself better than Venus, for it produces a difference of only one degree, which is remarkable, as its nature is never unambiguous. Certainly this is the one planet which most of all disgraces the reputation of the astrologers, and confounds the whole theory of things on high.

(3) And indeed in predicting winds (which it certainly stirs up, whenever it is in suitable positions) it is often off its course by such a constant number of days that at such times I can very nearly correct its circle, which is wrongly published in the Ephemeris. Thus if I were to see any astronomer devoting himself too intensely to

mium folicite rimandis planetæ huius erroribus incumbere, illum ego
monerem, vt tempus illud rectius collocaret, & Tellurem, atque hanc
ambientem Lunam, ἐν ἀρρέκτον fidus, quarum illam pedibus, hanc oculis proxime attingimus, hæc, inquam, fidera potius fpeculetur, quæque in eorum motibus inque Eclipfibus adhuc peccamus, limet; tum
demum operam ad Mercurium transferat. Interea fi venia digni funt
errores circa Telluris & Lunæ motus, multo magis id merebuntur errores in Mercurio, qui & remotior à nobis eft, & fere femper fub Sole
latet.

Atque hic rurfum vt priore capite, coronidis loco epiftolæ partem
afcribam, quam Mæftlinus ad me mifit; idque duabus de caufis, prima,
quia de re neceffaria te monet; altera, quia caput hoc paffim confirmat.
Sic ille:

*Tam mirabilis eft Mercurius, vt parum abfuerit, quin etiam me fefel-
liffet. Nec mirum, quia etiam Copernico & Rheinholdo admodum moleftum
fuiffe, animaduerto. Copernicus hoc de feipfo fatetur,* Multis (*inquit lib.* 5.
*cap.*30.) ambagibus & laboribus nos torfit hoc fidus, vt eius motus
fcrutaremur. *Vnde præterquam quod nullas fuas proprias recitat obferuationes in* ☿ *habitas, fed à Bernhardo VValtero Noribergico mutuatur: etiam in a-
pogæi ipfius loco ftatuendo, fibi non conftat. Nam quem (cap.* 26.*) in primis An-
tonini annis, circa annum* CHRISTI 140. *iuxta Ptolemæi obferuationes, inuenit in* 10.*grad.* ♎, *& fub ftellato orbe in* 183.*grad.* 20.*fcrup. à prima ftella* ♈: *eundem* 183.*grad.* 20.*fcrup. cap.* 29.*) reponit ad* 21.*annum Ptolemæi Philadelphi, perinde ac fi hoc* ☿ *apogæum intra* 400. *annos intermedios fub fphæra fixarum ftellarum immotum quieuiffet; cum tamen (cap.* 30.*in fine)* 63. *annis per vnum gradum
motum fuiffe ipfi videatur; addit autem: fi modo æqualis fuerit. Rheinholdum in
ijfdem difficultatibus hæfiffe, calculus Prutenicarum tabularum prodit, quo argui-
tur, Rheinholdum locum apogæi huius ad tempus illud Philadelphi affumpfiffe eundem quidem cum Copernico, vid.* 183.*gr.* 20.*fcrup. à prima ftella* ♈. *At ad Ptolemæi
tempus illud in locum longe alienum à manifeftis Ptolemæi obferuationibus & Copernici refumptionibus, cadit. Ibi enim locus eius computatur non* 183.20.*nec* 10.*gr.*
♎, *fed* 188.*gr.* 50.*fcr. fub orbe ftellato, &* 15.*gr.* 30.*fcr.* ♎. *Idcoq; numeri illi mei ad*

Hi nu-
meri
funt in
Tab. V.
quæ eft
cap.15.
ad ☿.

*Ptolemæi quidem feculu accommodati funt, non autem, vt cæteri per omnia calculo
Tabularum Prutenicarum, fed Ptolemæi obferuationibus conueniunt, eas enim
Copernicus quoque & retinuit, & fequutus eft, atque eofdem inde numeros produ-
xit. Ad noftram autem, fiue Copernici ætatem numeros hofce computare non
volui, propterea quod ij longe alij fierent, propter Eccentricitatem orbis magni
diminutam; & quod apud Copernicum nullis recentioribus obferuationibus in-
ueftigati & comprobati funt. Optarem autem (quemadmodum me coram dixiffe
meminiffe potes) Copernicum dimenfionum harum fundamenta non antiquas, fed
nouas obferuationes affumpfiffe. Grande enim & immane poftulatum illud eft (lib.* 5.
*cap.*30.*fol.*169.*b.lin.*7.*à fine) cum, concedendum, inquit, putamus, commenfuratio-
nes circulorum manfiffe à Ptolemæo etiam nunc.* (4) *Nam ipfa terrena Eccentricitas diminuta alios numeros poftulat. Nec enim verum eft, quod Rheticus in narratione dicit, quod in Mercurio nulla quoque, ficut in Ioue, fentiatur Eccentricitatis mutatio, nam non fimiliter folus Apogæi latus fuo Apogæo claudit. Huc accedit,
quod Ptolemaicæ obferuationes fatis craffæ & partiles funt: quas omnino præcifiori-
bus corrigere oportebat. Sed de his iam fruftra conqueri licet. In tuo autem propofito,
fi nu-*

scrutiny of the deviations of this planet, I should advise him to dispose of his time more fittingly, and that it should rather be the Earth, and the Moon which circles it, the clearest star — the former of which we touch most nearly with our feet, the latter with our eyes — these stars, I say, which he should examine, and that he should refine the mistakes which we still make in their motions and in the eclipses. Only after that should he transfer his effort to Mercury. Meanwhile if the errors in connection with the motions of the Earth and the Moon are pardonable, the errors in the case of Mercury deserve pardon all the more, as it is both more distant from us, and almost always hidden in the rays of the Sun.

And here again as in the previous chapter I shall append as a tailpiece part of a letter which Maestlin sent me, and for two reasons: the first that it communicates an important point to you, the second that it confirms this chapter throughout. It runs as follows.[5]

"So remarkable is Mercury that it very nearly defeated me as well, and not surprisingly, as I notice that it was also very troublesome even to Copernicus and Rheinhold. Copernicus admits this about himself: 'This star tormented me' (he says in Book V, Chapter 30) 'with many twistings and toilings, in trying to explore its motions.' Thus apart from the fact that he reports no observations of his own made on Mercury, but borrows from Bernhard Walter of Nuremberg, he also contradicts himself in determining the position of its apogee. For although (Chapter 26) in the first years of Antoninus, about A.D. 140, according to Ptolemy's observations, he found it was in 10° of Libra, and on the celestial sphere in 183°20′ from the first star of Aries, he places it (Chapter 29) in 183°20′ at the 21st year of Ptolemy Philadelphus, just as if the apogee of Mercury had remained motionless against the sphere of the fixed stars in the intervening 400 years — yet he thinks (Chapter 30, at the end) that its motion had been one degree in 63 years, adding however 'assuming it was regular.' That Rheinhold was caught up in the same difficulties is shown by the reckoning of the Prussian (Prutenic) Tables, from which it appears that Rheinhold assumed the position of the apogee at the time of Philadelphus to be the same as did Copernicus, that is 183°20′ from the first star of Aries; but at the time of Ptolemy it falls into a completely different position from the well-known observations of Ptolemy and Copernicus's restatements of them. For there the position is computed not as 183°20′ and 10° of Libra, but as 188°50′ (i.e., on the celestial sphere), and 15°30′ of Libra. Consequently my values have been adjusted for the time of Ptolemy, but not, as have the rest throughout, by the reckoning of the Prutenic Tables, but in conformity with the observations of Ptolemy, for Copernicus also retained and followed them.[6] However, I did not want to compute these values for our own time, or that of Copernicus, because they would be far different, on account of the decreased eccentricity of the Great Orbit, and because in Copernicus they have not been checked and verified by any more recent observations. However, I should prefer (as I can remember my saying in your presence) Copernicus to have taken as the bases of these measurements not ancient but fresh observations. For it is a huge and mighty assumption (Book V, Chapter 30, page 169b, 7 lines from the end) when he says, 'I think it must be accepted that the dimensions of the circles have remained the same from Ptolemy until now,' (4) as the very decrease of the Earth's eccentricity demands different values. For Rheticus's statement in his *Narratio*[7] that in the case of Mercury as well, just as in the case of Jupiter, no alteration is perceptible in the eccentricity, is not true because the relationship of the Sun's apogee to Mercury's apogee does not mask the alteration in the same way. In addition, the Ptolemaic observations are rather crude and isolated, and should be entirely corrected by more accurate ones. But on this subject at present we may complain in vain.[8] However, in the case of your scheme, if these values agree to any extent at all, you should consider that you have performed your task outstandingly, and warmly

These figures are in Plate V, Ch. 15, under Mercury.

ſi numeri hi vtcunque tibi reſpondeant, te putes officio tuo egregie functum, tibique quemadmodum Copernicus apud Rheticum in epiſtola, vehementer gratuleris (5) certiſſima ſpe fretus, propediem fore, vt occaſione horū, quæ à te ingenioſiſſime ſunt inuenta cætera quoque, quæ iam adhuc dubia ſunt, & Aſtronomorum cœtum non parum torquent, planiſſima ſint futura.

IN CAPVT DECIMVMNONVM
Notæ Auctoris.

(1) VAporum denſitate & phyſica parallaxi.] *Refractiones ſtellarum appellat Tycho Braheus, qui hanc doctrinæ Aſtronomicæ partem, conſtituit excoluitque, lib. Progymnaſmatum, qui ex eo tempore prodijt in lucem, quam etiam partem ſeci Aſtronomiæ Partis Opticæ ante 15. annos editæ, auxiꝗ in Epit. Aſtr. lib. 1. à fol. 52.*

(2) Neque mihi digna videtur eius motus diuerſitas.] *Ita creditum eſt hucuſque de Mercurio; nec nego, magnam eſſe verorum etiam eius motuum diuerſitatem, ſed quæ quantitatis eſt, non formæ ſeu principiorum, vt hactenus docebamur; his enim principiis ille nihil differt à cæteris.*

(3) Et in ventis quidem prædicendis.] *Sequebar id temporis communem opinionem; Mercurium ventos in ſpecie concitare, præ cæteris Planetis. At me multorum annorum docuit experientia, non eſſe diſtributas mutationum auræ formas inter Planetas; ſed generaliter incitari Naturam ſublunarem ab aſpectibus binorum, vel à ſtationibus ſingulorum; vt ita exſudet vapores, aut fumos ex montibus & officinis ſubterraneis, qui vapores & fumi, vel in pluuias, vel in niues, vel chaginata, vel ſalſuginæ, vel grandines, vel ventos degenereant, pro circumſtantiis locorum & temporum. Venti certe magni, vel nunquam, vel rariſſime ſunt ſoli: pluuia omnis ante ſe ventos agit, cum primum ingruit impetu acta; & cum plurimum ſurunt venti id indicium eſt humida conſtitutionis anni. Aut enim in montanis pluit, vnde venti ſpirant, aut nix ibi ſoluitur, aut vapor humidus impetu ſurſum latus alibi in guttas cogitur, alibi æſtuans in ſupernum frigus impingitur reſilitque, quæ quidem etiam lenis auræ geneſis eſt, cum ebullit vapor ex aliquo monte, repercutiturque & defluit in omnes circumiacta plagas. Eſt vbi omnis aer per totas Continentes extenſus, principio motus dato in montanis omniū altiſſimis, in fluxu conſtituitur. Ita omnis ventus, ab omnib. promiſcue cauſis, vel inueſtigationibus naturæ concitari poteſt; nec ſolum incuſare poteſt Mercurium, ortus Ventorum.*

(4) Nam ipſa terrena Eccentricitas diminuta.] *Supra dictum, id non eſſe probabile, nec tam accuratas veterum Obſeruationes ad hoc probandum requiſitas, vt demonſtratio efficiatur neceſſaria. Itaque amplector axioma Copernici hic poſitum,* Concedendum ſc. commenſurationes circulorum mantiſſe. *Id enim ſuadet cœli natura, & inductio à Planetis cæteris.*

(5) Certiſſima ſpe fretus, propediem fore.] *Ita tunc ille ſolebat has dictis animare ſpeque curas, qui etſi, quoad tempus, ſpe excidit, nec enim* propediem *eſt, quod viginti quatuor annis ſequitur; tandem tamen ſpei ſuæ compos eſt factus per meum Opus Harmonicum.*

K CAPVT

congratulate yourself like Copernicus according to Rheticus[9] in the letter, (5) relying on the sure hope that the day will soon come when by means of your own brilliant discoveries those points which are still doubtful, and which torment the company of astronomers considerably, will be made manifest."

AUTHOR'S NOTES ON CHAPTER NINETEEN

(1) *By the density of the atmosphere and the physical parallax.*] Tycho Brahe calls it the refractions of the stars, and established this part of the discipline of astronomy and elaborated it in his book *Progymnasmata*, which has brought it into the light from that time on. I have also made it part of my *Optical Part of Astronomy*, published 17 years ago, and have augmented it in my *Epitome of Astronomy*, Book I, from p. 52 on.

(2) *Nor do I think that the variation in its motion is worth.*] This has been the belief about Mercury up until now; and I do not deny that the variation in its true motions is also great. However, it is a variation in amount, not in form or in principles, as we have been stating up to the present; for in principles it does not differ at all from the rest.

(3) *And indeed in predicting winds.*] I was following at this time the common opinion that Mercury stirs up winds as a class, more than the other planets. But the experience of many years has taught me that the forms of changes in the atmosphere are not allocated among the planets, but that in general sublunar Nature is stirred by the aspects of pairs of planets, or the stationary points of individual planets, to discharge vapors or fumes from the mountains and underground workings; and these vapors and fumes degenerate either into rains, or into snows, or shooting stars, or lightnings, or hails, or winds, according to the circumstances of place or time. Certainly great winds are either never or very seldom unaccompanied. All rain drives winds before it, as soon as, driven by its onrush, it sets to; and when the winds rage most, that is a sign that the character of the year is wet. For either it is raining in the mountain country, from which the winds blow, or the snow is melting there, or a wet vapor carried up by its onrush is in some places forced into drops, in other places as it surges up strikes against the cold of the upper region and recoils. This indeed is the genesis of even a gentle breeze, when a vapor boils out from some mountain, and rebounds, and flows down on all the regions round about. There are places where all the air extending over whole continents, when movement has been started in the highest mountain country of all, is set in a state of flux. Thus every wind can be stirred up indiscriminately by all causes or searchings of Nature; and the origin of winds cannot be blamed on Mercury alone.

(4) *As the very decrease of the Earth's eccentricity.*] It has been said above that this is not probable, and the observations of the ancients required to prove it are not so accurate as to establish the demonstration which is necessary. I therefore adopt the axiom assumed by Copernicus on this point, in other words, "It must be conceded that the dimensions of the circles have remained the same." For that is supported by the nature of the heaven and induction from the other planets.

(5) *Relying on the sure hope that the day will soon come.*] In this way he used to stimulate my endeavors with his sayings and hope, although as far as time was concerned his hope was disappointed, for the day does not come soon; it follows twenty-four years later.

However his hope has at last been fulfilled by my work *Harmonice*.

CAPVT XX.

(I) *Quæ fit proportio motuum ad orbes.*

ATQVE hactenus quidem expeditum eft argumentum illud , quo ego plurimum roboris afferri puto nouatis hypothefibus: demôftratumque,quod proportione quinque regularium corporum vtantur ἀποσήματα orbium in hypothefibus Copernici. Videamus modo , vtrum altero etiam argumento ex motibus deducto poffint & nouæ hypothefes , & hæ ipfæ orbium dimenfiones Copernicanæ confirmari , atque in proportione motuum ad ἀποσήματα certior ratio ex Copernico,quam ex vfitatis hypothefibus,haberi. Qua in re *dum amplitudines orbium proximas Copernicanis ex motuum* περιοδικοῖς *temporibus bene cognitis extruo ,* faue facilis Vranie, pulcherrimo conatui; tuus iam honos agitur.

Primum omnes optant ; vt quo longius quilibet orbis abeft à medio,tanto tardiori motu incedat. Nihil enim rationi magis eft confentaneum,tefte Arift.lib. 2.de Cœlo cap.10. quam κατὰ λόγιν γίγνεθαι τὰς ἑκάςε κινήςις τοῖς ἀποσήμασι. Quo loco etfi Philofophus alienam affert ab inftituto noftro rationem alteram , fcilicet impedimentum ab occurfatione perniciffimi primi mobilis : tamen & altera ratione pro me adhuc , & tota fententia contra Ptolemæum,côtraque feipfum militat. Placet illi namque,motus æqualitatem à motoribus in omnes orbes venire; inæqualitatem reditus ab orbibus ipfis caufari : vt, Saturni quidem quælibet particula tam fit velox,quam eft infima Lunæ fphæra,vi motionis æqualis;fed illi iam accidat , vt amplius nacta fpacium , cum non citatior fit cæteris, tardius redeat. Atqui viliori hac æqualitate Philofophus in veterum traditione potiri non potuit ; quia neceffe erat, vt tribus Planetis inæqualium orbium,Soli,Veneri,Mercurio æquales reditus tribuerent,atq; fic femper fuperiorem in orbe fuo citatiorem efficerent inferiori.In Copernico prima fronte talis offert fefe proportio. Nam fex orbium mobilium femper qui anguftior eft,citius redit.Mercurij namque curfus trimeftris eft,Veneris fefquiocto menfium,Terrę annuus,Martis bimus,Iouis duodecim, Saturni triginta annorum. Verum fi ad calculos reuoces,ita vt quanta eft proportio motus Saturni ad ambitum orbis, fiue ad diftantiâ (eadem enim eft proportio circulorum,quæ femidiametrorum)tantam etiam facias proportionem cæterorum motuum cuiufque ad fuum orbem ; deprehendes eiufmodi fimplicem proportionem non habere locum.Cuius rei cape hanc tabellam indicem.

ħ Dies

CHAPTER XX.
(1) WHAT THE RATIO OF THE MOTIONS TO THE ORBITS IS

So far the argument by which I think a great deal of strength has been added to the novel hypotheses has run smoothly, and it has been shown that the distances of the orbits in the hypotheses of Copernicus use the ratios of the five regular solids. Let us now see whether also from a second argument drawn from the motions both the new hypotheses and the Copernican dimensions of the orbits themselves can be verified, and for the ratio of the motions to the distances a more accurate account can be obtained from Copernicus than from the customary hypotheses. In this affair, during my deduction of the sizes of the orbits, coming very close to the Copernican values, from the well-known periodic times of the motions, be gracious, kindly Urania, to this splendid endeavor: your good name is involved now.

First, everybody wants each planet to proceed with a slower motion the further its distance from the center. For nothing is more reasonable, witness Aristotle, *De Caelo*, Book II, Chapter 10,[1] than that "the motions of each should be in proportion to the distances." In that passage although the Philosopher is adducing one line of reasoning which is alien to our scheme, that is, resistance to the influence of the first moving sphere, which is the fastest moving, yet by another line of reasoning he is now fighting on my side, and by his whole notion against Ptolemy, and against himself. For he believes that equality of motion is imparted to all the orbits by their movers; but he takes the pretext for the inequality of the times in which they revolve from the orbits themselves. Thus each particle of Saturn is indeed as fast-moving as the lowest sphere of the Moon, by the force of their equal motion; but in fact as the former occupies a wider space, since it is no swifter than the rest, the result is that it revolves in a longer time. Yet the Philosopher could not achieve this rather paltry equality in the tradition of the ancients, because it was necessary for them to attribute equal times of revolution to three planets with unequal orbits, the Sun, Venus, and Mercury, and so they made the superior planet always swifter in its orbit than the inferior. In Copernicus such a ratio is apparent at first sight. For of the six moving orbits the narrower one always revolves faster. For Mercury passes round in three months, Venus in eight and a half months,[2] the Earth in a year, Mars in two years, Jupiter in twelve, and Saturn in thirty. Indeed if you compare it with the calculations, making the ratio of the motion of each of the other planets to its sphere the same as the ratio of the motion of Saturn to the circumference of its orbit, or to its distance (for the proportion of circles is always the same as that of their radii), you will discover that there is no room for a simple proportion of that kind. Of that fact take the following table as evidence.[3]

	♄ Dies scr.	♃ Dies scr.	♂ Dies scr.	Terra Dies scr.	♀ Dies scr.	☿ Dies scr.
♄	10759 12					
♃	6159	4332 37				
♂	1785	1282	686 59			
terr.	1174	843	452	365 15		
♀	844	606	325	262 30	224 42	
☿	474	312	167	135	115	87 58

Hic capita columellarum continent dies & dierum fcrupula,
quibus fuperinfcripti Planetæ fub orbe Stellato fuas periodos com-
plent: fequentes numeri indicant;quantum dierum quam proxime de-
beatur inferiori Planetæ , eadem proportione ad orbem , qua vtitur
ille,qui eft in capite columellæ. Vides igitur,veram periodum femper
minorem effe, quam eft illa , quæ illi attribuitur ad fimilitudinem fuperioris.

Interim tamen motuum binorum ad inuicem,non quidem eadem,
fimilis tamen femper eft proportio,quæ inter diftantias.

Dies fcr.

$$
\text{Naturali proportio} \left\{
\begin{matrix}
10759 & 12 & ♄ \\
4332 & 37 & ♃ \\
686 & 59 & ♂ \\
365 & 15 & \text{terræ} \\
224 & 42 & ♀
\end{matrix}
\right\}
\begin{matrix}
\text{accipiatur finus} \\
\text{totus 1000. Erit} \\
\text{in ea quantitate} \\
\text{motus periodi-} \\
\text{cus}
\end{matrix}
\left\{
\begin{matrix}
♃ & 403 \\
♂ & 159 \\
\text{terræ} & 532 \\
♀ & 615 \\
☿ & 392
\end{matrix}
\right\}
\begin{matrix}
\text{At fi fuperioris} \\
\text{media diftãtia} \\
\text{fit 1000. eft} \\
\text{inferioris in} \\
\text{Copernico}
\end{matrix}
\left\{
\begin{matrix}
♃ & 572 \\
♂ & 290 \\
\text{terræ} & 658 \\
♀ & 719 \\
☿ & 500
\end{matrix}
\right.
$$

Hic vide mihi in motibus medijs, fat certo cognitis,idque longe
prius atque de certa diftantiarum ratione Copernicus cogitaret , vide,
inquam,eandem diuerfitatem, quæ inter ipfas eft diftantias, ex περιοδα-
Caιρεται per Copernicum , & ex quinque corporibus per me extructas:
vtrinque fecus ♂ minima , inde fecus ☿, ♃, Terram,& maxima fecus ♀:
vtrinque fecus ♃ & ☿ æqualis penesitem & fecus terram, & ♀. Igitur vel
iam ftatim fatis explorata eft Copernico de mundo veteri victoria.

Quod fi tamen præcifius etiam ad veritatem accedere, & proportionum æqualitatem vllam fperare velimus ; duorum alterum ftatuen-
dum eft: aut (2) Motrices animas,quo funt à Sole remotiores,hoc effe imbecilliores : aut , (3) vnam effe motricem animam in orbium o-
mnium centro, fcilicet in Sole; quæ, vt quodlibet corpus eft vicinius,
ita vehementius incitet; in remotioribus propter elongationem & at-
tenuationem virtutis quodammodo languefcat. Sicut igitur fons Lucis in Sole eft, & principium circuli in loco Solis,fcilicet in centro; ita
nunc vita, motus & anima mundi in eundem Solem recidit; vt ita fixarum fit quies,Planetarum actus fecũdi motuum; Solis actus ipfe primus:
qui incomparabiliter nobilior eft actibus fecundis in rebus omnibus; nõ
fecus atque Sol ipfe & fpeciei pulchritudine,& virtutis efficacia, & lucis

K 2   fplen-

	Saturn		Jupiter		Mars		Earth		Venus		Mercury	
	Days	Sixtieths	Days	Sixtieths	Days	Sixtieths	Days	Sixtieths	Days	Sixtieths	Days	Sixtieths
Saturn	10759	12										
Jupiter	6159		4332	37								
Mars	1785		1282		686	59						
Earth	1174		843		452		365	15				
Venus	844		606		325		262	30	224	42		
Mercury	434		312		167		135		115		87	58

. Here the heads of the columns contain the days and sixtieths of days in which the planets shown above them complete their periods against the Sphere of the Stars. The numbers which follow show how many days, as nearly as possible, are due to the inferior planet, in the same ratio to its orbit as is taken by the one at the head of the column. You see, then, that the true period is always less than that which is appropriate for it by comparison with the superior planet.

Nevertheless, between pairs of motions there is, not indeed the same, but a similar ratio to that between the distances.[4]

days sixtieths

					for		But if the mean	for	
					for		distance of the		
	10759	12	Saturn	the whole sine is	Jupiter	403		Jupiter	572
For	4332	37	Jupiter	taken as 1000	Mars	159	superior planet is	Mars	290
if	686	59	Mars	units, the pro-	Earth	532	1000 units, that	Earth	658
for	365	15	Earth	portionate period-	Venus	615	of the inferior	Venus	719
	224	42	Venus	ic motion will be	Mercury	392	according to	Mercury	500
							Copernicus is		

Here please note that in the mean motions, which are accurately enough known, and that long before Copernicus thought about an accurate reckoning of the distance, note, I say, the same discrepancy as that between the distances deduced from the equations according to Copernicus and from the five solids according to me: in both cases it is smallest with Mars, then with Mercury, Jupiter, and Earth, and greatest with Venus; in both cases the discrepancy with Jupiter and Mercury is almost equal, and similarly with the Earth and Venus. Hence the victory of Copernicus over the ancient universe is straight away sufficiently confirmed.

But if, nevertheless, we wish to make an even more exact approach to the truth, and to hope for any regularity in the ratios, one of two conclusions must be reached: either (2) the moving souls are weaker the further they are from the Sun; or, there is (3) a single moving soul in the center of all the spheres, that is, in the Sun, and it impels each body more strongly in proportion to how near it is. In the more distant ones on account of their remoteness and the weakening of its power, it becomes faint, so to speak. Thus, just as the source of light is in the Sun, and the origin of the circle is at the position of the Sun, which is at the center, so in this case the life, the motion and the soul of the universe are assigned to that same Sun; so that to the fixed stars belongs rest, to the planets the secondary impulses of motions, but to the Sun the primary impulse. In the same way the Sun far excels all others in the beauty of his appearance, and the effectiveness of his power,

fplendore cæteris omnibus longe præftat. Hic iam longe rectius in Solem competunt illa nobilia epitheta, Cor mundi, Rex, Imperator ftellarum, Deus vifibilis, & reliqua. (4) Sed huius materiæ nobilitas longe aliud tempus locumque requirit,& iam antea fat clare apparet ex Narratione Rhetici.

Iam autem de modo conftituendæ huius qnæfitæ proportionis nobis cogitandum eft. Supra vifum eft, fi fola orbis amplitudo faceret ad augendum tempus πεϱιοδικὸν: quod motuum & diftantiarum mediarum eadem differentia futura fuiffet. Quæ nempe proportio 88. dierum periodicorum Mercurij, ad 225. dies Veneris: eadem foret femidiametri orbis Mercurialis ad Veneriam. Iam vero commifcet fe huic motuum proportioni debilitas motricis animæ in remotiori. Difpiciendum igitur, cum hac debilitate vt comparatum fit. Ponamus igitur, id quod valde verifimile eft, (5) eadem ratione motum à Sole difpenfari, qua lucem. Lucis autem ex centro prorogatæ debilitatio qua proportione fiat, docent Optici. Nam quantum lucis eft in paruo circulo, tantundem etiam lucis fiue radiorum folarium eft in magno. Hinc cum fit in paruo ftipatior, in magno tenuior, menfura huius attenuationis ex ipfa circulorum proportione petenda erit, idque tam in luce, quam in motrice virtute. Quare quanto amplior Venus Mercurio, tanto iftius, quam illius motus fortior, fiue citatior, fiue pernicior, fiue vigentior, feu quocunque verbo rem exprimere placet. At quanto orbis orbe amplior, tanto plus temporis etiam requirit ad ambitum, etfi vtrinque fit æqualis vis motus. Ergo hinc fequitur, vnam elongationem Planetæ à Sole maiorem bis facere ad augendam periodum: (6) & contra, incrementum periodi duplum effe ad ἀποϛημάτων differentiam.

Dimidium igitur incrementi additum periodo minori, exhibere debet proportionem veram diftantiarum, fic vt aggregatum fit, vt diftantia fuperioris,&, fimplex minor periodus repræfentet inferioris, fcilicet Planetæ fui diftantiam in eadem quantitate. Exemplum, ☿ motus periodicus eft 88. fere dierum, Veneris 224. cum beffe ferme, differentia 136. & bes, dimidium 68. & pars tertia. Hoc iunctum cum 88. efficit 156. & trientem. Ergo vt 88. ad 156. cum tertia, fic femidiameter circuli Mercurialis medij ad mediam Veneris. Hoc modo fi in fingulis opereris, atque prouenientes binas diftantias per numeros finuum explices, fic vt femper fuperioris femidiameter fit finus totus:

$$\text{proueniet femidiameter orbis}\begin{cases} ♃ & 574 \\ ♂ & 274 \\ terræ & 694 \\ ♀ & 762 \\ ☿ & 563 \end{cases} \text{At eft in Copern.} \begin{cases} 572 \\ 290 \\ 658 \\ 719 \\ 500 \end{cases}$$

(7) Propius, vt vides, ad veritatem acceffimus. Etfi vero dubito, an demonftratiua methodo, quod theorema inftituerat, praxis ifta diuifæ differentiæ affequuta fuerit per omnia: tamen non omnino nihil in hifce numeris latêre, credere me iubet alia numerandi methodus, qua ad eofdem numeros reuoluar. Quia enim probabile eft, fortitudinem

motus

and the brilliance of his light. Consequently the Sun has a far better claim to such noble epithets as heart of the universe, king, emperor of the stars, visible God, and so on. (4) But the nobility of this theme demands a far different time and place, and is already clearly apparent from the *Narratio* of Rheticus.[5]

Now, however, we must consider the means of establishing this ratio which we require. It has been seen above that if only the breadth of the sphere contributed to increasing the periodic time, there would have been the same difference between the motions and the mean distances. That is to say, the ratio of the 88 days of the period of Mercury to the 225 days of Venus would be the same as that of the radius of the sphere of Mercury to that of Venus. As it is, however, this ratio of the motions is compounded with the weakness of the moving spirit in the more distant planet. Therefore we must also discover what its relationship is with this weakness. Let us suppose, then, as is highly probable, that (5) motion is dispensed by the Sun in the same proportion as light. Now the ratio in which light spreading out from a center is weakened is stated by the opticians.[6] For the amount of light in a small circle is the same as the amount of light or of the solar rays in the great one. Hence, as it is more concentrated in the small circle, and more thinly spread in the great one, the measure of this thinning out must be sought in the actual ratio of the circles, both for light and for the moving power. Therefore in proportion as Venus is wider than Mercury, so Mercury's motion is stronger, or swifter, or brisker, or more vigorous than that of Venus, or whatever word is chosen to express the fact. But in proportion as one orbit is wider than another, it also requires more time to go round it, although the force of the motion is equal in both cases. Hence it follows that one excess in the distance of a planet from the Sun acts twice over in increasing the period: (6) and conversely, the increase in the period is double the difference in the distances.[7]

Therefore, adding half the increase to the smaller period should show the true ratio of the distances:[8] the sum is proportional to the distance of the superior planet, and the simple lesser period represents the distance of the inferior, that is, of its own planet, in the same proportion. For example: the periodic motion of Mercury takes about 88 days, that of Venus about 224⅔ days. The difference is 136⅔ days, and half that is 68⅓. Adding that to 88 makes 156⅓. Then as 88 is to 156⅓, so the radius of the mean circle of Mercury is to the mean distance of Venus. If you operate in this manner in the individual cases, and set out the resulting pairs of distances by sines, in such a way that the radius of the superior planet in each case is the whole sine:

then the	for	Jupiter	574		572
resulting		Mars	274	But in	290
radius of		Earth	694	Copernicus	658
the orbit		Venus	762	it is	719
will be		Mercury	563		500

We have arrived, (7) as you see, closer to the truth. * Although indeed I am doubtful whether by the demonstrative method this procedure of halving the difference has in all respects achieved what the theorem had proposed, yet I am led to suppose that there is some significance lurking in these values by another method of calculating which will bring me round to the same values. For as it is probable that the strength of motion is proportionate to the distances, it is also

motus cum diſtantiis eſſe in proportione ; erit & hoc probabile, quod quilibet Planeta, quantum ſuperat ſuperiorem fortitudine motus, tantum ſuperetur in diſtantia. Eſto igitur, exempli gratia, Martis & diſtantia & virtus vnitas. Igitur quota particula virtutis Martiæ Tellus Marte fortior eſt ; totam diſtantiæ Martiæ particulam amittet. Hoc facile fit per regulam Falſi : pono namque radium Telluris ad Martium eſſe vt 694.ad 1000.Ergo,inquio,ſi amplitudo circuli per 1000. notata perambulatur à vi motrice Martia 687. diebus: perambulabitur eadem vi Martia,circulus minor, per 694. notatus, diebus 477. Iam quia certum eſt terræ circuitum eſſe non 477. ſed 365. dierum : pergo per regulam inuerſam ſic: dies 477.conſumerentur à ſimplici vi Martia; quantum de vi Martia conſumit circuitum, 365. cum quadrante dierum per eundem ambitum, quem Mars conficeret 477. diebus? Nam dubium non eſt, quin fortior virtus requiratur quam eſt Martia. Prouenit igitur ſupra integram vim Martiam adhuc, $\frac{306}{1000}$ pars eiuſdem virtutis.Et tantum Tellus Marte fortior eſt : debet igitur & tanto propior eſſe Soli; nempe ſi Mars per 1000. à Sole receſſit (diſtantia enim ſuperioris ſemper eſt integrum quid) Tellus per 306 earum partium propior erit : & ſubtracto ſuperiori 306. ab inferiori 1000. debet prouenire numerus initio poſitus, videlicet 694.ſi vera fuit illa poſitio;ſin falſa foret;ergo operareris ſecundum præcepta regulæ,& eliceres veram poſitionem.

Vides hoc altero theoremate prouenire non alios,quam ſuperiores numeros ; vnde certum eſt duo iſta theoremata forma quidem differre, ſed reuera coincidere, & niti eodem fundamento, quod tamen quo pacto fiat,inueſtigare hactenus nunquam potui.

IN CAPVT VIGESIMVM
Notæ Auctoris.

(1) Q Væ ſit proportio motuum ad Orbes.] *Hæc eſt propria materia libri I V. Epitomes, tranſſumpta inde in lib.V.Harmonicorum. Nam illius libri cap. III. hæc ipſa quæſtio enodatur,& inter fundamenta aſſumitur, quibus demonſtratur, motus Planetarum extremos contineri proportionibus Harmonicis. Etſi vero in hoc capite nondum aſſequutus ſum, quod quærebam;pleraque tamen adhibita principia, quæ mihi iam tum naturæ rerum videbantur conſentanea, certiſſima, & totis his 25.annis vtiliſſima ſum expertus: præſertim in Commentariis de motibus Martis,parte IV.*

(2) Motrices animas.] *Quas nullas eſſe probaui in Comment.Martis.*

(3) Vnam eſſe motricem Animam.] *Si pro voce Anima,vocem, Vim,ſubſtituas,habes ipſiſſimum principium,ex quo Phyſica cæleſtis in Comment.Martis eſt conſtituta,& lib.IV.Epitomes Aſtr.exculta. Olim enim,cauſam mouentem Planetas abſolute Animam eſſe credebam,quippe imbutus dogmatibus I. C. Scaligeri,de Motricibus intelligentiis. At cum perpenderem,hanc cauſam motricem debilitari cum diſtantia,lumen Solis etiam attenuari cum diſtantia à Sole : hinc concluſi;Vim hanc eſſe corporeum aliquid, ſi non proprie,ſaltem æquiuoce; ſicut lumen dicimus eſſe aliquid corporeum,id eſt ſpeciem à corpore delapſum, ſed immateriatam.*

(4) Sed huius materiæ nobilitas longe aliud tempus locumque.] *Nimirum locum inuenit in Comment.Martis anno 1609.editus:inde tranſſumpta eſt ſumma rei, & repetita in Epit. Aſtron.lib.IV.*

(5) Eadem ratione motum à Sole.] *Hæc omnia ſine vlla mutatione valent etiam in Comment.Martis.*

(6) Et contra, incrementum periodi duplum.] *Hic error incipit. Hoc enim non*

K 3 *eſt idem*

probable that any planet will be exceeded in distance by the one superior to it by the same amount as it exceeds it in strength of motion. Then, for example, let both the distance and the power of Mars be unity. Then the Earth will lose the same fraction of Mars's distance as the fraction of Mars's power by which it is stronger than Mars. This is easily found by the rule of false assumption. For I take the ratio of the Earth's radius to that of Mars to be as 694 to 1000. Therefore, I say, if the width of the circle, denoted by 1000, is traversed by the moving force of Mars in 687 days, the lesser circle, denoted by 694, will be traversed by that same force of Mars in 477 days. As in fact it is accurately known that the Earth's circuit is not in 477 days but in 365, I proceed by the inverse rule as follows. 477 days would be taken up by the force of Mars on its own. What multiple of the force of Mars takes up 365¼ days for the same passage which Mars would complete in 477 days? For a stronger power than that of Mars is undoubtedly required. The result is, then, a further 306/1000 part of the same power over and above the complete force of Mars. Now this is the amount by which the Earth is stronger than Mars. Then it must be nearer to the Sun by the same amount. That is, if Mars is 1000 units away from the Sun (for the distance of the superior planet is always a round number), the Earth will be nearer by 306 of the same units; and on subtracting from the 1000 for the superior planet 306 for the distance from the inferior, the result should be the value assumed at the start, namely 694, if that assumption was true. But if it was false, then you would operate as the rule directs, and extract the true assumption.

You see that from this alternative theorem the values which result are no different from those above.[9] Hence it is certain that both theorems differ indeed in form, but in actuality are equivalent, and rest on the same basis. (8) However, by what means that comes about I have never so far been able to discover.

AUTHOR'S NOTES ON CHAPTER TWENTY

(1) *What the ratio of the motions to the orbits is.*] This is the proper subject matter of Book IV of the *Epitome*, transferred from there to Book V of the *Harmonice*. For in Chapter 3 of that book this very question is unraveled, and it is included among the basic assumptions by which it is demonstrated that the extreme motions of the planets are defined by the harmonic proportions. Although in fact in this chapter I had not yet attained what I was seeking, yet a number of principles were introduced which then seemed to me in agreement with the nature of things, and quite certain, and which I have found very useful throughout the last 25 years, especially in the *Commentaries on the motions of Mars*, Part IV.[10]

(2) *The moving souls.*] Of which I have proved there are none in the *Commentaries on Mars*.

(3) *There is a single moving soul.*] If for the word "soul" you substitute the word "force," you have the very same principle on which the Celestial Physics[11] is established in the *Commentaries on Mars*, and elaborated in Book IV of the *Epitome of Astronomy*. For once I believed that the cause which moves the planets was precisely a soul, as I was of course imbued with the doctrines of J.C. Scaliger on moving intelligences. But when I pondered that this moving cause grows weaker with distance, and that the Sun's light also grows thinner with distance from the Sun, from that I concluded, that this force is something corporeal, that is, an emanation which a body emits, but an immaterial one.

(4) *But the nobility of this theme (demands) a far different time and place.*] Naturally it finds a place in the *Commentaries on Mars* published in the year 1609, and a summary of the matter was transferred from there and repeated in the *Epitome of Astronomy*, Book IV.

(5) *Motion (is dispensed) by the Sun in the same proportion.*] All this is also valid without any alteration in the *Commentaries on Mars*.[12]

(6) *And conversely, the increase in the period is double.*] Here the mistake begins.[13] For this is not the

est idem in contrarium, cum eo quod præmittitur, scil. elongationem à Sole bis facere ad augendam periodum. Sic autem debui colligere, & contra, proportionem periodorum duplam esse ἀποστημάτων proportionis, non quod hoc verum esse teneam, est enim eius tantummodo sesquialtera, vt audiemus: sed quia ex hac argumentatione hoc legitime sequebatur: Vides vti hic medium arithmeticum sit sumptum, per dimidiationem differentiæ, cum debuisset medium Geometricum sumi.

(7) Propius, vt vides ad veritatem.] *Propius sane per talem mediationem arithmeticam, quam per Geometricam, quamuis Geometrica legitime concludebatur ex assumptis principiis: quia cum reuera sit proportio proportionis non dupla, sed tantum sesquialtera: accidit hic, vt medium arithmeticum appropinquaret medio proportionis sesquialteræ, plus quam medium Geometricum, seu proportionis duplæ: quia medium arithmeticum semper propius est maiori termino, quam medium Geometricum: vt in Exemplo 6.9.12. & 6.8.12. hic medium arithmeticum 9. maius est Geometrico 8.*

* Etsi vero dubito.] *Citra dubium, praxis ista non fuit assecuta Theorematis scopum, vt iam est explicatum: Medium enim arithmeticum non est idem cum Geometrico.*

(8) Quod tamen, quo pacto fiat, inuestigare hactenus nunquam potui.] *Quia nimirum incedebam vagis gressibus flexiloquorum verborum, non lege arithmetica. Vide hic iam vtrumque processum: Prior sic erat:*

Periodus Martis 687.
Periodus Terræ $365\frac{1}{4}$.

Differentia $321\frac{3}{4}$.
Dimidium $160\frac{7}{8}$.

Medium arithmeticum $526\frac{1}{8}$.
$526\frac{1}{4}$ *dat distantiam Martis* 1000. *quid* $365\frac{1}{4}$?
Sequitur, distantia Telluris 694.

Posterior sic erat. Posito distantiam Telluris esse 694. Dico sic: Distantia Martis 1000. dat periodum 687. quid distantia Terræ 694? sequitur tanquam periodus Terræ 477. Pergo igitur per euersam proportionem.

Vera periodus $365\frac{1}{4}$ *dat falsam 477. tanquā ex Marte quid 1000. tanquā vis Martis? sequitur 1306. tanquā vis Telluris. Excessus igitur virtutis telluris 306. supra Martiā 1000. est idē, qui excessus Martiæ distantiæ 1000. supra Telluris assumptā 694. Hoc sit ideo, quia Marti applico numerū 1000. tam periodi indicem, quam virtutis, quam etiam distantiæ. Atqui hoc non est, reuolui per necessitatem regulæ Falsi ad eosdem numeros, qui erant in processu priori; sed est inuenire iterum, quod initio posueras. Cum enim in primo processu fiat mediatio arithmetica inter 687. &* $365\frac{1}{4}$ *per* $526\frac{1}{8}$; *duæ igitur diuersæ constituuntur proportiones, vt in omni tali mediatione, superior quidem & minor 687.* $526\frac{1}{8}$. *inferior vero & maior* $526\frac{1}{8}$. $365\frac{1}{4}$. *quæ per regulam Detri translata fuit in distantias 1000.694.*

In secundo processu, dum ponitur distantia Martis 1000. Terræ 694. ponitur igitur inter distantias Martis & Terræ proportionis periodorum arithmetice bisectæ pars inferior, scilicet $526\frac{1}{8}$. $365\frac{1}{4}$. *Illa vero transfertur in alios numeros, sc. 687.477. per regulam Detri. Si ergo à proportione 687.* $365\frac{1}{4}$ *auferas partem diuisæ arithmetice inferiorem, applicatam tamen termino superiori 687. relinqui necesse est eiusdem partem superiorem, apud terminum inferiorem, scilicet 477.* $365\frac{1}{4}$. *Quali transpositione, vt obiter moneam, vsus sum etiam in digressione politica ad finem libri 3. Harmonicorum. Atqui per Detri translata fuit hæc proportio in numeros alios, 1306.1000. Quare cum idem numerus 1000. sit in vtraque parte proportionis; sequitur igitur, vt inter duos terminos eiusdem socios, inter sc. 694. primo assumptum, & 1306. vltimo constitutum, facta sit mediatio arithmetica per 1000. Quia quæ prius inter 687.* $365\frac{1}{4}$ *erat pars inferior, sc.* $526\frac{1}{8}$. $365\frac{1}{4}$. *ea hic rursum assumpta fuit pars inferior 1000.694. quæ vero ibi pars superior, sc. 687.* $526\frac{1}{8}$ *(eadem enim est, quæ 477.* $365\frac{1}{4}$*) ea hîc rursum superior constituta fuit, scil. 1306.1000. Si inter 1306. & 694. constitutum fuit medium arithmeticum 1000. necesse est differentias æquales prodire, sc. vtrinque 306. Sufficiebat igitur, proposuisse facere vt* $526\frac{1}{8}$ *ad 687. &* $365\frac{1}{4}$ *sic 1000. ad duos alios: id per simplicem Detri fieret an per Falsi, perinde erat. Certum enim erat, minimum terminum proditurum 694. quia etiam in primo processu fiebat vt* $526\frac{1}{8}$ *ad* $365\frac{1}{4}$ *sic 1000. ad 694.*

Interim animaduerte, quod hoc imaginario concursu turbatus (veluti qui dextra sinistram nescius

exact converse of what precedes, that is, that the distance of the Sun makes a double contribution to the increase of the period. Now what I ought to have inferred, together with its converse, is that the ratio of the periods is the square of the ratio of the distances, not because I hold it to be true, for it is only the 3/2th power, as we shall hear, but because it was the legitimate conclusion from this line of argument. You see how at this point the arithmetic mean was taken, by halving the difference, when the geometrical mean should have been taken.

(7) *As you see, closer to the truth.*] Closer, to be sure, by taking the arithmetic mean in that way than by taking the geometrical mean, though from the principles adopted the geometrical mean was the legitimate conclusion. The reason is that actually the ratio is not the square of the ratio, but only the 3/2th power. The result of that was that the arithmetic mean came closer to the mean according to the 3/2th power than the geometrical mean, or that according to the square, because the arithmetic mean is always closer to the greater term than the geometrical mean, as in the example 6:9:12 and 6:8:12. Here the arithmetic mean 9 is greater than the geometric mean 8.

**Although indeed I am doubtful.*] It is beyond doubt that this procedure has not achieved the aim of the theorem, as has already been explained; for the arithmetic mean is not the same as the geometric.

(8) *However, by what means that comes about I have never so far been able to discover.*] Naturally, because I was proceeding by the wandering steps of ambiguous words, not by the law of arithmetic. Now consider both processes here. The former was as follows:

Period of Mars	687		
Period of Earth	365¼		
		Difference	321¾
		Half Diff.	160⅞
Arithmetic mean	526⅛		

526⅛ corresponds with a distance for Mars of 1000. What corresponds with 365¼?

It is found that the distance of the Earth is 694.

The latter process was as follows: Take the Earth's distance as 694. I argue as follows. Taking the distance of Mars as 1000 gives a period of 687. What does taking the Earth's distance as 694 give? It is found that the period of the Earth is 477. Proceed, then, by inverse proportion.

The true period 365¼ gives a false period of 477 as taken from Mars. What does taking 1000 as the force of Mars give? It is found that the force of the Earth is 1306. Then the excess of the power of the Earth over that of Mars, taken as 1000, which is 306, is the same as the excess of the distance of Mars, 1000, over that of the Earth taken as 694. This comes about because I allot to Mars the number 1000 both to represent its period, and its power, and also its distance. But this is not a case of returning by the necessity of the *regula Falsi* to the same numbers as those in the former process; but it is finding again what you had assumed at the start. For when in the first process taking the arithmetic mean between 687 and 365¼ comes out to 526⅛, then, as in every such taking of the mean, two different ratios are established, the upper and smaller one being 687:526⅛, and the lower and greater 526⅛:365¼, which by the rule of three was converted into the distances of 1000:694.

In the second process, when the distance of Mars is taken as 1000, and of Earth as 694, we are therefore taking as the ratio of the distances between Mars and the Earth the lower part of the ratio of the periods divided arithmetically, that is 526⅛:365¼. This is converted into the other numbers, that is 687:477 by the rule of three. If therefore from the ratio 687:365¼ you remove the part of the arithmetical division which as referred to the upper term, 687, is the lower, what is left must necessarily be the upper part of it, with respect to the lower term, that is 477:365¼. I also used a similar transposition, to take a comment in passing, in the political digression at the end of Book III of the *Harmonice.*[14] But by the rule of three this ratio was converted into other numbers, 1306:1000. Consequently since the same number, 1000, is in both parts of the proportion, it therefore follows that between two terms which are related to the same term, that is, between 694 which was taken originally and 1306 which was eventually established, the arithmetical mean comes out to 1000. For the ratio which was previously taken as the lower part of the arithmetic division between 687 and 365¼, that is, 526⅛:365¼, has again been taken here as the lower part, 1000:694; but the ratio which was in that case the upper part, that is, 687:526⅛ (for this is the same as 477:365¼) has again been established here as the upper part, that is, 1306:1000. If 1000 has been established as the arithmetic mean between 1306 and 694, it must necessarily produce equal differences, that is 306 in each direction. For it was enough to have proposed that the ratio of 1000 to the other two terms should be made the same as that of 526⅛ to 687 and 365¼; whether it was done by the simple rule of three or the *regula Falsi,* it would be the same. For it was certain that the smallest term would yield 694, because in the first process also 1000:694 was found to be as 526⅛:365¼.

Meanwhile take note that being confused by this imaginary coincidence (like someone who touches his

nescius in tenebris contingit & horrescit) aberrauerim à proposito, volens eandem virtutum propor-
tionem probare, quæ esset distantiarum; cum tamen vt tutum hic proportionem minorem statuam,
Martis scil. 1000. *distantiarum maiorem, Martis* 1000. *Terræ* 694. *Fuisset vero*
eadem vtrinque proportio si non arithmetice, sed Geometrice mediassem.

Nimis multa de hoc processit, sepeliendus enim est non erran. tantum, sed si etiam plane legi-
time procedat; quia proportio periodorum non est dupla proportionis distantiarum mediarum, sed
perfectissime & absolutissime, eiusdem sesquialtera: hoc est, si quarantur radices cubicæ ex Planeta-
rum temporibus periodicis vt 687. *&* 365¼. *& hæ radices multiplicentur quadrate: tunc in*
quadratis his numeris inest certissima proportio semidiametrorum Orbium. Perfici vero possunt ope-
rationes istæ facile, vel per Tabulam Cuborum Clauij, quæ adiecta est eius Geometriæ Practicæ, vel
longe facilius per Logarithmos Neperi Baronis Scoti sic: Prolongentur nostri numeri pro necessitate
& commoditate, vt sint 68700. *&* 36525. *nec iam sequemur summam subtilitatem: Logarithmi*
eorum sunt ex Canone Neperi 37543. *&* 100715. *circiter.*

Horum partes tertiæ sunt 12514. *&* 33572.
Et harum dupla, illarum bessis 25029. *&* 67144. *quæ exhibent, inter sinus, numeros hosce* 77858.
& 51097. *Inter hos est proportio orbium Martis & Telluris. Transponatur enim proportio in alios*
numeros, & sint vt 51097. *ad* 100000. *sic* 77858. *ad* 152373. *quæ plane est quantitas mediocris*
distantiæ Martis, qualium Terra à Sole distat 100000.

Causam cur non sit dupla proportio periodorum, ad proportionem Orbium, sed saltem sesqui-
altera, inuenies explicatam in Epit. Astr. lib. 4. fol. 530.

Hoc igitur alterum & præstantissimum quidem secretum auctarij loco nunc accedat Myste-
riis hisce Cosmographicis: quo in vulgus enunciato, lubet nunc vniuersos, tam Theologos, quam Phi-
losophos clara voce ad censuram dogmatis Aristarchici conuocare: **Attendite viri Religiosis-**
simi, Profundissimi, doctissimi:

Si verum dicit Ptolemæus de motu corporum Mundanorum, & dispositione ,,
Orbium: tunc nulla est constans & identica per omnes Planetas proportio Motuum, ,,
seu periodicorum temporum ad Orbes. ,,

Si verum dicit Tycho Braheus, Solem quidem esse centrum Planetarum quin- ,,
que, veluti quinque Epicyclorum: Terram vero esse centrum orbis Solis, vt Terra ,,
quiescente, Sol ei cumeat, portans & luxans systema totum Planetarium: tunc est ,,
quidem eadem proportio periodicorum temporum ad orbes, per omnes Planetas; sci- ,,
licet proportio periodorum, (verbi causa, Solis & Martis) est sesquialtera propor- ,,
tionis orbium suorum, sed motus nõ ab eodem centro dispensatur, Motus enim quin- ,,
que planetarum circa Solem dispensatur à Sole, motus vero Solis circa terram dispē- ,,
satur à terra; at sic Sol planetarum, Terra vero Solis motor constituitur. ,,

Si denique verum dicit Aristarchus Solem esse centrum & quinque Planeta- ,,
riorum Orbium, & sexti etiam, qui Tellurem vehit, vt Sole quiescente, Tellus in- ,,
ter Planetas cæteros circa Solem vehatur; tũc binorum quorumcunque Planetarum ,,
orbes inter se proportionem talem habent, quæ duas tertias complectatur proportio- ,,
nis periodorum, vel, proportio periodorum est perfectissime sesquialtera proportio- ,,
nis orbium; & motus tam Telluris quam cæterorum quinque ex vnico fonte Solaris ,,
corporis dispensatur. ,,

Hic nulla plane est exceptio, proportio est munitissima ex vtroque latere; ex ,,
parte quidem sensus attestatur Astronomorum obseruationes quotidianæ, cum omni ,,
subtilitate sua: ex parte vero rationis, astipulatur nobis Arist. in generalib. in specie ve- ,,
ro causæ suppetunt euidentissimæ, posita specie immateriata corporis Solaris, cur pro- ,,
portio debeat esse, nec simpla, nec dupla, sed plane sesquialtera: causæ et suppetũt cur Sol ,,
potius Terræ vt Planetarum cæterorum, quam Terra Solis motor esse possit; denique ,,
naturale rationis lumen dictat, digniorem & magis Archetypam esse speciem Ope- ,,
rum Dei

left hand with his right hand unawares in the dark and is scared), I have wandered away from my intention, as I was meaning to prove that the ratio of the powers was the same as that of the distances, whereas I here establish that the ratio of the powers is smaller, that is 1000 for Mars:1306 for the Earth, and that of the distances greater, 1000 for Mars:694 for the Earth. Now the ratio would have been the same in each case if I had taken not the arithmetic, but the geometrical mean.

I have said too much about this process; for it should be buried not only as mistaken but even if it were plainly carried out legitimately, because the ratio of the periods is not the square of the ratio of the mean distances, but quite perfectly and precisely the 3/2th power of that ratio. That is, if the cube roots of the periodic times of the planets are found, such as 687 and 365¼, and these cube roots are squared, then in these squares the ratio is exactly that of the radii of the orbits. These operations can in fact easily be carried out either by Clavius's Table of Cubes, which is appended to his *Practical Geometry,* or much more easily by the logarithms of Napier the Scots baron, as follows: let our numbers be lengthened, for necessity and convenience, to 68700 and 36525. We shall not now aim for the greatest accuracy. Their logarithms, from Napier's table, are 37543 and 100715, approximately.

The third parts of these are 12514 and 33572; and twice these, two thirds of the former numbers, is 25029 and 67144. The numbers shown for these in the sine tables are 77858 and 51097.[15] The ratio of these is the ratio between the orbits of Mars and the Earth. For if the ratio is converted into other numbers, it turns out that 51097:100000 as 77858:152373, which is clearly the amount of the mean distance of Mars, in units in which the distance of the Earth from the Sun is 100000.

The reason why the ratio of the periods is not as the square of that of the orbits, but in fact as the 3/2th power, you will find explained in the *Epitome of Astronomy,* Book IV, page 530.

This, then, is another and an outstanding secret which now comes as an addition to these Secrets of the Universe; and now that it has been announced publicly, it is our pleasure to call together both theologians and philosophers one and all with uplifted voice to pass judgment on the Aristarchan doctrine. Attend, most religious, profound, and learned men.

"If what Ptolemy says about the motion of earthly bodies and the arrangement of the orbits is true, then there is no ratio of the motions or of the periodic times to the orbits which is permanent and constant for all the planets.

"If it is true as Tycho Brahe says that the Sun is indeed the center of the five planets, as if of five epicycles, but the Earth is the center of the Sun's orbit, so that with the Earth at rest the Sun goes round, carrying and illuminating the whole planetary system, then the ratio of the periodic times to the orbits is indeed the same for all the planets, that is the ratio of the periods (for instance, of the Sun and Mars) is as the 3/2th power of the ratio of their orbits, but the motion is not controlled by the same center. For the motion of the five planets round the Sun is controlled by the Sun, whereas the motion of the Sun round the Earth is controlled by the Earth; and in this way the Sun is established as mover of the planets, but the Earth as mover of the Sun.

"Lastly, if it is true as Aristarchus says that the Sun is the center both of the five planetary orbits and also of the sixth, which carries the Earth, so that with the Sun at rest the Earth is carried round the Sun among the other planets, then the orbits of each pair of planets are in the same ratio to each other as the 2/3rds power of the ratio of their periods, or the ratio of the periods is quite precisely as the 3/2th power of the ratio of the orbits, and the motion both of the Earth and of the other five is controlled from the single source of the solar body.

"In this case there is plainly no exception, the ratio is completely secured on both sides. On the side of the senses the daily observations of the astronomers attest it with all their accuracy, and on the side of reason Aristarchus agrees with us in general, and in particular very clear reasons are available, assuming the immaterial emanation of the solar body, why the ratio should not be either simple, nor as the square, but plainly as the 3/2th power; and also reasons are available why the Sun can rather be the mover of the Earth as of the other planets, than the Earth the mover of the Sun. Lastly the natural light of reason

,, *rum Dei, ſi motus omnes ab vno fonte fluant, quam ſi plerique quidem ab vno illo*
,, *fonte, fontis vero ipſius ab alio ignobiliore fonte.*

,,   *Accedat vero formatio ipſa proportionis orbium ſeorſim ante motus facta,*
,, *per quinque figuras & per Harmonias. Nam ſi Braheus verum dicit; locum iſta*
,, *non habent, niſi aſcito circulo aliquo Telluris inter orbis Martis & Veneris per*
,, *imaginationem circumducto: & Deus non rei ipſius, ſed imaginationis potius cu-*
,, *ram habuit diſtorquens opus ipſum Mundanum, vt operis imaginatio pulchra eſſe*
,, *poſſet: cum tamen infinitæ aliæ ſimiles imaginariæ ſpecies, (vt ſtationum & retro-*
,, *gradationum) careant tali ornatu: at ſi verum dicit Ariſtarchus; tunc ornatus iſte*
,, *inuenitur in re; Species vero imaginariæ omnes, nulla excepta, permittuntur neceſſi-*
,, *tatibus legum opticarum.*

,,   *Hiſce perpenſis ſpero vos æquos dogmatum cenſores fore; nec hoſtes vos geſturos*
ornatus Operum diuinorum exquiſitiſſimi. Valete.

CAPVT XXI.

(I) *Quid ex defectu colligendum.*

SIc igitur hoc alterum argumentum habet: quo probatum eſt Ariſtotelis auctoritate, potiores eſſe nouas hypotheſes, propterea quod per eas motus duplici nomine, &
virtutis intentione, & celeritate reditus fiant proportio-
nales ἀπ ϛἰμαϭι Copernicanis, quod in veterum de mundo traditione fieri nullo pacto potuit. Atque hæc quidem
huius de motu tractatus intentio ſola debebat eſſe. Verum non difficile
mihi eſt conijcere; extituros, qui optauerint, vt hanc vltimam opuſculi
partem omiſiſſem. Etenim (dicent) ſi veram per corpora proportionem cœlorum conſtituiſſes: vtique motus illam confirmarent. Veritas enim à ſeipſa non diſſidet. Atqui vides ipſe, KEPLERE, quantum inter ſe diſſideant motus & corpora, hoc eſt diſtantiæ vtrinque extructæ.
Quare nudum hoſti latus obijcis, imo teipſum feris, nec opus alieno iu-
gulere gladio.

    His igitur vt reſpondeam, primum inuerto rationem, & ipſorum,
imo omnium appello iudicium & conſcientiam; vtrum argumentum
putent veriſimilius eſſe, num alterum de corporibus, an hoc de motu.
Neque mihi probabile eſt, quenquam aliter dicturum, quam hanc motuum ad orbes accommodationem admodum concinnam eſſe, atque
admirabile Dei opificis χειρούργημα. Proinde ſi alterutri argumento fides
habenda ſit, huic præ corporibus, aſtipulaturos, tanquam rei magis euidenti; quamuis numeri adhuc aliquantum à Copernicanis diſcrepent.
Quod ſi obtinui Lectoris confeſſione, vtar pro confirmatione corporum, & excuſatione diſcordiæ illius, vt quæ multis partibus minor eſt,
quam hæc in motu diſſonantia. Nam ſi Lector hîc propter concinnita-
tem inuenti magnum errorem libenter diſſimulat; paruum illic errorem longe facilius tolerabit. Diuerſitas enim illa penes corpora, calculum

declares that it is a more worthy and archetypal emanation of the works of God, if all the motions flow from one source, than if most indeed flow from that one source but those of the source itself from another, more ignoble source.

"To this is added the actual design of the proportion of the orbits which was made separately before the motions from the five figures and the harmonies. For if what Brahe says is true, there is no place for these things, except by the introduction of some circle for the Earth drawn round in the imagination between the orbits of Mars and Venus, and unless God paid attention to imagination rather than to reality, distorting the earthly work itself, so that the imagined work could be beautiful, whereas an infinite number of other similar imaginary appearances (such as the stations and retrogressions) lack such a display; but if what Aristarchus says is true, then that display is found in the reality, while all the imaginary appearances, without exception, are permitted by the requirements of the laws of optics.

"After pondering these things I hope you will judge the doctrines fairly, and not conduct yourselves as enemies of the most excellent display of the divine works.

"J. Kepler."

CHAPTER XXI.
(1) WHAT IS TO BE INFERRED FROM THE DEFICIENCY

That, then, is the position of this alternative argument. It has been proved by it on the authority of Aristotle that the new hypotheses are preferable, because by them the motions are under two headings, both from the extent of the power, and the speed of revolution, made proportional to the Copernican distances, which could not be done by any means within the tradition of the ancients on the universe. That indeed should have been the sole intention of this treatise on motion. Moreover it is not difficult for me to guess that there will be those who wish I had omitted this last part of my little work. "For" (they will say) "if you had established the true proportions of the heavens by means of the solids, the motions would assuredly confirm it. For the truth does not disagree with itself. But you see for yourself, Kepler, the extent to which the motions and the solids, that is, the distances based on them on each side, disagree with each other. So you are exposing your flank to the enemy, or rather you are striking at yourself, and there is no need for you to be slaughtered by someone else's sword."

To answer them, then, I first invert the argument, and call on their judgment and fairness to say which argument they think more probable, the other about the solids, or this one about the motion. It seems to me unlikely that anyone will give any other answer than that this fitting of the motions to the spheres is very neat, a wonderful piece of handiwork by God the craftsman. Consequently, if one or other argument must be accepted, they will assent to the second argument rather than to the one from the solids, as being the more obviously acceptable, even though the values still have a slight discrepancy from the Copernican ones. But if I have obtained the reader's agreement, I shall use it to reinforce the solids, and to excuse the discrepancy in them, since it is much smaller than the conflict in the motion. For if the reader in the latter case willingly overlooks a large error because the discovery fits so neatly, he will far more easily tolerate the small error in the former. For the difference with respect to the solids does not disturb astronomical calculation in the least; but the difference with respect to the motions has a little greater effect. This is the first point: the blow has been parried.

cu!um Aftronomicum nihil admodum turbat: ifta vero penes motus paulo quid maius infert. A tq; hoc primum eft;plaga nempe repofita.

Deinde (2) cum corpora diffentiant à motibus, vt vere mihi obij-citur;fateri vtique cogor,alterutros in errore verfari. Veruntamen erro-rê ita demonftrari poffe exiftimo, (3) vt neutrum inuentum (neque de motuum,neque de orbium proportione) penitus relinquere neceffe fit. Vtrum autem inuentorum in culpa fit, ex fuperioribus facile eft coijce-re. Primum diftantiæ motoriæ longius à Copernicanis recedunt, quam figurales. Deinde, fi motorias cum Copernicanis conferas,fingulas cum fingulis,defectusq; afcribas, videbis aliquam defectuum cum ipfis nume-ris,atque adeo cum corporibus cognationem,præterquam in Mercurio. Ecce:

Copern. Motoriæ Differentiæ

♄	♃	572	574	✢	2	Cubus.
♃	♂	290	274	---	16	Tetraedron.
♂	Terræ	658	690	✢	26	Dodecaedron.
Terræ	♀	719	762	✢	43	Icofaedron.
♀	☿	500	563	✢	63	Octaedron.
	vel	559		✢	4	

Plus fcilicet in quatuor, minus in quinto. Nam ex quatuor, bina femper corpora funt fimilia,quintum folitarium eft. Deinde Mercuriú, vt eft varius,in ordinem redige, & cogita, debere aliquid altius media or-bis fpiffitudine pro media diftantia cenferi, (4) tantum nempe,quantus eft orbis Octaedri,(quod fupra audiuifti media fpiffitudine amplius effe) &obtinebit pro media diftantia 559. non 500. Erit igitur hic ordo eius nu-merorum ♀ ☿ 559 | 563 | ✢ 4.Ecce in ♄ ♃,& ♀ ☿ differentias minores, fc. 2. 4. in ♂ terra, terra ♀ maiores, fc. 26.43. ficut interiecta corpora illic Cu-bus & Octaedron, hîc Dodecaedron & Icofaedron funt fimilia. Et a-nimaduerte, quod illic, vbi magna differentia eft infcriptorum & circú-fcriptorum, parua eft differentia diftantiarum: viciffim vbi propemo-dum æquales afcripti, magno interuallo diffident diftantiæ motoriæ à Copernicanis.

Cum igitur in defectu hoc fit quædam æqualitas, & vero nihil or-dinatum fortuito accidat:ideo cogitandum numeros hofce ad veritatem quidê alludere;nondum tamen eam penitus affecutos. (5) Nepe in ipfo theoremate adhuc limari quid poteft; aut theorema quidê recte habet, (6) fed eius fenfum neutra operatio affecuta eft. Quod quamuis initio ftatim fufpicari potui,nolui tamen, Lectorem hac occafione,&veluti fti-mulo plura tentandi,carere. (7) Quid fi namque aliquando diem illum videamus,quo ambo hæc inuenta conciliata erunt? (8) Quid fi hinc ra-tio eccentricitatum elici poffit? Nam quo pertinacius retineam etiam hoc de motibus theorema,illud inter cætera in caufa eft,quod vnius mo-toriæ diftantiæ ad alteram proportio, nunquam à toto orbe Copernica-no aberrat, fed femper ad aliquid digitum intendit, quod pertinet ad or-bium fpiffitudinem. Eftq; in hoc, quod mirari poffis aliqua etiam æqua-litas. Quam vt videas, explico tibi ordinem diftantiarum motoriarum in partibus, quarum media Telluris remotio eft 1000. & appono diftan-tias Copernicanas:

Secondly, (2) since the solids disagree with the motions, an objection which is truly made against me, I am certainly forced to admit that one or the other is subject to error. Nevertheless I think the error can be explained in such a way (3) that it is not necessary to relinquish either discovery altogether (about the ratio of either the motions or the spheres). However it is easy to conjecture from the foregoing which of the discoveries is at fault. First, the distances according to the motions are further away from the Copernican distances than those according to the figures. Secondly, if you compare the distances according to the motions with the Copernican distances individually, and make a list of the discrepancies, you will see that the discrepancies are related to the actual values, and therefore to the solids, except in the case of Mercury. Take note:

		Copernican distances	Distances from motion	Differences	
Saturn	Jupiter	572	574	+ 2	Cube
Jupiter	Mars	290	274	− 16	Tetrahedron
Mars	Earth	658	694	+ 36	Dodecahedron
Earth	Venus	719	762	+ 43	Icosahedron
Venus	Mercury	500	563	+ 63	Octahedron
		or			
		559		+ 4	

Plainly the difference is positive in four cases, negative in the fifth. For among the four, all the pairs of bodies are alike; the fifth is on its own. Next bring Mercury back into the pattern, as it varies from it, and consider that some height greater than halfway through the thickness of the sphere ought to be taken instead of the mean distance, (4) that is, the radius of the inscribed sphere of the octahedron (which as you have heard above extends beyond halfway through the thickness), and it will achieve for its mean distance 559, not 500.[1] Therefore the pattern for its values will be:

Venus	Mercury	559	563	+ 4

Notice that for Saturn and Jupiter, and for Venus and Mercury, the differences are smaller, that is 2 and 4; for Mars and the Earth, and the Earth and Venus, they are larger, that is 36 and 43. Similarly the solids interposed, the cube and octahedron in the former cases, the dodecahedron and icosahedron in the latter, are alike. And observe, that in the former, where the difference between the inscribed and circumscribed spheres is great, the difference in the distances is small; but on the other hand where the related spheres are almost equal, the distances according to the motions differ by a wide margin from the Copernican ones.

Since, then, there is a certain regularity in this deficiency, and indeed no pattern occurs accidentally, we must therefore consider that these values hint at the truth, but have not yet completely achieved it. (5) That is to say, there is something in the theorem itself which can still be improved, or else the actual theorem is correct, (6) but neither procedure has carried through its intention. Although I could suspect that at the start, yet I did not want the reader to be without this opportunity, and so to speak stimulus for making further attempts. (7) For what if at some future time we should see the day when both these discoveries are reconciled? (8) What if the rationale of the eccentricities can be deduced from it? For among the reasons which make me cling more tenaciously to this theorem about the motions is the fact that the ratio of one distance, according to the motions, to another never strays outside the complete Copernican sphere, but always points to something which relates to the thickness of the spheres. In this fact, at which you can wonder, there is also a regularity. To show it to you, I set out for you a table of the distances according to the motions, in units in which the mean distance of the Earth is 1000, and I append the Copernican distances:[2]

(9) Copernici Motoriæ

Summa		9987					
Media	♄	9164	9163				
Ima		8341		vt	1000	ad	577
				ſic 9163		ad	5290
Summa		5492		proximus			5261
Media	♃	5246	5261				
Ima		5000a		vt	1000	ad	333
				ſic a 5000		ad	1666
Summa		1648b		proximus			1648b
Media	♂	1520	1440				
Ima		1393c		vt	1000	ad	795
				ſic c 1393		ad	1107
Sum.terræ1042 terræ		1102d		proximus			1102
Med.ſim- 1000 cum		1000	1000				d
Ima plicis.858 e ☽		898		vt	1000	ad	795
				ſic e 958		ad	762
Summa		741h		proximus			762 f
Media	♀	719	762f				
Ima		696		vt	1000	ad	577
				ſic 741		ad	429 g
Summa		489		proximus			741 h
Media	☿	360	429g				
Ima		231					

Æqualitas hæc eſt, quod in remotis à terra ad medias diſtantias proxime acceditur: in vicinis Marte & Venere, motoria diſtantia vtrinq; vicinior eſt terræ, quam Copernicana media.

Vides etiã nuſquam, nec excludi loco ſuo corpus, neq; ordinẽ turbari, ſed ad minimũ, hiatum tantũ inter medias diſtãtias patêre, qui corpus recipiat. Vt ſi quis maxime motorias haſce pro optime demõſtratis acceptare velit(quo de dubitatur tamẽ)is (10) modũ fortaſſis interpoſitionis corporũ tollat, interpoſitionẽipſam nõ tollat. Fere n.indicantmotoriæ, quaſi (11) duo exteriora ſimilia ſimiliter inter medias interſint, duo interiora ſimilia inter mediã &extremã, nẽpe dodecaedron ab imaMartis ad mediã Terræ, Icoſaedron à media Terræ ad ſummã Veneris. Tetraedron vero ẽt ſuis fruatur priuilegijs, atq; inter vtrãq; extremã interſit. Verũ hẽc omnia ſuo loco cẽſeantur, nẽpe ex incertis extructa numeris motoriarũ, nec in aliũ finẽ, quã vt extimulẽtur alij ad cõciliationẽ:ad quã viã præiui.

In Cap. XXI. Notæ Auctoris.

(1) Q Vid ex defectu colligendum.] *Superuacua iam porro eſt hæc coniectatio. Vera enim proportione inuenta, in qua defectus plane nullus, quid mihi opus eſt falſæ defectu?*

(2) Cum corpora diſſentiant à motibus.] *Quia nec corpora ſu figuræ, ſolæ formant interualla Planetarũ, nec motuum talis in indiuiduo eſt proportio. Ita vtrumq, in errore verſabatur.*

(3) Vt neutrum inuentum penitus relinquere cogamur.] *Conciliata ſunt inter ſe libro 5. Harmonicorum.*

(4) Tantum nempe, quantus eſt orbis Octaedri.] *Poſito orbe perihelio Veneris, cui*

				(9) Copernican	From motions			
Greatest distance				9987				
Mean distance		of Saturn		9164	9163			
Least distance				8341			1000 :	577
						as	9163 :	5290
Greatest distance				5492			closest 5261	
Mean distance		of Jupiter		5246	5261			
Least distance				5000a			1000 :	333
						as a	5000 :	1666
Greatest distance				1648b			closest 1648b	
Mean distance		of Mars		1520	1440			
Least distance				1393c			1000 :	795
						as c	1393 :	1107
Greatest dist.	of	1042	of Earth	1102d			closest 1102d	
Mean distance	Earth	1000	with	1000	1000			
Least distance	alone	958e	Moon	898			1000 :	795
						as e	958 :	762
Greatest distance				741h			closest 762f	
Mean distance		of Venus		719	762f			
Least distance				696			1000 :	577
						as	741 :	429g
Greatest distance				489			closest 741h	
Mean distance		of Mercury		360	429g			
Least distance				231				

The regularity is this, that in the cases of those far from the Earth the values are very close to the mean distances: in the cases of its neighbors Mars and Venus, the distance according to the motions is for both planets closer to the Earth than the Copernican mean distance.

Also you see that never is a solid shut out of its position, or the arrangement disturbed, but at the least, a large enough gap is open between the mean distances to accept the solid. So that anyone who is willing to accept these values according to the motions particularly as very well established (about which, however, there is doubt), (10) may perhaps discard the method of interposing the solids, but not the interpolation itself. For that is almost implied by the distances according to the motions, as if (11) the two outer similar solids were interposed in a similar way between the mean distances, the two inner similar solids between the mean and the extreme distance, in other words, the dodecahedron from the least distance of Mars to the mean distance of the Earth, the icosahedron from the mean distance of the Earth to the greatest distance of Venus. The tetrahedron indeed enjoys its own privileges, and is interposed between two extreme distances.[3] Yet all these points should be assessed at their proper value, namely as having been based on the inaccurate values of the distances derived from the motions, and for the sole purpose of stimulating others to reconcile them. On that road I have been the forerunner.

AUTHOR'S NOTES ON CHAPTER TWENTY-ONE

(1) *What is to be inferred from the deficiency.*] This conjecture is from now on completely pointless. For as I have found the true proportion, in which there is plainly no deficiency, what need have I for a false deficiency?

(2) *Since the solids disagree with the motions.*] Because neither do the solids or the figures alone regulate the intervals between the planets, nor is there any such proportion between the motions depending on the individual case. Thus both were erroneous.

(3) *That it is not necessary to relinquish either discovery altogether.*] They have been reconciled with each other in Book V of the *Harmonice*.

(4) *That is, the radius of the inscribed sphere of the octahedron.*] If the distance of the perihelion of the orbit of Venus, in which the octahedron is inscribed, is taken as 1000 units, the distance of the centers

cui Octaedron inscribatur, partium 1000. centra Octaedri distabunt à centro systematis partibus 559. cum Mercurij summa distantia ex Copernico promatur 723. media 500. itaque punctum, vbi terminantur partes 559. est in ipso spacio, seu spissitudine orbis; at non in medio, sed inter medium 500. & summum 723.

(5) Nempe in ipso theoremate.[Hoc nimirum limandum erat, Proportionem alteram esse alterius non duplam, sed sesquialteram.

(6) Sed eius sensum neutra.] Vt clarum feci priori cap. in annotationibus.

(7) Quid si namque aliquando diem illum videamus.] Vidimus post 22. annos, & gauisi sumus, saltem ego, puto & Mæstlinus, & plurimi alii qui lib. 5. Harmon. sunt lecturi, participes erunt gaudii.

(8) Quid si hinc ratio Eccentricitatum.] Ita somniabam de veritate, opinor bono Deo inspirante. Elicita est, non hinc quidem, sed ex Harmoniis, ratio Eccentricitatum, sed tamen mediante hoc inuento; nec illud ante fieri potuit, quam hoc emendatum haberetur. Nam lib. 5. Harmon. cap. 3. ponitur inter principia demonstrationis hæc sesquialtera proportio.

(9) Copernici summa &c.] Pro his non perfectis interuallis ex Copernico habes Harm. lib. 5. perfectissima ex Astronomia per Obseruationes Braheanas restaurata.

(10) Modum fortassis interpositionis corporum tollat.] Rursum somniabam de veritate. Vide emendatum modum hunc lib. 5. Harm. cap. 9. Prop. 46. 47. 48. 49.

(11) Duo exteriora similia similiter.] Cubus exteriorum & Octaedron interiorum vltimæ, similiter id est, penetratiue inter sunt, at non intermedias distantias nimium hoc. Duo vero interiora, Dodecaedron & Icosaedron, similia, rursum similiter, id est, defectiue, at non inter extremam & mediam, rursum hoc nimium est: Tetraedron vero omnino suo fruitur etiam hic priuilegio interestque inter extremas distantias: imam Iouis summam Martis. Hoc sic esse debere, demonstraui propositionibus iam allegatis.

Cætera errantium numerorum ad veritatem allusiones, quas passim allego, fortuitæ sunt, nec dignæ, quæ excutiantur; iucundæ tamen mihi recognitu; quia monent, quibus mæandris, quorum parietum palpatione, per tenebras ignorantiæ, ad pellucens ostium veritatis deuenerim.

CAPVT XXII.

Planeta cur super æquantis centro æqualiter moueatur.

IDICISTI modo, Lector, etiam imperfecta cognoscere, quo minus metuo, te vltimam hanc & frigidam catastrophen explosurum. Vltimo autem referre volui, cû quia vltimo loco habeo; tum quia cum motibus cohæret, nec expediri sine XX. capite potest, quamuis ad 14. proprie pertineat, vti ibi monitus es.

Cum hanc figuralem cœlorum proportionem Mæstlini censuræ Vide Ta IV. Cap XIV. subiecissem: is me de superiorum epicyclijs monuit, quos Copernicus loco æquantium introduxit, quique duplo maiorem efficiant orbi spissitudinem, quam Planetæ ascensus descensusque requirit. Et in inferioribus quidem alij motus sunt, quibus Planeta ad omnem illius epicycli altitudinem euehitur, ad omnem eius humilitatem descendit, vnde in illis pro eccentrepicyclo eccentrus eccentri à Copernico assumptus est: in Mercurio vero peculiaris quædam diameter, per quam accedit & recedit à Sole, similiter longe remotius à Sole interdum exportigitur, quam Stella vnquam. Existimauit igitur, eam orbibus relinquendam esse spissitudinem, quæ motibus demonstrandis sufficiat. Cui respondi,

L 2

of the octahedron from the center of the system will be 559 units, while the greatest distance of Mercury according to Copernicus comes out at 723 units, the mean distance at 500. Thus the point to which 559 units extend is within the space, or thickness, of the orbit; yet not at the mean distance, but in between the mean distance of 500 units and the greatest, 723 units.

(5) *That is to say . . . in the theorem itself.*] Obviously the improvement which should have been made was that the ratio of the one to the other is not the square, but the 3/2th power.

(6) *But neither . . . its intention.*] As I have made clear in the notes on the previous chapter.

(7) *What if at some future time we should see the day.*] We have seen it, 22 years later, and we have rejoiced; at least I have, and I believe both Maestlin and a great many others, who are going to read Book V of the *Harmonice*, will share my joy.

(8) *What if the rationale of the eccentricities (can be deduced) from it?*] In this way I dreamed of the truth, I think inspired by the good Lord. The rationale of the eccentricities was deduced, not indeed from this, but from the harmonies, yet with this discovery as the intermedium; and the deduction could not have been made before the discovery had been amended. For in Book V of the *Harmonice*, Chapter 3, this 3/2th power of the ratio is taken among the bases of the demonstration.

(9) *Copernican greatest (distance), etc.*] Instead of these imprecise intervals taken from Copernicus you can find them in the *Harmonice*, Book V, taken absolutely precise from astronomy restored by means of the Brahean Observations.[4]

(10) *May perhaps discard the method of interposing the solids.*] Again I was dreaming of the truth. See this method emended in Book V of the *Harmonice*, Chapter 9, Propositions 46, 47, 48, and 49.

(11) *The two outer similar solids (were interposed) in a similar way.*] The cube is interposed within the last of the outer, and the octahedron within the last of the inner, *in a similar way*, that is, so that they penetrate, but not *between the mean* distances; that is too much. On the other hand the two inner similar solids, the dodecahedron and icosahedron, are again interposed *in a similar way*, that is, so that they fall short, but not between an extreme and the mean: again that is too much. The tetrahedron, however, here also absolutely *enjoys its own privilege*, and is interposed between extreme distances, the least of Jupiter, the greatest of Mars. I have shown that this must be so in the propositions already quoted.

The remaining hints at the truth which are offered by erroneous values, and which I quote everywhere, are fortuitous, but do not deserve to be deleted; yet I enjoy recognizing them, because they tell me by what meanders, and by feeling along what walls through the darkness of ignorance, I have reached the shining gateway of truth.

CHAPTER XXII.
WHY A PLANET MOVES UNIFORMLY ABOUT THE CENTER OF THE EQUANT

You have just learnt, reader, to take cognizance even of imperfect ideas, and so I am less afraid that you will jeer off the stage this last feeble dénouement. However, I wanted to keep it till last, both because I discovered it last in order, and also because it is connected with the motions and cannot be expounded without Chapter 20, though it properly belongs to 14, as you were informed there.

As I had submitted this proportion of the heavens based on the figures to Maestlin's criticism, he mentioned to me the epicycles of the superior planets, which Copernicus introduced in place of equants, and which make the thickness of the sphere twice as great as the upward and downward movement of the planet requires.[1] And in the case of the inferior planets indeed, there are other motions, by which the planet is lifted up to the full height of its epicycle, and goes down to its lowest level; so in their cases instead of an epicycle on an eccentric, an eccentric on an eccentric was postulated by Copernicus. In fact in the case of Mercury a certain special diameter, along which it moves towards and away from the Sun, is similarly extended to a much greater distance from the Sun than the planet ever is. He therefore considered that a thickness sufficient for deriving the motions should be left for the spheres. To that I have replied, first, that the whole undertaking should be abandoned, if the spheres were made twice as fat, for the equa-

See Plate IV, Chap. 14.

respondi,primum,deserendum esse totum negotium,si duplo crassiores
fiant orbes:nam nimiũ ϖϱϑαφαμέϛᾳν ademptum iri : Deinde nihil de-
cedere nobilitati miraculosæ huius machinationis,si modo viæ ipsæ;pla-
netarum descriptæ globulis,retineant hanc proportionem;quibuscun-
que illi agitentur orbib.magnis an paruis. Et addidi, qnæ cap.16.habes,
de materia figurarum,quæ nulla sit;atq; inde non absurdum esse,corpo-
ra cum orbibus eodem loco includere. Imo vero vel sine orbibus hâc viæ
inæqualitatem defendi posse.In qua sententia video Nobilem & excel-
lentiss. Mathematicum Tychonem Brahe,Danum,versari. Causam tamen & modum hæc nostra disertius indicant. (1) Nempe si eadem sit
causa tarditatis & velocitatis in singulorum orbibus,quæ supra cap.20.
fuit in vniuerso mundo,hoc modo : Via Planetæ eccentrica,tarda superius est,inferius velox. Ad hoc enim demonstrandũ assumpta (2) Copernico epicyclia,Ptolemæo æquantes. Describatur igitur concentricus æ-
qualis viæ Planetariæ eccentricæ; cuius motus vndiquaque æqualis erit,
quia æqualiter ab origine motus distat. Ergo in medietate viæ eccétricæ
supra concentricũ eminenti tardior erit Planeta, quia longius à Sole recedit,& à virtute debiliori mouetur:in reliqua celerior,quia Soli vicini-
or,& in fortiori virtute.Atq; hanc variationem motus non secus per cir-
cellum demonstrari , ac si vere in eo circello Planeta moueretur æquali
motu,cuilibet facile est colligere.Habes causam tarditatis huius,videa-
mus nũc & mensuram:A sit fons animæ mo-

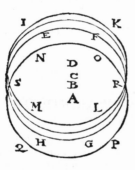

uentis,sc. Sol. B centrum viæ E F G H, quã Planeta,sed inæquali passu,incedit,B D sit vt B A,
& C B eius dimidium. Cũ igitur E F sit remotior ab A,quam N O quantitate A B : coueuie-
bat vt Planeta in E F tam tardus esset,ac si duplo longius ab A recessisset,quátitate sc. A D,
& super centro D curreret. Et econtra,cum
H G sit propior ipsi A quam P Q, eadem A B
quantitate,conueniebat,vt Planeta in G H tã
velox esset,ac si duplo propius ad A accessis-
set,nimirum itidem quantitate A D.V trobiq;
ergo tantundem est,ac si super D centro incederet. (3) Supra enim cap.
20.ea motuum ad orbes fuit proportio. Quare cogita , quæ ibi loci duæ
causæ per totum circulum concurrerunt, eas hîc inuersas & permixtas
esse. Illic orbis eiusdem integer ambitus maior & remotior periodũ au-
xit,& minor atq; propior diminuit:Hîc autem circuli N O P Q & E F G H æ-
quales sunt,& huius pars altera remotior, altera propior est cétro A So-
li.Quapropter motrix virtus in A agit in E F, & in G H, tanquam planeta
illic esset in I K,hîc in L M. V triusque autem, tarditatis illius, & velocitatis huius communis mensura inuenitur in D. Itaque Planeta in E F G H via
progrediens,tardus veloxque,nec non mediocris circa R & S fit,perinde
tanquam in I K L M,super D centrũ æqualiter iret. Iá vide artifices,qui penitus idé statuerunt. Népe Ptolemæus D centrũ æquantis, & B centrum
viæ planetariæ fecit.Copern.vero circa C centrum,medium inter D & B,
eccentrum eccentri vel eccétrepicyclum circumducit. Ei ergo fit,vt via
planetæ sit quã proxime E F G H,sed motus æqualitas,sicut ipsius orbis in-
termedij inter E F G H & I K L M circa C,ita planetæ circa D,reguletur.

(4) Cau-

tion to be subtracted would be too large; second, there is no detraction from the miraculous nobility of this mechanism if only the paths themselves traced out by the planets' globes retain this proportion, by whatever spheres they are impelled, large or small. And I have added what you have found in Chapter 16, on the material of the figures, which is non-existent; and hence that it is not absurd to include the solids in the same location as the spheres, or rather that even without the spheres this irregularity in their path can be defended.

I see that the noble and excellent Danish mathematician Tycho Brahe is of the same opinion. However, the reason and the means are shown more clearly by these writings of our own, (1) that is, if the cause of the retardation and acceleration is the same for the spheres of the individual planets as it was above in Chapter 20 for the whole universe, in the following manner. The path of the planet is eccentric, and it is slower when it is further out, and swift when it is further in. For it was to explain this that (2) Copernicus postulated epicycles, Ptolemy equants. Then describe a concentric circle equal to the eccentric path of the planet, of which the motion will be equal at all points, since it is equally distant from the source of the motion. Therefore at the middle part of the eccentric path where it projects above the concentric circle, the planet will be slower, because it moves further away from the Sun and is moved by a weaker power; and in the remaining part it will be faster, because it is closer to the Sun and subject to a stronger power. And it is easy for anyone to understand that this variation in motion can be explained by a little circle exactly as if the planet were really moving on that circle with a uniform motion. You know the cause of this slowness; let us now look at the measure of it as well. Let A be the source of this moving spirit, namely the Sun; let B be the center of the path EFGH, on which the planet moves, but at an irregular pace; and let BD be as BA, and CB half of it. Then since EF is further away from A than NO by the extent of AB, it turns out that the planet on EF would be slow by the same amount as if it had moved twice as far from A, that is, by the extent of AD, and its course was about D as center. And on the other hand, since HG is nearer to A than PQ, again by the extent of AB, it turns out that the planet on GH would be fast by the same amount as if it had moved twice as near to A, naturally by the extent of AD once more. Then in both cases it is just the same as if the planet were moving about D as center. (3) For in Chapter 20 above, that was the ratio of the motions to the orbits. Consider, therefore, that the two causes which in the former position act in combination all round the circle, are in the latter position inverted and have different effects.[2] In the former case the complete circuit of a given planet had an increased period when it was larger and further away, and a diminished period when it was smaller and nearer: in the latter case the circles NOPQ and EFGH are equal, and one part of the latter is further away from the center, A, the Sun, and the other part is nearer to it. Consequently the motive power at A acts along EF, and along GH, as if the planets were on IK in the former case, on LM in the latter. However the common measure for each case, of the retardation in the former and of the acceleration in the latter, is found at D. Thus a planet passing along the path EFGH becomes slow and fast, and also of medium speed in the region of R and S, exactly as if it were moving uniformly on IKLM, about D as center.[3] Now notice the practitioners, who came to exactly the same conclusion. Ptolemy of course made D the center of the equant, and B the center of the planetary path. Copernicus, to be sure, constructed an eccentric on an eccentric, or an epicycle on an eccentric about C as center, halfway between D and B. For him therefore the path of the planet is as nearly as possible EFGH, but the uniformity of the motion of the planet is referred to D as center, just as that of the orbit intermediate between EFGH and IKLM is to C as center.

(4) Caufam habes, cur æquantis centrum parte tertia eccentrici-tatis totius â centro eccentrici diftet. (5) Nempe mundus totus anima plenus efto, quæ rapiat, quicquid adipifcitur ftellarum fiue cometarum, idque ea pernicitate, quam requirit locia Sole diftantia & ibi fortitudo virtutis. Deinde efto in quolibet Planeta peculiaris anima, cuius remigio ftella afcendat in fuo ambitu: Et orbibus remotis eadem fequentur.

Atque hæc de Æquâte, vbi legerint aliqui, fcio geftient. Nam fi mi-rantur Aftronomi Ptolemæum indemonftratam fumpfiffe hanc eâdem menfuram centri Æquantis: multo magis iam mirabuntur quidam, fuif-fe caufam huius rei, neque tamen de ea Ptolemæo fuboluiffe, cum ipfam rem ita, vti habet, fumeret; & quafi diuino nutu cæcus ad locum debitum perueniret.

Sed tamen eos admonitos velim, nihil effe ex omni parte beatum. (6) Nam in Venere & Mercurio ifta tarditas & velocitas non ad plane-tæ à Sole digreffionem, fed ad folum Terræ motum accommodatur. Et fi quis huic rei prætexat diuerfam motus conditionem à motu fuperiorû: quam denique in (7) Terræ annuo motu caufam afferet? Is enim neque apud Ptolemæum, neque apud Copernicum Æquante indiguit. Qua-re & hæc incerta lis fub Aftronomo iudice pendeat.

In Cap. XXII. Notæ Auctoris.

(1) NEmpe fi eadem fit caufa.] *Si quæ caufa efficit, vt Saturnus altus, fit tardior Ioue humi-liori & foli viciniori, eadem efficiat, vt Saturnus altus & apogæus, fit tardior feipfo perigæo & humili. Cæla vtriufque rei eft, elongatio Planetæ à Sole rectilinea, maior vel minor, quia longe diftans à Sole verfatur in virtute Solari tenuiore & imbecilliore.*

(2) Copernico Epicyclia, Ptolemæo æquantes.] *Quam æquipollentiam hypothe-fium docui in Comment. Martis part. 1.*

(3) Supra enim Cap. XX. ea motuum ad Orbes.] *Hoc vero in annotationibus e-mendauimus. Non dupla erat periodorum, & fic tarditatum proportio ad proportionem orbium, fed fefquialtera faltem. At in Planetæ vnius motibus, ex fole apparentibus, Aphelio & perihelio, regnat proportio diftantiarum præcife dupla, in motibus ipfis diurnis, vt funt arcus eccentricorum, proportio ipfiffima diftantiarum fimpla, vide Comment. Martis, part. 3. & 4. Caufam diuerfitatis euidentiffi-mam habet lib. 4. Epit. Aftron. fol. 533.*

(4) Caufam habes, cur æquantis centrum parte tertia.] *Hoc de Copernico verum eft, cui C centrum eft æquantis, feu potius eccentrieccentrici, B centrum viæ Planetæ, & ipfius A C pars tertia B C. At in Ptolemæo ratio eft alia. Ille enim D eft centrum æquantis, B Eccentrici, quare ipfius A D femiffis eft B D.*

(5) Nempe mundus totus anima plenus.] *Rurfum pro anima intellige Solis fpeciem immateriatam, extenfam vt lumen: & habebis hic breuibus verbis fummam meæ phyficæ cœleftis, traditam in Comment. Martis part. 3. & 4. & repetitam lib. 4. Epit. Aftron.*

(6) Nam in Venere & Mercurio.] *Nihil opus exceptione: vero verius eft etiam de Ve-nere & Mercurio. Nam quod Copernicus aliquas horum Planetarum inæqualitates alligat ad mo-tum orbis annui, id eft errore eft.*

(7) Terræ annuus motus æquante non indiguit.] *Apud Ptolemæum quidem & Copernicum. At ego in Comment. Martis, præcipuorum libri membrorum hoc vnum feci, & velut angularem lapidem in fundamento pofui; imo clauem Aftronomiæ merito appellaui: quod ex ipfis mo-tibus Martis liquido demonftraui, feu Solis feu Terræ motum annuum regulari circa alienum centrû æquantis, eiufque eccentricitatem orbita, dimidium folum habere, Eccentricitatis ab auctoribus cre-ditæ.*

Vides itaque, Lector ftudiofe, libello hoc femina fparfa effe omnium & fingulorum, quæ ex eo

Chapter XXII

(4) You now know the reason why the distance of the center of the equant from the center of the eccentric is one-third of the whole eccentricity.[4] (5) Then, naturally, let the whole universe be full of a spirit which whirls along any stars or comets it reaches, and that with the speed which is required by the distance from the Sun of their positions and the strength of its power there. Next let there be in each planet a peculiar spirit, by the impulsion of which the star goes up in its circuit; and even without the spheres the same results will follow.

Anyone who reads this passage on the equant will, I know, rejoice. For if the astronomers are surprised that Ptolemy assumed this same measure of the center of the equant without proof, some people will now be all the more surprised that there was an explanation for it, and Ptolemy did not suspect it, since he assumed the fact to be as it is, and as if by divine guidance[5] arrived blind at the proper destination.

Yet I should like them to take warning that nothing is pleasing in every way. (6) For in the cases of Venus and Mercury this slowness and quickness fits in, not with the planet's distance from the Sun, but only with the Earth's motion. And if anyone elaborates this question with a law of their motion different from that of the superior planets, what explanation will he eventually put forward for (7) the annual motion of the Earth? For it did not need an equant either in Ptolemy's theory or in Copernicus's. Consequently, this is also a doubtful case awaiting the judgment of astronomy.

AUTHOR'S NOTES ON CHAPTER TWENTY-TWO

(1) *That is, if the cause . . . is the same.*] If any cause has the effect that Saturn when it is high is slower than Jupiter when it is lower and closer to the Sun, the same cause would have the effect that Saturn when it is high and at apogee would be slower than it is itself at perigee and low. The cause of both effects is the greater or smaller distance of the planet from the Sun in a straight line, because when it is far distant from the Sun the power of the Sun which it experiences is thinner and weaker.[6]

(2) *Copernicus (postulated) epicycles, Ptolemy equants.*] I have explained this equivalence of the hypotheses in my *Commentaries on Mars*, Part I.

(3) *For in Chapter 20 above that (was the ratio) of the motions of the orbits.*] However we have emended it in the notes. The ratio of the periods, and so of the slownesses, was not the square of the ratio of the orbits, but in fact the 3/2th power. But in the apparent motions of a single planet as seen from the Sun, at aphelion and perihelion, the reigning ratio is precisely the square of that of the distances: in the daily motions themselves, as they are arcs of the eccentrics, the actual ratio of the distances is simple. See the *Commentaries on Mars*, Parts III and IV. The quite obvious cause of the difference is found in Book IV of the *Epitome of Astronomy*, page 533.

(4) *You now know the reason why (the distance of) the center of the equant . . . is one third.*] This is true for Copernicus, for whom C is the center of the equant, or rather of the eccentric on an eccentric, B the center of the path of the planet, and BC one-third of AC. But in Ptolemy the reasoning is different. For him D is the center of the equant, B of the eccentric, so that BD is half AD.

(5) *Then, naturally, (let) the whole universe (be) full of a spirit.*] Again, instead of "spirit" understand the immaterial emanation of the Sun, spreading out like light: and you will have here in a few words a summary of my celestial physics, expounded in my *Commentaries on Mars*, Parts III and IV, and repeated in Book IV of the *Epitome of Astronomy*.

(6) *For in the cases of Venus and Mercury.*] There is no need of the exception: indeed it is even more true of Venus and Mercury. For in Copernicus's linking of certain irregularities of the planets to the motion of the annual orbit is where the error lies.

(7) *The annual motion of the Earth . . . did not need an equant.*] In Ptolemy's theory indeed and in Copernicus's. But in my *Commentaries on Mars* I have made this one of the chief features of the book, and have laid it like a cornerstone at the foundation. Indeed I deservedly called the key to astronomy the fact, which I have demonstrated clearly from the actual motions of Mars, that the annual motion either of the Sun or of the Earth is controlled by a different center from the equant, and that the eccentricity of its orbit is only half the eccentricity believed by the authorities.[7]

You see, then, assiduous reader, that in this book there were scattered the seeds of each and every one

tempore in Astronomia noua & vulgo absurda, ex certissimis Brahei obseruationibus à me constituta & demonstrata sunt: itaq; spero te iocum meum lib. 4. Harm. de meis Imaginibus, ex Procli Paradigmatibus delapsis, non iniqua censura flagellaturum.

CAPUT XXIII.

De initio & fine Mundi Astronomico & anno Platonico.

POST epulas, post fastidium ex saturitate, veniamus ad bellaria. Problemata duo pono nobilia. Primum est de principio motus; alterum de fine. (1) Certe non temere Deus instituit motus, sed ab vno quodam certo principio & illustri stellarum coniunctione, & in initio Zodiaci, quod creator per inclinationem Telluris domicilij nostri effinxit, quia omnia propter hominē. (2) Annus igitur Christi 1595. si referatur in 5572. mundi (qui communiter & à probatissimis 5557. censetur) veniet creatio in illustrem constellationem in principio ♈. Nam anno primo assumpti numeri, die Aprilis 27. Iuliano retro computato, feria prima, qui dies Creationis omnium est, hora vndecima meridici Borussiæ, quæ est sexta vespertina in India, talis exhibetur cœli facies à Prutenico calculo.

☉	3	♈
☽	3	♎
♄	15	♈
♃	10	♈
♂	24	♊
♀	10	♉
☿	3	♈
☊	18	♍

Motus ♂ ♀ & ☊ paulisper morare, aut promoue, & venient in loca cognata, & forte ☊ in o. ♎ ad ☽. Scaliger male Nouilunium vult. Nam Luna in potestatem noctis condita, nocte vtiq; prima fulsit. Verisimilius initium calculus multis retro porroque annis non suppeditat. (3) Sed si rationes sequamur, oportet hoc initium, ☉ in ♎ versante, quærere, nempe hac cœli facie.

♄	0	♈
♃	0	♈
♂	0	♈
☋	0	♈
☽	0	♑
☊	0	♎
♀	0	♎
☿	0	♎
☉	0	♎

Vult hoc veterum auctoritas, Mundum in Autumno creatum, & ratio ipsa ex Copernico, vt Tellus sub eodem initio stet, quo reliqui. Apparebunt igitur superiores in ♈, inferiores & ☉ in ♎, Luna cum circa terram sit neque in ♈, neque in ♎ competit, ne turbet numerum terna-

of the things which since that time in this new and, to the masses, absurd astronomy I have established and demonstrated from the thoroughly exact observations of Brahe; and I therefore hope that you will not lash with an unfair judgment my joke in Book IV of the *Harmonice*[8] about my Images, which the Paradigms of Proclus let fall.

CHAPTER XXIII.
ON THE ASTRONOMICAL BEGINNING AND END OF THE UNIVERSE AND THE PLATONIC YEAR

After the feasting, after the weariness of repletion, let us come to the dessert. I pose two noble problems. The first concerns the start of motion, the other its end. (1) Certainly God did not start the motions at random, but from some single definite starting point, some illustrious conjunction of stars, and at the beginning of the zodiac, which the Creator formed according to the inclination of the Earth, our dwelling, since everything is for the sake of Man. (2) Then if the year of Christ 1595 is taken as the year 5572 of the universe (though it is generally and by very sound authority reckoned as 5557), the creation will come to an illustrious combination of stars at the start of Aries. For, if we take that number of years, in the first year, on the 27th April by the Julian calendar, counted backwards, on the first day of the week, which is the day of the creation of all things, at the eleventh hour at the Prussian meridian, which is the sixth hour of the evening in India, this is the appearance of the heaven calculated according to the Prutenic Tables.

Sun	3°	Aries
Moon	3°	Libra
Saturn	15°	Aries
Jupiter	10°	Aries
Mars	24°	Gemini
Venus	10°	Taurus
Mercury	3°	Aries
Ascending node	18°	Virgo

Delay the motions of Mars, Venus, and the ascending node a little, or hasten them, and they will come to associated positions, and perhaps the ascending node will be in 0° of Libra with the Moon. Scaliger prefers the New Moon, wrongly. For the Moon was created to have power in the night, and undoubtedly shone in the first night. Calculation for many years backwards and forwards does not afford a more likely beginning. (3) But if we follow reason, we should look for a beginning, with the Sun in Libra, that is with the following appearance of the heaven.

Saturn	0°	Aries
Jupiter	0°	Aries
Mars	0°	Aries
Descending node	0°	Aries
Moon	0°	Capricorn
Ascending node	0°	Libra
Venus	0°	Libra
Mercury	0°	Libra
Sun	0°	Libra

The conclusion of the ancient authorities is that the universe was created in the autumn, and the inference from Copernicus is that the Earth is located at the same starting point as the other planets. Therefore the superior planets will appear in Aries, the inferior planets and the Sun in Libra, and the Moon, since it is a satellite of the Earth, does not belong either in Aries or in Libra, in case it should upset the threefold number of the superior and inferior planets. When the Sun is setting (for

ternarium superiorum & inferiorum. Et sole occidente (sic enim conditus mundus est) nocti nullibi rectius dominatur, quâ ex medio cœli, quod est o. ♄ Sicque poterit in epicycli summa absside consistere. Et quia orbis eius aduentitius est, sortiatur & ipsa aduentitium & peculiarem situm principij. Lunationes etiam eius nobilitas & fama inter homines, lunationúque potissima quadrans. Caput autem in libram, & caudam in Arietem refero, vt sit in rationali situ cum Luna, absq; Eclipsi tamen: & vt Luna sit in maximo limite horeo. Erit igitur terra oculari etiam positu media inter stellas; sicut orbis eius inter orbes medium locum certo Dei consilio obtinuit; quia omnia propter homiem Quod si Solem etiam hic in ♈ loces: erit in ♄ ♎ & ☽ in 69. & reliqua similiter. (4) Sumendi autem motus medij, nam hos in principio cursus, veros esse conuenit, nempe ab absidibus. Hæc palma in medio posita, quam aut simile si quis aut ex calculo, aut ex restauratione Astronomiæ adeptus fuerit, is Phyllida solus habebit. Hæc de initio.

(5) Finem motui nullum cum ratione statui, nullumque fore Platonicum annum ex postulato vno probabo. Detur namque eccentricitaté esse cum orbe in proportione rationali: erunt igitur orbium radij inuicem irrationales, quia habent se, vt inscripti & circumscripti corporibus, qui irrationales sunt, quia sequuntur ex ratione subtensæ in quadrato, & sectionis secundum extremum & mediam rationem; quæ duo sunt exempla irrationalium in Geometria. Iam autem motus cum radijs in proportione sunt; Ergo motus inter se irrationales, & sic nunquam ad idé redibunt initium, etsi durarent infinitis seculis: quia nunquam, ne in infinita quidem sectione temporis, occurreret communis mensura, qua sæpius repetita, motuum omnium vnus terminus, & meta anni Platonici constituatur. Etiam vel tandem cum diuino Copernico libet exclamare: *Tanta nimirum diuina hæc est Opt. Max. fabrica:* & cum Plinio: *Sacer est (mundus) immensus, totus in toto, imo vero ipse totum, finitus & infinito similis.*

In Caput XXIII. Notæ Auctoris.

(1) **C**Erte non temere Deus instituit motus.] *Non tamen statim de coniunctione omniũ Planetarum sub eodem Zodiaci gradu concludere possumus: sufficit; si saltem in genere fuerit aliqua Harmonica dispositio, & Zodiaci per planetas diuisio, si nõ ex Terra, at saltem ex Solis centro. Vide Harm. lib. 4. cap. 2. & 3.*

(2) Annus igitur 1595. si referatur.] *Non tolerat Astronomia, supposita periodorum æquabilitate, vt constellatio hæc perficiatur, adque meram Harmonicam dispositionem redigatur.*

(3) Sed si rationes sequamur oportet.] *Nec hoc necessarium; nec auctoritas veterum rigide vrgenda de Creatione: potuit enim frugum prouentus (non creationis memoria) causam dare, cur anni finis autumnus haberetur.*

(4) Sumendi autem motus medij.] *Quid si ne hoc quidem? quid si non in absidibus erati planetæ, vt in Extremis, vbi æquatio nulla, sed in interuallo medio, vbi æquatio maxima? Itaq; superet exercitatio ista proposita omnibus calculatoribus Astronomis, & plena quidem piæ persuasione de ortu temporis. Mæstlinus aliqua tentauit. Accipe & à me aliam, vbi ex centro Solis omnia in locis oppositis & quadratis, & punctis quidem Cardinalibus.*

Currente ante æram nostram vulgarem Anno 3993. *Iuliano retro extenso, die* 24. *Iulij ad vesperam incipiente in Chaldæa feria secunda, Sol & Luna in principio Cancri prope cor Leonis, omnes Lunæ motus in quadrantibus sunt, vt & omnes reliqui: Saturnus & Mercurius versus libra initium;* Iupiter,

that was how the universe was created) there is no place where the Moon more fittingly holds sway over the night than from the middle of the heaven, which is 0° of Capricorn. In that way it will be able to stand at the upper apsis[1] of its epicycle. And since its orbit is an extraordinary one, for itself also it is allotted an extraordinary and peculiar position for its starting point. Also its phases are its nobility and its renown among men, and the most important of the phases is when it is in quadrature. Now I place the head in Libra, and the tail in Aries,[2] so that it is in a logical relationship to the Moon, though without involving an eclipse; and so that the Moon is at its furthest northern limit. Then the Earth's position as judged by eye will be midway among the stars, just as among the spheres its sphere has been given the middle place by God's express intention, since all things are for the sake of Man. But if on the other hand you locate the Sun in Aries, Saturn will be in Libra, and the Moon in Cancer, and the other things similarly. (4) Furthermore the mean motions must be adopted, for it is agreed that at the start the paths were the true ones, that is to say, away from the apsides. This prize is open to all comers, and if anyone attains it, either by calculation or by a reform of astronomy, he alone shall have the princess. So much for the beginning.

(5) I have not established any end to the motion by argument, and I shall prove that there will be no Platonic[3] year by means of a single assumption. For let it be granted that the eccentricity stands in rational proportion to the sphere. Then the radii of the spheres will in turn be irrational, because they are in the same ratio as the inscribed and circumscribed circles of the solids; and those circles are irrational, because they follow the ratio of the diagonal in a square, and the division in the extreme and mean ratio, which are two examples of irrationals in geometry. But in fact the motions are proportionate to the radii. Therefore the motions are in irrational proportions to each other, and thus they will never return to the same starting point, even if they were to last for infinite ages, since there would never, even indeed in an infinite division of time, occur a common measure, by the more frequent repetition of which a single endpoint for all the motions, and the goal of the Platonic year, would be established. And now indeed we are at last happy to exclaim with the divine Copernicus,[4] "Such truly is the size of this structure of the Almighty's," and with Pliny, "Holy is the boundless (universe), whole in a whole, nay truly itself a whole, finite and like to the infinite."

AUTHOR'S NOTES ON CHAPTER TWENTY-THREE

(1) *Certainly God did not start the motions at random.*] However we cannot immediately come to a conclusion about the conjunction of all the planets under the same degree of the zodiac. It is enough, if at least in a general way there was some harmonic arrangement, and a division of the zodiac among the planets, if not from the Earth's, then at least from the Sun's center. See the *Harmonice*, Book IV, Chapters 2 and 3.

(2) *Then if the year . . . 1595 is taken.*] Astronomy does not tolerate, assuming regularity of the periodic times, the achievement of this arrangement of stars, and its reduction to a pure harmonic pattern.

(3) *But if we follow reason, we should.*] This is not necessary, and the authority of the ancients should not be too rigorously pressed on the subject of the Creation; for the coming forth of the fruits (not the memory of Creation) might provide the reason why autumn was held to be the end of the year.

(4) *Furthermore the mean motions must be adopted.*] What if not even that? What if the planets were not created at their apsides, as extreme points, where there is no equation, but in the middle of the intervening space, where the equation is a maximum? This exercise, then, remains as a problem for all astronomical calculators, and indeed one which is full of incitement to piety on the subject of the origin of time. Maestlin has attempted it somewhere. You shall have from me another attempt, in which from the center of the Sun everything is in positions which are opposite or at quadrants, and indeed at cardinal points.

In the course of the Julian year 3993, reckoned backwards from our standard era, on the 24th day of July at evening, at the beginning of the second day of the week in Chaldea, the Sun and Moon are at the beginning of Cancer near the heart of Leo, all the motions of the Moon are at the quadrants, as are all the remaining motions: Saturn and Mercury are towards the beginning of Libra, Jupiter and the Earth towards Capricorn, the Moon, Mars, and Venus towards Cancer. In the case of Mercury there are some degrees

Iupiter, Tellus, versus Capricornum, Luna, Mars, Venus versus Cancrū. In Mercurio abūdant gradus aliquot, sed qui consumi possunt eius æquatione maxima ablatiua, si modo satis cognitus est eius motus medius, vt non per huius correctionem consumantur. In Venere etiam abundat aliquid, quod æquatione tolli non potest. Feria secunda est Firmamenti, seu expansionis inter aquas & aquas; quasi Orbes seu Planetæ, per hanc expansionem ire iussi, statim in ipso ortu expansi, ceperint ire; feria vero quarta demum exornatum cœlum extimum fixis, & Sol, & Luna, &c. vltima manu imposita.

(5) Finem motui nullum cum ratione statui.] Dogma innitebatur huic vt primario fundamento: quod inter Orbes cœlestes sit proportio, illa quæ est Orbium Geometricorum cuiuslibet, ex quinque figuris. Illarum enim quatuor proportiones sunt ineffabiles, seu vt hic cum vulgo appellant, irrationales. Iam vero fundamentum hoc refutauimus: quia proportio cœlestium orbium non est ex solis quinque figuris. Quæritur, quid iam porro de hoc dogmate tenendum, & num detur aliqua perfecta Apocatastasis motuum omnium? Dico, quamuis hoc fundamento subruto, nullam tamen dari Apocatastasin. Id probabo. Certum igitur est, si proportiones saltem periodicorum temporum sunt effabiles, dari ἀποκαταστασιν: si ineffabiles non dari. Iam effabiles dentur an ineffabiles, sic diiudicandum. Omnes motuum Apogæorum & Perigæorum proportiones, tam binorum, quam singulorum, sunt effabiles; sunt enim desumptæ ex Harmoniis, & illæ sunt omnes effabiles, vt & Concinna & concinnis inseruientia interualla omnia. Itaque lib. V. Harmonicorum cap. IX. pro. XLVIII. Omnes hi motus suis numeris expressi & estati sunt: Numeri enim illi præcisi sunt intelligendi. Iam vero periodicorum temporum inter se proportio est eadem quātitate, quæ est & motuum mediorum. Motus vero medii participant de medio arithmetico inter extremas, aphelium & perihelium; quod medium est inter effabiles hos terminos, effabile: participant & de medio inter eosdem Geometrico. At inter effabiles terminos, non est semper effabile medium Geometricum. Sunt igitur motus planetarum medii ineffabiles, & incommensurabiles motibus extremis Planetarum omnium. Vide Harmon. lib. V. Cap. IX. Prop. XLVIII. Cum autem à priori nulla sit ratio, quæ formet motus medios, sed cum resiliant singuli ex suis motibus extremis: non erunt medij motus ne inter se quidem commensurabiles; nullum enim ordinatum, vt effabilitas, casu existere solet. Quare neq; periodi temporum inter se commensurabiles erunt. Nulla igitur data perfecta motuum Apocatastasis, quæ pro fine motuum formali, seu rationali haberi possit.

Habes igitur, Lector, examen Libelli mei, cui titulus à Mysterio Cosmographico, promissum ante annos X. in Comm. Martis Part. III. Verum ante Harmonicorum editionem locus huic examini non fuit. Quare fine commentationi imposito, conuertamur ad hymnum, qui librum claudit.

CONCLVSIO LIBRI.

Tu nunc, amice Lector, finem omnium horum ne obliuiscare, qui est, Cognitio, admiratio & veneratio Sapientissimi Opificis. Nihil enim est ab oculis ad mentem, à visu ad contemplationem, a cursu aspectabili ad profundissimum Creatoris consilium processisse: si hic quiescere velis; & non vno impetu, totaque animi deuotione sursum in Creatoris notitiam, amorem cultumque efferare. Quare casta mente, & grato animo mecum perfectissimi operis architecto sequentem Hymnum accine.

IOVA Sator Mundi, nostrúmque æterna potestas,
Quanta tua est omnem terrarum fama per orbe?
Gloria quanta tua est? Cœli quæ didita supra
Mœnia, concussis volat admirabilis alis.
Agnoscit puer & spreto satur vbere, balbis
Te dictante struit valida argumenta labellis:
Argumenta, quibus tumidus confunditur hostis
Contemptorq; tui, & contemptor iuris & æqui:
Ast ego, quo credam spacioso Numen in orbe?
Suspiciam attonitus vasti molimina cœli.
Magniopus Artificis, valida miracula dextræ;
Quinq; vti siderios normis distinxeris orbes,
Quos intra medius Lucisq; animæq; Minister
Qua lege æterni cursus moderetur habenas,
Quas capiat variata vices, quos Luna labores,
Sparseris immenso quam plurima Sidera campo.

Maxime mundi Opifex, qua te ratione coegit
Paruus, inops, humilis, tamq; exigua Incola glebæ
Adamides rerum curas agitare suarum?
Respicis immeritum, vehis in sublime, Deorum
Tantum non genus est, tantos largiris honores,
Magnificumq; caput cingis diademate, Regem
Constituisq; super manuum monumenta tuarum.
Quod supra caput est, magnos cum motibus orbes,
Subijcis ingenio: quicquid Tellure creatur,
Natum operis pecus, atq; aris fumantibus aptum.
Quaq; habitant siluas reliquarum sæcla ferarum,
Quodq; genus, volucres, leuibus ferit aera pennis,
Quiq; maris tractus tranant & flumina, pisces.
Omne iubes premere imperio, dextraq; potenti.

Ioua sator Mundi, nostrúmq; æterna potestas
Quanta tua est omnem terrarum fama per orbem!

over; but they can be disposed of by its maximum subtractive equation, if only its mean motion is adequately known, so that they are not used up in correcting it. In the case of Venus there is also something over, which cannot be removed by the equation. The second day of the week is that of the Firmament, and of its spreading out between the waters and the waters; as if the orbits or planets, given the order to move by this spreading out, began to move immediately at the origin of this spreading out. The fourth day of the week, however, is that of the embellishment of the outermost heaven with the fixed stars, and the setting of the Sun and Moon, etc., in it as the finishing touch.

(5) *I have not established any end to the motion by argument.*] The doctrine leant on the following fact as its chief foundation: that the ratio between the celestial spheres is that between any of the geometrical spheres derived from the five figures. For four of those ratios are inexpressible, or as I have called them in the common way, irrational. Now, however, we have refuted this basis, because the ratios of the celestial spheres are not derived solely from the five figures. The question is, what belief must we hold about this doctrine from now on, and is some exact return of all the motions to their starting point to be found? I say that although this foundation has collapsed under us, yet no such return is to be found. I shall prove it. It is certain, then, that if the ratios of the periodic times at least are expressible, a return to the starting point is to be found: if they are inexpressible, it is not. Now whether they are expressible or inexpressible must be decided in the following way. All the ratios of the motions of the apogees and perigees, both in pairs and singly, are expressible; for they have been taken from the harmonies, and they are all expressible, as all the intervals are either melodic or subsidiary to melodic intervals.[5] Thus in Book V of the *Harmonice*, Chapter 9, Proposition 48, all these motions have been set out and expressed by their own numbers. For those numbers are to be understood as precise. Now in fact the ratio of the periodic times to each other is the same quantity as that of the mean motions. However, the mean motions are formed from the arithmetic mean between the extremes, the aphelion and the perihelion, and that mean between these expressible terms is expressible. They are also formed from the geometric mean between the same terms. But the geometric mean between expressible terms is not always expressible. Therefore the mean motions of the planets are inexpressible, and incommensurable with the extreme motions of all the planets. See the *Harmonice*, Book V, Chapter 9, Proposition 48. However, since *a priori* there is no proportion which controls the mean motions, but they spring individually from their own extreme motions, the mean motions will not be commensur..ble even among themselves; for no regular property, such as expressibility, normally exists by accident. Therefore no exact return of the motions to their starting point is to be found, which can be taken as an end to the motions in accordance with form and reason.

You now have, then, reader, the revision of my little book, entitled *The Secret of the Universe*, which was promised ten years ago in Part III of my *Commentaries on Mars*. However, before the publication of the *Harmonice*, there was no room for this revision. Therefore, having made an end of the commentary, let us turn to the hymn which closes the book.[6]

CONCLUSION OF THE BOOK

Now, friendly reader, do not forget the end of all this, which is the conception, admiration and veneration of the Most Wise Maker. For it is nothing to have progressed from the eyes to the mind, from sight to contemplation, from the visible motion to the Creator's most profound plan, if you are willing to rest there, and do not soar in a single bound and with complete dedication of spirit to knowledge, love, and worship of the Creator. Therefore with pure mind and thankful spirit sing with me the following hymn to the Architect of this most perfect work:

Great God, Creator of the Universe,
And our eternal power, how great thy fame
In every corner of the whole wide world!
How great thy glory, which flies wondrously
Above the far-flung ramparts of the heavens
With rushing wings! The babe salutes it, spurning
The breast, replete, and with his halting lips
Bears powerful witness — witness which confounds
The haughty enemy, who shows contempt
For thee, and shows contempt for law and justice.
Yet, to believe thy Godhead is within
This spacious sphere, let me look up astonished
At thy achievement of this mighty heaven,
The work of the great Craftsman, miracles
Of thy strong hand; see how thou hast marked out
The five-fold pattern of the starry spheres,
Dispensing light and spirit from their midst;
See by what law thou dost control the reins
Of their eternal course; see how the Moon
Varies her path, her toils, how many stars
Thy hand has scattered over that boundless field.

Great Builder of the Universe, what plea
Of the poor, humble, small inhabitant
Of this so tiny plot compelled thy care
For his harsh troubles? Yet thou dost look down
On his unworthiness, carry him up
On high, a little lower than the Gods,
Bestow great honors on him, crown his head
Nobly with diadem, appoint him king
Over the tokens of thy handiwork.
Thou makest all that is above his head,
The great spheres with their motions, bow before
His genius. All creatures of the Earth,
The herds bred for his works, and fitted for
The smoking altars, and the generation
Of wild beasts which remain to dwell in woods,
The birds, which with light feathers strike the air,
The fish, which swim through rivers and through seas,
Over all these by thy command he rules
By his dominion and his strong right hand.

Great God, Creator of the Universe,
And our eternal power, how great thy fame
In every corner of the whole wide world!

225

TRANSLATION OF THE ANNOTATIONS TO THE PLATES

PLATE I. Showing the order of the moving celestial spheres, and at the same time the true proportion of their size according to their mean distances, also the angles of the corrections for the same on the Earth's Great Orbit, according to Copernicus's theory.

In the center or near it is the Sun, motionless.

EF the smallest circle round the Sun is that of Mercury, which returns in about 88 days.

That is followed by CD the circle of Venus, which revolves about the same Sun in 224⅔ days.

AB the circle which follows is that of the Earth, which revolves in 365¼ days. It is called the Great Orbit because it has many applications.

Round the Earth is the little orbit, like an epicycle, of the Sphere of the Moon, at A, which returns with the same motion in the space of a year along with the Earth to the same Fixed Star. But its own revolution referred to the Sun occupies 29½ days.

After that is the orbit of Mars, GH, which completes one passage under the Fixed Stars, or referred to the Sun, in 687 days.

That is enclosed after a large gap by the Sphere of Jupiter, IK, which has a round of 4332 days plus about ⅝.

LM, the furthest and largest circle, is that of Saturn: its periodic time is 10,759⅕ days.

The Fixed Stars, however, are still higher by an interval so immeasurable that in comparison the gap between the Sun and the Earth is insensible. Also they at the edge, like the Sun in the center, are completely motionless.

Angle TGV, or arc TV, is the correction, or parallax, of the Earth's Great Orbit with respect to the Sphere of Mars.

Similarly PIN is the parallax of the same Great Orbit with respect to the Sphere of Jupiter, and PLN or RLS, or arc RS, with respect to the Sphere of Saturn.

Also XAY, or arc XY is the parallax of the Sphere of Venus; and in the same way ZAÆ, or ZÆ, is the parallax of the Sphere of Mars with respect to the Great Orbit.

PLATE II. Showing the order of the celestial spheres, and in each case the proportion of the orbits and epicycles, and the angles or arcs of the equations for the same, according to their mean distances, in accordance with the theory of the ancients.

In the center is the Earth, which alone is motionless.

The innermost small orbit round the Earth represents the Sphere of the Moon, of which the motion is monthly.

(continued next page)

(Plate II continued)

The next round that is the orbit of Mercury; it is followed by that of Venus and after it is the Sphere of the Sun, which all go round in an annual revolution. The orbits of the other three, the superior planets, Mars, Jupiter, and Saturn, as well as the Sphere of the Fixed Stars, are indicated by arcs, which anyone can complete by describing the whole of them about the Earth as the center.

The orbit of Mars makes a turn in two years.

That of Jupiter requires twelve years, as nearly as possible, and that of Saturn about thirty years. The Fixed Stars complete a period in 49,000 years, according to the tenets of the Alfonsine tables. The amounts of the equations for each of them (except the Moon) produced by the epicycles on the concentric circle at their mean distances are shown by the arcs intercepted by straight lines drawn from the Earth and touching each of the epicycles, that number of degrees being added.

To follow Chapter I, page 85.

PLATE III. Showing the dimensions of the spheres of the planets, and their separations according to the five regular geometrical solids.

(left column)

You are wondering, spectator, at the work of Kepler, a diagram of heaven which you never saw before. For the five solids of Euclid tell the distance between the orbits of the planets. How well the doctrine which Copernicus once declared agrees, the author's work now reveals to you. Naturally the author has shown himself grateful for his great benefaction to the Duke of Teck, not without praise.

> Drawn by
> Christopher Leibfried.
> Tübingen: 1597.

(right column)

α Sphere of Saturn.

β Cube, the first regular geometrical solid, showing distance from Sphere of Saturn to Jupiter.

γ Sphere of Jupiter.

δ Tetrahedron or pyramid, touching Sphere of Jupiter outside, and Mars inside, producing greatest distance between planets.

ϵ Sphere of Mars.

ζ Dodecahedron, the third solid, showing distance from Sphere of Mars up to the Great Sphere which carries the Earth along with the Moon.

η Great Sphere of the Earth.

θ Icosahedron indicating true distance from Earth's Great Sphere to Sphere of Venus.

ι Sphere of Venus.

κ Octahedron, showing distance from Sphere of Venus to Sphere of Mercury.

λ Sphere of Mercury.

μ The Sun, the unmoving midpoint or center of the Universe.

Plate to be placed at page 101.

PLATE IV. Showing the true breadth of the heavenly spheres, and of the intervening spaces, according to the calculations and theory of Copernicus.

At Chap. 14,
page 157.

The outermost circle represents the zodiac on the sphere of the stars, and is drawn about the center of the universe or of the Earth's orbit, or even from the Earth's globe, as the whole of the Earth's orbit is insensibly small compared with it.

A System of Saturn, drawn concentrically about G, the center of the Earth's orbit.

B System of Jupiter.

C That of Mars.

D Circle or path of the center of the Earth's globe, drawn concentrically about the center G, together with the Moon's small sphere shown on it in two places. The two dotted circular lines mark the thickness of the Earth's sphere with the Moon included.

E Two small circles delineating the thickness of the system of Venus, within which the whole of the variations of its motions are performed.

F The space between the two small circles, within which the whole of the variations of the motions of the star Mercury are performed.

G The center of all things, and near it the Solar body.

The circle passing through O and P (of which only two arcs appear here) is the epicycle on the eccentric of Saturn.

The curved line through Q, through the perigee of the epicycle situated at the apogee O of the eccentric, and through the apogee of the same epicycle at the perigee P of the eccentric, is the eccentric path of the planet. It is not in fact a circle, though it does not sensibly differ from a circular line.

HI is the thickness included between the two concentric circles which the eccentric path of Saturn occupies.

The curved line or virtual circle passing through M, through the apogee of the epicycle at O, and through the perigee of the same epicycle at P, is the eccentric which Ptolemy calls the equant.

KL is the thickness intercepted by the two dotted circles, which is required by the whole epicycle and the aforesaid equant. The planet, however, never goes outside H or inside I. The remaining spheres are understood to be marked off by similar circles in each case, though they are omitted here in case the multiplicity of lines should obscure the point rather than make it clear. Therefore in the cases of Jupiter and Mars it is sufficient to draw their eccentric path, and the two concentric circles which contain it, and in the other cases only the concentric circles.

Intermediate spaces.   R: position of the Cube.   S: of the Tetrahedron T: of the Dodecahedron.   V: of the Icosahedron.   X: of the Octahedron. Z is the space between Saturn and the fixed stars, which is virtually infinite.

PLATE V.   Showing the positions of the centers of the eccentric spheres of the universe according to the theory of Copernicus, and the values in the Prutenic Tables.

<table>
<tr><td>(over left diagram)
At the time of Ptolemy, about
140 A.D. At Chapter 15, p. 161.</td><td>(over right diagram)
At the time of Copernicus, about
1525 A.D. At Chapter 15. p. 161.</td></tr>
</table>

At A is the Sun, the center of the universe. The very small circle at B is the circle of the eccentricity of the Earth's Great Orbit. The center of the eccentric Great Orbit stood in the time of Ptolemy at its edge or in a position further from the Sun, but in the time of Copernicus in a nearer position. That is, the eccentricity of the Great Orbit was in the former case nearly at its maximum; in the latter case almost at minimum. The former instance may be seen in the earlier, or left-hand diagram, the latter in the later, or right-hand diagram.

AB in the earlier diagram is 4170, where the semidiameter of the Great Orbit is 100,000. Hence the greatest separation of the Earth from the Sun is 104,170 and the least is 95,830. But in the other diagram the eccentricity, which is almost at its minimum, is 32,195.

AC is the small circle of the eccentricity of Venus. Its semidiameter (where the semidiameter of the Great Orbit is 100,000 units) is 1040, and BC (in the right-hand figure), the eccentricity of the center of the small circle from the center of the Great Orbit B, is 3120. But AC, the eccentricity of the same center from the Sun A, is 1262. Hence the maximum distance of Venus from the Sun is 74,232 units, and the minimum is 69,628.

D is the center of the small circle of eccentricity of Mercury. Its semidiameter, in the same units as above, is 2114½, and its eccentricity from the center of the Great Orbit, DB, is 7345½; but DA, its eccentricity from the Sun, is 10,270. Whence the maximum distance of Mercury from the Sun is found to be 48,114½, and the minimum 23,345½.

E is the center of the small circle of eccentricity of Mars. Its semidiameter is 7602½, and BE, its eccentricity from the center of the Great Orbit, is 22,807½. But AE, its eccentricity from the Sun, is 20,342. Whence the maximum distance of Mars from the Sun is 164,780, the minimum 139,300.

F is the center of the small circle of eccentricity of Jupiter. Its semidiameter is 12,000, and BF, its eccentricity from B, is 36,000. But AF from the Sun is 36,656. The maximum distance of Jupiter from the Sun is 549,256, the minimum 499,944.

G is the center of the small circle of eccentricity of Saturn. Its semidiameter is 26,075. BG is 78,225, and AG, its eccentricity from the Sun, is 82,290. The maximum separation of Saturn from the Sun is 998,740, and the minimum 834,160.

The straight line HBT is the equator with respect to the Earth; but IAS is the equator with respect to the Sun. Thus the straight line NBβ is the line of the solstices with respect to the Earth, and MAγ with respect to the Sun.

(continued next page)

(Plate V continued)

			At time of Ptolemy			Copernicus	
	Saturn	BGY	23		Scorpio	27 42	Sagittarius
	Jupiter	BFQ	11		Virgo	6 21	Libra
At	Mars	BEO	25 30	Cancer	27	Leo
apogee	Venus	BCK	25		Taurus	15 44	Gemini
	Mercury	BDV	10		Libra	28 30	Scorpio
	Sun	BAL	6 8	Cancer	6 40	Cancer

			At time of Ptolemy			Copernicus	
	Saturn	AGZ	23 40	Scorpio	28 3	Sagittarius
	Jupiter	AFR	17 31	Virgo	11 30	Libra
At	Mars	AEP	4 27	Leo	4 21	Virgo
aphelion	Venus	ACδ	4 39	Capricorn	19 48	Aquarius
	Mercury	ADX	29 42	Libra	13 40	Sagittarius
	Earth	ABα	6 8	Sagittarius	6 40	Capricorn

COMMENTARY NOTES

TITLE PAGE AND DEDICATORY LETTERS

1. The verse on the verso of the title page is an epigram attributed to Ptolemy with Kepler's translation of it. In the first edition, the translation only is given on the title page itself. On the sources for the epigram and other Latin translations, see F. Seck, Johannes Kepler als Dichter, *Internationales Kepler-Symposium* (see Introduction, note 4), pp. 427-450, especially p. 440 and pp. 449-450.

2. This is an allusion to the conical columns set in the ground at the limits of the Roman Circus to serve as turning posts in the chariot races.

3. Kepler follows the Vulgate. The references here are to the King James version (with the Vulgate locations in parentheses where these differ). Ps. 19 (18), v. 1; Ps. 8, v. 3 (4); Ps. 147 (146), vv. 4-5; Ps. 148, v. 1 and v. 3.

4. Ovid, *Fasti* 1, 297-298.

5. In the last two years of his life the Emperor Charles V retired to apartments adjacent to the monastery of San Geromino de Yuste in Estremadura, where he died on 21 September 1558. As curator of his clocks Charles invited to Estremadura Giovanni (Juanelo) Turriano of Cremona, who had accepted a commission to construct for him a clock with planetarium. This took three and a half years to construct and was still incomplete when Charles died. From Yuste Juanelo went on to Toledo where he achieved fame by his construction of the aqueduct. On Juanelo Turriano (Iannellus Turrianus Cremonensis) and his works, see Ambrosio de Morales, *Las Antigüedades de las ciudades de España,* f. 91r-f. 93v. (This work is published as part of Florian de Ocampo, *De la Corónica de España,* continued by Ambrosio de Morales, Alcalá de Henares, 1574-1586, vol. 3). See also William Stirling, *The Cloister Life of the Emperor Charles the Fifth,* 2nd edition London, 1853, pp. 60, 97-98 and 268-270.

6. These letters are printed in KGW 13, nos. 73, 69, 96 and 92 respectively. There is a facsimile and a German translation of Galileo's letter in Walther Gerlach and Martha List, *Johannes Kepler, Dokumente zu Leben und Werk,* Munich, 1971, pp. 70-73.

7. Seneca, *Naturales questiones,* vii, 31.

8. According to custom, Kepler expected to receive an honorarium for the dedication. This was granted (250 Gulden) in 1600, when he evidently found it useful in moving his family to Prague. (See KGW 19, p. 29).

ORIGINAL PREFACE TO THE READER

1. Kepler knew the edition of Euclid's *Elementa* by François de Fois, Comte de Candale (Paris, 1566 and 1578), which contains extensive commentaries on the five regular solids. He cites this edition in the *Mysterium cosmographicum,* chapter 13.

2. According to Kepler's testimony in this passage, Maestlin was openly committed to the Copernican system, teaching this system and discussing its advantages over that of Ptolemy in his lectures to students. In his *Epitome astronomiae* (Heidelberg, 1582; Tübingen, 1588, 1593, 1597, 1598, 1610 and 1614) Maestlin restricted his treatment to the Ptolemaic system, though in the later editions he added some remarks supporting the Copernican system. His position is made quite clear in his letter to Friedrich von Württemberg recommending the work of Kepler, where he explains that, while the familiar ancient hypothesis was easier

for beginners to understand, and therefore more suitable in an elementary text-book, all practitioners (*artifices*) agreed with the demonstrations of Copernicus. (KGW 13, p. 68. Cf. Introduction, p. 22.)

3. As we remarked in the introduction, this comment indicates that Kepler made his first comparison of the two systems before reading the *Narratio prima*. In Graz, however, he made use of both the *Narratio prima* and *De revolutionibus* itself.

4. Virgil, *Aeneid*, 4, 175.

5. The analogy is to be seen in terms of the Christian interpretation of the Platonic doctrine of participation—symbolized in the idea of the Book of Nature—whereby God is revealed through the creation. Kepler's promised amplification of the symbolism of the Trinity is given in the *Epitome astronomiae copernicanae* (KGW 7, p. 47, p. 51 and p. 258.). See also W. Petri, Die betrachtende Kreatur im trinitarischen Kosmos, in *Kepler Festschrift 1971* (see Introduction, note 4), pp. 64-98.

6. The inspiration for this hypothesis probably came from Plato's construction of the World-Soul in the *Timaeus,* 35B-36A, from the geometric series 1, 2, 4, 8 and 1, 3, 9, 27.

7. The idea of postulating invisible bodies in the search for what we may call archetypal causes (a term used by Kepler himself in the *Harmonice mundi* and the *Epitome astronomiae copernicanae*) had been introduced by the Pythagoreans. As related by Aristotle (*Metaphysics*, 986 a 10-15), since they considered 10 to be the essence of the numerical system, they asserted that 10 bodies must revolve in the heavens, "and there being only nine that are visible, they make the counter-earth [an invisible planet revolving round the central fire in opposition to the earth] the tenth."

8. See E. Rosen, *Three Copernican treatises*, New York, 1971, p. 147. As Kepler explained to Maestlin in his letter of 3 October 1595 (KGW 13, p. 34), the principles underlying the world's construction were to be sought in geometrical relations and not in pure numbers, whose properties were accidental.

9. This hypothesis may have been inspired by reading Regiomontanus, *De triangulis omnimodis*, where the sine function is introduced. For a facsimile of this work with English translation, see *Regiomontanus on triangles*, translated by Barnabas Hughes, Madison, 1967.

10. Here can be seen the beginnings of Kepler's physical theory in which the moving virtues of the individual planets will be replaced by a single moving virtue located in the sun.

11. Chance or Providence, in leading Kepler to illustrate the pattern of Jupiter-Saturn conjunctions in a certain way, simply provided the initial insight for the invention of the polyhedral hypothesis. As he explains, it was only when he recognized the need for a pattern of 3-dimensional figures (after unsuccessful trials with polygons) that he achieved his goal.

12. An allusion to Virgil, *Aeneid*, 10, 652.

13. Horace, *Ars poetica*, 388.

14. Cicero, *De amicitia* (*Laelius*), 23.

15. This is an allusion to Terence, *Heauton timorumenos*, 4, 3, 41.

16. See *Epitome*, Book 1, part 5 (KGW 7, pp. 80-100). The original disputation is not extant.

17. See *Astronomia nova* (KGW 3, pp. 22-24) and *Epitome*, Book 4, part 2, especially chapter 5 (KGW 7, pp. 312-316).

18. *Epitome,* Book 1, part 2 (KGW 7, p. 47 and p. 51) and Book 4, part 1 (KGW 7, p. 258).

19. Kepler explains the distinction between *numeri numerantes* (counting numbers) and *numeri numerati* (counted numbers) in *Harmonice mundi*, Book 5,

appendix (KGW 6, p. 370). The former are abstract numbers (whose properties are accidental), the latter concrete numbers or numbers embodied in real things; that is, for Kepler, numbers embodied in geometrical objects such as regular polygons and the regular and semi-regular solids.

CHAPTER I

1. In his report to the University, transmitted to Kepler by Hafenreffer (KGW 13, pp. 86-87), Maestlin recommended the addition of a preface explaining the Copernican system (KGW 13, pp. 84-86). Kepler responded by suggesting the addition of an extract from either Copernicus himself or Rheticus, but Hafenreffer did not favor this idea. By placing the new material in this chapter, Kepler reveals his acceptance of Hafenreffer's specific advice, given in a further letter (KGW 13, p. 90), to insert a brief description of the Copernican system, illustrated with diagrams, "after the preface to the reader."

2. Hafenreffer's letter to Kepler of 12 April 1598 (KGW 13, p. 203: 35-37) seems to indicate that Kepler had his approval to retain this brief statement concerning the reconciliation of the Copernican hypothesis with the Bible. On the Lutheran attitude to the Copernican hypothesis, see E. Rosen, Kepler and the Lutheran attitude towards Copernicanism, in *Johannes Kepler, Werk und Leistung*, Linz 1971, pp. 137-158 and R. S. Westman, The Melanchthon circle, Rheticus and the Wittenberg interpretation of the Copernican theory, *Isis*, 66 (1975), 165-193.

3. Later, in the preface to his *Rudolphine tables*, which effect an improvement of nearly two orders of magnitude in the prediction of the planetary positions, Kepler had to acknowledge that Reinhold's *Prutenic tables*, based on Copernicus, were no more accurate than the *Alfonsine tables*. See J. Kepler, Preface to the *Rudolphine tables*, translated by O. Gingerich and W. Walderman, *Quarterly Journal of the Royal Astronomical Society*, 13 (1972), 367.

4. See E. Rosen, *Three Copernican treatises*, pp. 136-153. Introducing his principal arguments in support of the Copernican system, Rheticus remarks, "there is something divine in the circumstance that a sure understanding of celestial phenomena must depend on the regular and uniform motions of the terrestrial globe alone."

5. The true diurnal appearances are demonstrated from opposite premises (namely the immobile and rotating earth) because the true cause of the appearances is a difference with respect to motion between the heavens and the earth. This difference (or relativity) is here the genus and the immobile and rotating earth two species of it. Properties inferred from the genus (that is, from the nature of a subject) are called by Aristotle *per se* or essential. Although the same properties may be inferred from a species, it is clearly the genus and not the species which is the cause or logical basis of these properties. On the *kat' auto* or *per se* rule, see Aristotle, *Posterior analytics*, 73 a 21-74 a 44. See also R. S. Westman, Kepler's theory of hypothesis and the realist dilemma, in *Internationales Kepler-Symposium* (see Introduction, note 4), pp. 32-33.

6. For Kepler, the quality of simplicity provides the ground for the physical truth of the Copernican hypothesis. Since nature loves simplicity, physical truth is to be sought in choosing what is simple or natural in preference to what is contrived or seemingly miraculous. For example, the Copernican hypothesis explains the agreement of the retrogressions with the position and apparent motion of the sun, whereas in the old hypotheses, such coincidences could only provoke astonishment.

7. For example, the polyhedral hypothesis was the *a priori* reason for the number, dimensions and arrangement of the planetary orbits.

8. For an extensive account of Maestlin's treatise on the comet of 1577, see C. Doris Hellman, *The comet of 1577: its place in the history of astronomy*, New York 1971, pp. 145-159 and p. 384. See also R. S. Westman, The comet and the cosmos: Kepler, Mästlin and the Copernican hypothesis, in J. Dobrzycki (editor), *The reception of Copernicus' heliocentric theory*, Dordrecht, 1972, pp. 7-30.

9. Here the term *prosthaphaeresis* means, in relation to the Ptolemaic system, the angle subtended at the earth by the epicycle of the planetary orbit. Interpreted in relation to the Copernican system, the *prosthaphaeresis*, in the case of the inferior planets, is the angle subtended at the earth by their orbits, and in the case of the superior planets, the angle subtended at the planet by the earth's orbit.

10. Copernicus's term *orbis magnus* for the earth's annual orbit came to be commonly used in the sixteenth and seventeenth centuries. See E. Rosen, *Three Copernican treatises*, pp. 16-17, note 45. See also KGW 7, p. 403.

11. This conical motion of the earth's axis about the axis of the ecliptic serves to neutralize the rotation in the opposite sense arising from the earth's annual motion; as envisaged by Copernicus, the earth's annual motion would, of itself, keep the earth's axis inclined at a fixed angle to the line joining the earth to the sun. A slight inequality in the two motions is postulated by Copernicus to explain the precession of the equinoxes (*De revolutionibus*, Book 3, chapter 1). In Medieval and Renaissance astronomy, this phenomenon had been described by Copernicus, as a motion of the eighth sphere. The *Alfonsine tables* combined two previous hypotheses; namely, a uniform motion in consequence, transmitted by a ninth sphere, and a trepidation or oscillation, arising from the motion of the eighth sphere itself. A tenth sphere was needed to transmit the daily rotation of the heavens. (See P. Duhem, *Le système du monde*, Paris 1913-1959, vol. 2, pp. 261-262.) On the significance of the conceptual distinction between a motion of the eighth sphere in consequence and a motion of the equinoctial points in precedence, see J. R. Ravetz, *Astronomy and cosmology in the achievement of Nicolaus Copernicus*, Warsaw, 1965, pp. 15-20.

12. These motions in the solsticial and equinoctial colures (great circles through the poles of the mean equator and the solsticial and equinoctial points respectively) were introduced by Copernicus (*De revolutionibus*, Book 3, chapter 3) to explain a supposed variation in the rate of precession of the equinoxes (for which observations seemed to offer some evidence). See the account of Rheticus (Rosen, *Three Copernican treatises*, pp. 153-162), Maestlin's commentary in his édition of the *Narratio prima* (KGW 1, pp. 111-112) and the modern analysis of K. P. Moesgaard, The 1717 Egyptian years and the Copernican theory of precession, *Centaurus*, 13 (1968), 120-138.

13. An allusion to the harmony of the three classes in the just state and the corresponding harmony of the three elements of the soul of the just man, in Plato's *Republic* (πολιτεία), 433A and 443D-E. Kepler was familiar with the writings on the ideal state of Campanella, More, Erasmus and especially Jean Bodin, whose *Les six livres de la république*, published in Paris in 1583, provided the starting point for his own Digressio politica, appended to Book 3 of the *Harmonice mundi*. In opposition to Bodin and Aristotle, who advocated a mean between a democracy and an aristocracy, represented by arithmetic and geometric series, Kepler proposed that the harmonious state, represented by his harmonic series (which was based neither on the arithmetic nor the geometric series), was different in kind from these extremes. In Kepler's ideal state, the citizens are inspired to cooperative activity under the influence of harmony. See A. Nitschke, Keplers Staats- und Rechtslehre, in *Internationales Kepler-Symposium* (see Introduction, note 4), pp. 409-424.

14. In *Astronomia nova*, chapter 21, Kepler explains how an equant-type theory, though false in its representation of the distances of the planet from the sun, may give longitudes within the limits of observational error. The reason is that the longitudes are more sensitive to errors in the position of the center of the eccentric than they are to errors in the position of the equant point. See C. Wilson, Kepler's derivation of the elliptical path, *Isis*, 59 (1968), 8-9. It was clearly his recognition of the possibility of representing longitudes accurately by this type of theory that led Kepler to persevere with the further development of his vicarious hypothesis even when he knew that it could not represent the true orbit. See O. Gingerich, Kepler's treatment of redundant observations, in *Internationales Kepler-Symposium* (see Introduction, note 4), pp. 307-314.

15. Kepler made this remark in chapter 14 of the *Astronomia nova* (KGW 3, p. 141) after having shown that the plane of the orbit of Mars passes through the sun. Copernicus, seeking to correct Ptolemy instead of accepting the natural truth, as Kepler put it, had taken the plane of the planetary orbit to pass through the center of the earth's orbit and in consequence had missed the discovery that the inclination of the orbit is constant.

16. *De mundi aetherei recentioribus phaenomenis*, Book 2, chapter 10 (Tycho Brahe, *Opera omnia*, ed. J. L. E. Dreyer, 1913-1929, vol. 4, pp. 259-367). See also C. D. Hellman, *The comet of 1577* (see note 8 above), pp. 337-338.

17. *De cometis libelli tres* (KGW 8, pp. 131-262).

18. See the letter of Tycho Brahe to Kepler, 1 April 1598 (KGW 13, pp. 187-200).

19. In order to have a fixed reference system, Kepler introduced the concept of the *via regia* or mean ecliptic, which he identified with the plane of the equator of the rotating sun. See *Astronomia nova*, chapter 68.

CHAPTER II

1. The primary source for Kepler's comparison of God with the curved and the created universe with the straight may be located in Nicholas of Cusa, *Complementum theologicum*, chapter 3. See D. Mahnke, *Unendliche Sphäre und Allmittelpunkt*, Halle 1937 (reprint Stuttgart-Bad Cannstatt 1966), p. 141. Kepler's own concrete geometrical image of the Trinity is a modification of the symbolism described by Cusanus in this work (chapter 6). In effect, Cusanus sees the center of the circle as symbolizing God the Father, the radius, "a principle issuing from a principle and therefore concerning the supreme equality of the source," as symbolizing God the Son, and the circumference, which is a "union or synthesis," as symbolizing the Third Person of the Trinity. The traditional abstract symbolism of unity, equality and synthesis, going back to St. Augustine, is described by Cusanus in his major work *De docta ignorantia*, Book 1, chapter 9, known to Kepler only through hearsay.

2. *Timaeus Platonis, sive de universitate, interpretibus M. Tullio Cicerone & Chalcidio, una cum eius docta explanatione*, Paris, 1563, pp. 19-20. Cf. *Timaeus*, 30A.

3. It follows that the plan of creation (which, as Kepler writes to Herwart, KGW 13, p. 309, God intended that we should discover by sharing in his thoughts) will possess qualities of beauty and simplicity reflecting the divine attributes of perfection.

4. In seeking to comprehend God's thoughts through human thoughts, certainty was unattainable, so that Kepler's *a priori* reasons (or archetypal causes) remained hypothetical.

5. *De caelo*, 286 b 10 - 287 a 5.

6. See Genesis, chapter 1, vv. 6-7.

7. Psalm 147 (146 in the Vulgate), v. 4: "He telleth the number of the stars; he calleth them by their names."

8. Kepler argues against the infinity of the universe in *De stella nova* (KGW 1, pp. 251-257) and *Epitome astronomiae copernicanae* (KGW 7, pp. 45-46). See also A. Koyré, *From the closed world to the infinite universe*, Baltimore, 1957, pp. 58-87.

9. Kepler's *a priori* reasons, as we have remarked, remained hypothetical. Writing to Herwart on 12 July 1600 (KGW 14, p. 130) Kepler states that "these *a priori* speculations must not contradict manifest experience but rather be in agreement with it."

10. Plato's association of the solids with the elements is given in *Timaeus*, 53C-56C, and also in Campanus de Novara's commentary on Euclid's *Elementa*, which Kepler cites in the *Mysterium cosmographicum*, chapter 13. From *De placitis philosophorum*, believed at that time to be authentic Plutarch, Kepler learnt that Plato had imitated Pythagoras in associating the solids with the elements. Kepler saw the doctrine of Pythagoras as an attempt to solve the mystery of the cosmos that necessarily failed because Pythagoras had no knowledge of the true, Copernican system.

11. In both the first and second editions, the heavens are assigned the icosahedron instead of the dodecahedron. Kepler correctly assigns the dodecahedron to the heavens in the *Harmonice mundi* (KGW 6, p. 79, diagram).

12. Here is an illustration of the part played by the empirical data in the construction of hypotheses. Writing to Fabricius on 4 July 1603, Kepler emphasizes that a hypothesis "is built upon and confirmed by observations" (KGW 14, p. 412). In the following chapters Kepler will establish with *a priori* reaons the order of the solids suggested by the observations.

13. In the first edition Kepler dedicated this diagram to the Duke, who had authorized the construction of a model in the form of a "Kredenzbecher." See Kepler's letter to the Duke of 17 February 1596 and the marginal note in the Duke's hand (KGW 13, pp. 50-51). This idea developed into a design for a planetarium but the technical difficulties proved to be too great for the craftsmen. See F. D. Prager, Kepler als Erfinder, in *Internationales Kepler-Symposium* (see Introduction, note 4), pp. 386-392. According to Maestlin (KGW 13, p. 151), Kepler's dedication of this key diagram to the Duke had deterred the theologians in Tübingen from expressing their criticism openly.

The dedication, "ILLUSTRISS: PRINCIPI, AC DÑO. DÑO. FRIDERICO, DUCI WIRTENBERGICO, ET TECCIO, COMITI MONTIS BELGARUM, ETC. CONSECRATA," was omitted in the second edition. Also the diagram itself was inverted laterally. (A facsimile of the diagram with the dedication, as it appeared in the first edition, is shown as a frontispiece).

14. These annotations were added to the first edition at the suggestion of Maestlin (see KGW 13, p. 85).

CHAPTER III

1. Simple because three faces is the minimum number needed to form a solid angle.

2. This geocentrism of importance (as we may call it) in the Copernican universe of Kepler may be compared with the heliocentrism of importance, which attached special significance to the median position of the sun between the earth and the fixed stars in the geocentric universe of Renaissance Platonism. See, for example, M. Ficino, *Théologie platonicienne*, translated by R. Marcel, Paris. 1964-1970, Book 18, chapter 3, p. 191.

Commentary Notes

CHAPTER V

1. The *Monobiblos* to which Simplicius refers is a lost work of Ptolemy (in one book) entitled *Peri Diastaseos* (*On Dimension*). Simplicius writes: "The admirable Ptolemy in his monobiblos *On Dimension* showed well that there are not more than three dimensions; for it is necessary for dimensions to be determinate, determinate dimensions are found along perpendicular straight lines, and not more than three mutually perpendicular lines can be found, two of them determining a plane and the third measuring depth; therefore, if another were added after the third dimension, it would be completely without measure and indeterminate." Simplicius, *In Aristotelis de caelo commentaria* (*Commentaria in Aristotelem Graeca*, vol. 7), edited by J. L. Heiberg, Berlin, 1894, p. 9, lines 21-27.

CHAPTER IX

1. By physics Kepler here means physical astrology, and the *vires naturales* of the planets are their astrological powers.
2. For example, a pentagonal section of the icosahedron is related to the vertex directly above its center in the same way as the leaves of a plant and its umbilicus, a projection standing in the middle.
3. See Virgil, *Aeneid*, 4, 569.
4. What Kepler seems to mean is that when the octahedron is rotated about an axis joining two opposite vertices, four successive edges (forming a square) move entirely in the plane through the center of the solid and perpendicular to the axis. The disposition of these edges therefore facilitates their smooth movement. By contrast, however other solids are rotated, all the edges are inclined (some one way and some another) to the plane through the center of the solid and perpendicular to the axis of rotation, so that they rotate awkwardly.
5. See Ptolemy, *Tetrabiblos*, Book 1, chapters 4-7 (Loeb edition, 1971). For editions of Ptolemy's *Harmonica*, see U. Klein, Johannes Keplers Bemühungen um die Harmonieschriften des Ptolemaios und Porphyrios, in *Johannes Kepler, Werk und Leistung*, Linz, 1971, pp. 51-60. Kepler himself made a Latin translation of Book 3 of Ptolemy's *Harmonica*, which was first published in the nineteenth century by C. Frisch, *Kepleri opera omnia*, Frankfurt and Erlangen, 1858-1871, vol 5, pp. 335-412. However, Kepler's notes on Ptolemy's *Harmonica* were published by him as an appendix to Book V of the *Harmonice mundi*.
6. On Kepler's astrology, see F. Hammer, Die Astrologie des Johannes Kepler, *Sudhoffs Archiv*, 55 (1971), 113-135 and G. Simon, Kepler's astrology: the direction of a reform, *Vistas in Astronomy*, 18 (1975), 439-448.

CHAPTER X

1. See notes for Original Preface to the Reader, note 19.

CHAPTER XI

1. Kepler means that the solid is laid out flat in such a way that the octahedron-square is opened out in a straight line.
2. Following Plato, Kepler assigns to mathematical forms an existence prior to sense objects. This view, which underlies his central theme of a divine harmony based on geometry, is expounded in detail in the *Harmonice mundi* (Book 4,

chapter 1), where he quotes a long extract from Proclus's commentary on Euclid's *Elementa* in support.

3. Aristotle, *De caelo*, 291 b 28 - 292 a 9.

4. See *Epitome astronomiae copernicanae*, Book 1, part 2. On Kepler's ideas concerning infinity see also W. Petri (*loc. cit.*, notes for Original Preface to the Reader, note 5), pp. 69-72, but note that, on p. 72, line 6, the first word should be 'endlich'. See also A. Koyré (*loc. cit.*, notes for chapter 2, note 8).

5. Aristotle, *De caelo*, 287 b 22-32.

6. Aristotle, *De caelo*, 284 b 6 - 286 a 2.

7. See notes for chapter 1, note 19.

CHAPTER XII

1. The aspects are the angular separations of planets on the celestial sphere corresponding to certain fractional parts of the circle. Traditional astrology, as represented by the teachings of Ptolemy in the *Tetrabiblos*, recognized five aspects, namely conjunction and opposition (0° and 180°), trine, quartile and sextile (120°, 90° and 60°). In his definitive account of the origin of the aspects, given in the *Harmonice mundi*, Book 4, chapter 5, Kepler added a number of others, notably the quintile and biquintile (72° and 144°), to the set of effective astrological aspects. See KGW 6, pp. 250-251.

2. Kepler here uses the nomenclature of the hexachordal system, which had been used by singers since the time of Guido d'Arezzo in the eleventh century to keep their tonal bearings when sight-reading. In this system, the vocal range, starting on the G shown in Kepler's diagrams, was represented by a set of overlapping hexachords, each consisting of six diatonic notes, with a single semitone (between the middle pair), called ut, re, mi, fa, sol, la. These names were taken from the opening syllables of six lines of a Latin hymn. The hexachords were classified as hard, natural or soft according to whether they contained B, no B or B flat. Thus the lowest hexachord, starting on G, is hard; the next, starting on C, is natural, and the third, starting on F, is soft. The note described by Kepler as F (fa, ut) is the F below middle C, which is fa of the second hexachord and ut of the third. Similarly C (sol, fa, ut) is middle C, which is sol, fa and ut respectively of the third, fourth and fifth hexachords.

Kepler extended the characterization of the thirds as 'hard' (*dura*) and 'soft' (*mollis*), to be found in the medieval hexachord system, also to the sixths. In the sixteenth century, Gioseffo Zarlino used the terms major and minor in relation to both intervals and these terms were adopted in Germany by Johannes Lippius. Although Kepler occasionally uses the terms major and minor, he generally describes thirds and sixths as hard and soft. In the *Harmonice mundi* (Book 3, chapter 5, KGW 6, p. 135), Kepler explains the origin of the terminology. The minor, he remarks, sounds softer and more soothing (*mollior et blandior*) to the ear, while the major sounds hard or harsh (*dura sive aspera*).

On the concept of tonality in Kepler's music theory (which is definitively set out in *Harmonice mundi*, Book 3), see M. Dickreiter, Dur und Moll in Keplers Musiktheorie, in Johannes Kepler, *Werk und Leistung*, Linz, 1971, pp. 41-50; M. Dickreiter, *Der Musiktheoretiker Johannes Kepler*, Bern and Munich, 1973, especially pp. 160-187; D. P. Walker, Kepler's celestial music, *Journal of the Warburg and Courtauld Institutes*, 30 (1967), 228-250.

3. Thus, a minor third added to a major sixth, or a major third added to a minor sixth, produces an octave; a minor third added to a major third produces a perfect fifth; a minor sixth added to a major sixth produces an octave plus a perfect fifth.

4. Each edge of the octahedron is a chord of a quadrant of a great circle of the circumscribing sphere.

Commentary Notes

5. By B flat Kepler means the minor third, the base note G being understood.

6. Consider the diagrams:

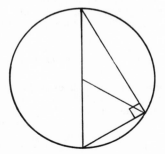 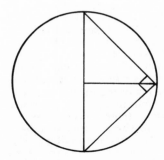

In the first, the right angle is constructed from a trine and a sextile; in the second, from two quartiles.

7. The imperfect harmony B flat again means the minor third with base note G.

8. Here Kepler implies that effective aspects only arise from regular polygons which tessellate in a plane.

9. Regiomontanus issued a prospectus of books he intended to edit and publish, among them the *Harmonica* of Ptolemy with the commentary of Porphyry, but his early death prevented him from carrying out this scheme. The prospectus is reproduced in E. Zinner, *Leben und Wirken des Johannes Müller von Königsberg gennant Regiomontanus*, Munich, 1938, Tafel 26. An imperfect Latin translation of Ptolemy's *Harmonica* by Antonius Gogavinus had been published in Venice in 1562. Preparatory to the writing of his *Harmonice mundi*, Kepler made a study of this Latin version and also of manuscript copies of the Greek texts of both Ptolemy and Porphyry, loaned to him by Herwart (KGW 14, p. 137 and KGW 15, p. 408). On Kepler's study of Ptolemy and Porphyry, see U. Klein (*loc. cit.*, notes for chapter 9, note 5), pp. 51-60. The texts of Ptolemy's *Harmonica* and Porphyry's commentary, together with a German translation of the *Harmonica*, have been published by I. Düring, *Göteborgs högskolas Arsskrift*, 36 (1930) No 1, 38 (1932) No 2 and 40 (1934) No 1.

10. Cardanus, *De rerum varietate*, Basel 1557, Book 17, chapter 98, p. 680.

11. In *De stella nova*, as Kepler remarks in a letter to Herwart (KGW 15, p. 453), he had rejected almost all of judicial astrology, except for the aspects. Kepler clarified his views on astrology in two German works, *Antwort auf Röslini Discurs* (KGW 4, pp. 99-144) and *Tertius Interveniens* (KGW 4, pp. 145-258), where he took a middle course between the astrologer Röslin and the anti-astrologer Feselius. Kepler's definitive account of the efficacy of the aspects and their relation to the musical harmonies is given in *Harmonice mundi*, Book 4, chapters 6 and 7. Whereas the division of the zodiac into twelve signs was arbitrary, so that these signs (and similarly, the twelve houses tied to the observer's horizon) could have no natural effects, the aspects or configurations reflected the divine harmony, and in addition to producing meteorological effects by a kind of resonance with the earth-soul, could also invoke an instinctive response from the soul of the newly-born infant, whose disposition was therefore influenced, to some extent, by the birth-constellation; this was the justification for horoscopes. For an account of Kepler's "Persönlichkeitslehre," see B. Sticker, Johannes Kepler - homo iste, in *Internationales Kepler-Symposium* (see Introduction, note 4), pp. 463-467.

12. See notes for chapter 11, note 2.

13. See *Epitome astronomiae copernicanae*, Book 4, part 2, chapter 6 (KGW 7, p. 316).

14. Kepler explains this terminology in the *Harmonice mundi*, Book 3, chapter 5. The Greeks called the octave διαπασῶν (*Diapason*). The expression *Diapason epidiapente* or διὰ πέντε ἐπι διὰ πασῶν means 'the octave over the fifth' or 'the fifth over the octave'.

15. As Kepler explains, the practical musician uses tempered intervals. See, for example, *Harmonice mundi*, Book 3, chapter 9. See also M. Dickreiter. *Der Musiktheoretiker Johannes Kepler,* Bern and Munich, 1973, p. 158.

16. Following Euclid, Kepler describes incommensurable lines whose squares are commensurable as 'expressible' (*effabiles*). The actual Greek term is ῥητὴ δυνάμει (potentially expressible). For the opposite (inexpressible) Euclid uses the term ἄλογοι , translated by Kepler as *'ineffabiles'*. In the *Harmonice mundi* (Book 1, definition 15) Kepler warns the reader of the ambiguity contained in the usual Latin translations of ἄλογοι as 'irrational'. In the cases of the equilateral triangle, square and regular hexagon, the ratios of the sides to the radius of the circumscribing circle are respectively √3:1, √2:1 and 1:1, so that the sides are 'expressible' (they can be constructed using the ruler-and-compasses construction of square roots). In the case of the hexagon, the sides are also, of course, commensurable.

17. That is, neither 1:11 nor 11:12 and neither 5:7 nor 7:12 represent consonances.

18. See *Harmonice mundi*, Book 4, chapter 5, proposition 14.

19. By correlating the aspects with harmonic intervals extending over several octaves, Kepler was able to show that the octile, trioctile, decile and tridecile also represented musical consonances. While the consonances within an octave are represented by the complement of the fraction of the circle defining the aspect, consonances greater than an octave are represented by the fraction itself. For example, the trioctile represents the minor sixth (5:8) and also the combined interval of a fourth and an octave (3:8). In the case of the octile itself, the ratio 7:8 does not represent a consonance, but the ratio 1:8 corresponds to three octaves, which is, of course, a consonance. See *Harmonice mundi*, Book 4, chapter 6 (KGW 6, p. 261).

20. KGW 6, p. 260.

21. For example, although regular dodecagons do not tessellate, a suitable combination of regular dodecagons and equilateral triangles will tessellate.

22. In his explanation of the efficacy of the aspects, Kepler made use of a reciprocal figure placed at the center, such that the angle between adjacent sides was equal to the angle subtended at the center of the circle by a side of the circumferential polygon. Thus the central figure was formed from the angle between the light rays marking the termini of the aspect. When the soul recognized the harmony of the light rays, it was at first concerned with the central figure. But as the efficacy of the aspect was primary, and the way in which it was perceived by the soul a secondary consideration, the circumferential polygon had greater importance. See *Harmonice mundi*, Book 4, chapter 5, proposition 6 (KGW 6, p. 247).

.23. The melodic intervals (*concinna*) are differences between pairs of neighboring consonant intervals smaller than an octave. See *Harmonice mundi*, Book 3, chapter 4 (KGW 6, p. 128). Kepler remarks that Ptolemy considered the thirds and sixths not to be consonances, but divided the interval between ut and fa into two intervals, each held to be melodic, whereas singers recognized three melodic intervals between these notes: ut, re, mi, fa. See *Harmonice mundi*, Book 3, introduction (KGW 6, p. 99).

24. Konrad Dasypodius appended the pseudo-Euclidean *Harmonica* to his edition of Euclid (Strasbourg, 1571). On the relation between Kepler and Dasypodius, see H. Balmer, Keplers Beziehungen zu Jost Bürgi und anderen Schweizern, in Johannes Kepler, *Werk und Leistung*, Linz, 1971, pp. 123-124.

CHAPTER XIII

1. Here Kepler expresses quite clearly that his *a priori* reasons were only probable and needed to be tested against the empircal data.

2. Euclid, Book 15, proposition 13. The Latin translation of Campanus de Novara from the Arabic was first printed in 1482. Kepler possessed the edition published in Basel in 1537, a reprint of the edition prepared by Jacques Lefèvre d'Etaples and published in Paris in 1514.

3. The expression *sinus totus* (whole sine) means sin 90°, here taken as 1000 units. The sine of any arc was the perpendicular from one extremity to the diameter through the other extremity. Taking the sine of the quadrant, that is sin 90° or the radius of the circle, as any convenient number of units, the sines of other arcs could be expressed in terms of these units, without the introduction of fractions.

4. See notes for Original Preface to the Reader, note 1.

5. In his calculation of the greatest distance of Mercury according to the polyhedral hypothesis, Kepler used the value 707, that is, the radius of the circle inscribed in the square formed by four middle edges of the octahedron, instead of the radius of the inscribed sphere.

CHAPTER XIV

1. The distances on which the comparison given in this table is based are measured from the center of the earth's orbit, except in the cases of the ratios Mars-earth and earth-Venus, where the greatest and least distances of the earth are measured from the sun, so that the earth's sphere is given a thickness in accordance with the eccentricity of the orbit. Thus the distances used in the calculation are those given by Kepler in the first column of the table on page 162, except in the case of the earth, where the distances used are those given in the second column. In the case of Saturn-Jupiter, there is a slight arithmetical error, for the Copernican data imply a ratio of 1000:631, which is nearer the value 1000:630 used by Kepler in his letter to Maestlin of 2 August 1595 (KGW 13, p. 28).

2. The values used here by Kepler for the radii of the inner and outer surfaces of the earth's sphere are those given in the fourth column of the table on page 162, so that the radius of the lunar orbit is taken to be 3' 36".

3. Here we see the idea of simplicity (again in the sense of a preference for the natural over the seemingly miraculous) used as a justification for the hypothesis. For the agreement of the hypothesis with the empirical data would be unthinkable were it not a consequence of God's plan of creation.

4. *Epitome astronomiae copernicanae* (KGW 7, p. 280). Archetypal reasons led Kepler to equate the ratio of the distances of the sun and moon from the earth to the ratio of the lunar distance to the radius of the earth. Observational techniques had not thus far permitted a more accurate determination.

Commentary Notes

CHAPTER XV

1. Kepler here took the step which converted the Copernican system into a truly heliocentric system. Again the principle of simplicity provided justification for the step. For Kepler's innovation brought the earth's orbit into line with those of the other planets.

2. See Introduction, pp. 19-20.

3. Plate V. The basis of these figures, prepared by Maestlin, is the representation of planetary motion called by Copernicus eccentric-on-eccentric (see *De revolutionibus*, Book 5, chapter 4). The point A represents the sun, B the center of the earth's orbit and the lines BC, BD, BE, BF and BG the eccentricities of the respective planetary orbits as defined by Copernicus; that is, they are the eccentricities of the deferent in the epicycle-on-eccentric representation or three quarters of the eccentricities of the equant in the simple eccentric representation. The directions of these lines in the two diagrams show the positions of the lines of apsides in the times of Ptolemy and Copernicus respectively. Despite his reference to *dextrae figurae* in introducing his values for the planetary distances, Maestlin calculates these on the basis of the data for the time of Ptolemy. The diagrams show, for each planet, the small eccentric, a circle of radius ⅓ ϵ, where ϵ is the eccentricity as defined above; it may be noted that, if 2e represents the eccentricity of the equant, then ϵ = 3/2e. The planet moves on a large eccentric of radius a (not shown in the diagram) whose center moves on the small eccentric in such a way that (except in the case of Mercury, for which a more complicated combination of circles is needed) the greatest and least distances of the planet from the center of the earth's orbit are respectively a + ⅔ ϵ and a − ⅔ ϵ. (For a detailed description of the various representations, see the Appendix.)

4. In Kepler's table the numbers are given in sexagesimal form, the radius of the earth's eccentric being taken as 1°. There are mistakes of various kinds, and in particular, discrepancies between the distances given by Kepler in the first column (which may be compared with those he gave in his first detailed communication of the polyhedral hypothesis to Maestlin, KGW 13, p. 44) and those implicit in the data accompanying Maestlin's diagrams. The principal reason for these differences is that, whereas Maestlin derived his data by new calculations from the *Prutenic tables* (see KGW 13, p. 65), Kepler simply accepted the values given by Copernicus. In the first column (giving the distances from the center of the earth's eccentric), the values for Saturn, Jupiter and Mars are taken from *De revolutionibus*, Book 5, chapters 9, 14 and 19 respectively, though in the case of Mars, Kepler gave the greatest distance correctly as 1° 39′ 56″, correcting a misprint in Copernicus, where this distance is given as 1° 38′ 57″. In the case of Venus, although Maestlin's value of the mean distance agrees with that of Copernicus, the consequent values of 0° 44′ 25″ and 0° 41′ 55″ for the greatest and least distances from the center of the earth's orbit differ from those given by Kepler, because Kepler has inadvertently calculated a±4/3 ϵ instead of a±2/3 ϵ. Again, in the case of Mercury, Kepler's values differ appreciably from those of Maestlin. Both Maestlin and Kepler calculate the distances according to the formula (which applies only to Mercury):

(radius of large eccentric) ± (ϵ + radius of small eccentric).

Using Maestlin's values, this gives 35730 ± (7345½ + 2114½), where the radius of the earth's eccentric is taken as 100,000. This is equivalent, in the sexagesimal notation of Kepler's table, to 0° 27′ 7″ and 0° 15′ 46″ for the greatest and least distances respectively. (Cf. KGW 1, p. 145). The Mercury theory is complicated, however, by a variation in the radius of the eccentric according to the position of the earth in relation to the line of apsides (see *De revolutionibus*, Book 5, chapters 25 and 27 and also the Appendix). Whereas Maestlin used the minimum

244

value of 35730 for the radius of the eccentric, Kepler used the maximum value of 39530, and this accounts for the differences. In the case of the sun, Kepler's values for the greatest and least distances are simply the values given by Copernicus (*De revolutionibus*, Book 3, chapter 21. Cf. KGW 1, p. 92) for the eccentricities of the earth's orbit in the time of Ptolemy and in the sixteenth century. Copernicus had supposed that the eccentricity oscillated in the same period as the obliquity of the ecliptic. (See notes for chapter 1, note 12.)

In the second column Kepler gives the greatest and least distances of the planets from the sun. Here he relies on the values of Maestlin, though there is a slight discrepancy in the case of Venus (Maestlin's value for the greatest distance being 0° 44′ 32″) and a larger difference in the case of Mercury. For Mercury Kepler used the values originally communicated to him by Maestlin but revised without Kepler's knowledge during the printing of the *Mysterium cosmographicum* (see Introduction, p. 20).

The method of calculation of the distances from the sun is described by Maestlin in a letter of 11 April 1596, where he discusses the distances of Mercury (KGW 13, p. 78). The lines AC, AD,...are calculated from the triangles ACB, ADB,...using the known values of BC, BD,...and the angles at B (derived from the known directions of the apogees). Clearly, the angles at A (or the directions of the aphelions) can also be calculated. In the letter, Maestlin calculates the distances of Mercury on the basis that, in the time of Ptolemy, the apogee of Mercury was in 15° 30′ of Libra (computed from the *Prutenic tables*). However, the distances given in his table are based on Ptolemy's own position of 10° of Libra for the apogee of Mercury. (See notes for chapter 19, note 6.) Indeed, all the distances in Maestlin's tables (accompanying the diagrams) are based on the data for the time of Ptolemy. Apart from the case of Jupiter, where the eccentricity in relation to the sun is nearer 36600 than the value of 36656 given by Maestlin, his solutions of the triangles (giving the eccentricities in relation to the sun and the directions of the aphelions) are in almost perfect agreement with the data. There are, however, major errors in the calculation of the distances of Venus and Saturn. In the case of Venus, the greatest and least distances from the sun are taken to be

(greatest distance from center of earth's orbit) ± (AC − radius of small eccentric)

instead of

(mean distance from center of earth's orbit) ± (AC − radius of small eccentric)

The greatest distance should be 72152 or 0° 43′ 17″ and the least distance 71708 or 0° 43′ 1″. In the case of Saturn, the radius of the small eccentric has been neglected. When this is taken into account, the greatest and least distances become 9° 43′ 36″ and 8° 36′ 7″ respectively. For the greatest and least distances of the earth from the sun, Kepler takes the values accompanying the diagrams, corresponding to an eccentricity of 4170. There is a misprint in the table of apogees and aphelions. The direction of BAL in the time of Ptolemy should be in Gemini (not Cancer). (Cf. KGW 13, p. 78.) In calculating the directions of aphelion in the time of Copernicus (of which no application is made), Maestlin took into account the changes in the eccentricities that had been noted by Copernicus.

In the third column, Kepler calculates the distances according to the polyhedral hypothesis. He starts with the earth's sphere, whose inner and outer surfaces have radii equal to the least and greatest distances of the earth given in the second column. The outer surface of the earth's sphere is taken as the inscribed sphere of the

dodecahedron and the circumscribed sphere of this dodecahedron then becomes the inner surface of the sphere of Mars. The radius of the outer sphere of Mars (that is, the theoretical greatest distance of Mars) is then calculated from the known radius of the inner surface, using the ratio of distances of Mars given in the second column. This process is continued upwards to Saturn and downwards to Mercury. The values of the fourth column are calculated in the same way as those of the third, except that the thickness of the earth's sphere is increased to include the moon's orbit.

Kepler's results in the third and fourth columns reflect the errors in those of the second. A significant factor contributing to the errors was no doubt the fact that Kepler was not himself able to check the manuscript being prepared for the printer. This was undertaken by Maestlin in Tübingen, who therefore had responsibility for the final corrections. Kepler made no attempt to correct these errors in the second edition. (See Introduction, p. 29).

The true relation between the polyhedral hypothesis and the Copernican data may be seen in Table I of the Introduction, where the distances given by Kepler in the second, third and fourth columns of his table are compared with the corrected distances (in parentheses).

5. Kepler's values are compared with the corrected angles (in parentheses) in Table II of the Introduction. The angle 29° 19′ given by Kepler for Mercury does not correspond to his value for the greatest distance of this planet from the sun. For agreement with this distance (0° 29′ 19″), the angle should be 29° 15′. It seems likely that Kepler has mistakenly written the 'distance' in place of the angle.

6. *Astronomia nova*, chapter 6.

CHAPTER XVI

1. Kepler's reason for the existence of the lunar sphere would seem less convincing after the discovery of the satellites of Jupiter, though he makes no comment on this point in the second edition.

2. Kepler makes clear that the polyhedra and spheres are purely geometrical concepts without material reality.

3. Diogenes Laertius, *De vitis philosophorum*, ii, 8. For Kepler's quotation from Plutarch's *De facie in orbe lunae*, xvii, see KGW 2, p. 203.

4. *Kepler's conversation with Galileo's sidereal messenger*, translated by E. Rosen, New York, 1965.

5. See *Astronomia nova*, introduction and chapters 33-34.

6. Kepler here uses the expression *inertia materiae* to denote the inclination of matter to remain at rest. Although the concept is embodied by Kepler in the axiom "Every corporeal substance, as corporeal, will rest in any place in which it is found isolated, outside the reach of bodies of the same kind," printed in the introduction to the *Astronomia nova*, it is in the *Epitome astronomiae copernicanae*, Book 1, part 5 and Book 4, parts 2 and 3, that he introduces the term 'inertia'. For example, Kepler attributes to the planets a "natural and material resistance or inertia to leaving a place, once occupied." (KGW 7, p. 333. Cf. p. 339). See I. B. Cohen, Dynamics, the key to the new science of the seventeenth century, *Acta historiae rerum naturalium necnon technicarum*, Special Issue No. 3 (Prague, 1967), pp. 83-100. Cf. I. B. Cohen, Kepler's century, *Vistas in astronomy*, 18 (1975), 3-36, especially 21-22.

7. See, for example, *Epitome astronomiae copernicanae*, Book 4, (KGW 7, p. 332).

8. See chapter 1, appendix (KGW 2, pp. 39-42).

CHAPTER XVII

1. The orbit of Mercury has a large eccentricity compared with those of the other planets. But Kepler will later admit (see his note 1 in the second edition) that the archetypal reason for this peculiarity of Mercury is not to be found in the octahedron.

2. Kepler's diagram is visually misleading. The figure XQHS represents the dotted rhombus in the diagram of the octahedron shown below.

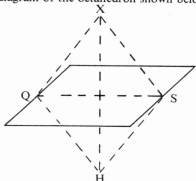

3. For a description of the Mercury theory of Copernicus see the Appendix. What Kepler here describes as a variation in the radius of the large eccentric is represented by Copernicus as an oscillation along the diameter of an epicycle.

4. The values for the greatest and least distances of Mercury used here by Kepler are those communicated to him by Maestlin in the letter of 11 April 1596 (KGW 13, p. 37, lines 37 and 43) and not the revised values accompanying Maestlin's diagrams. (Cf. notes for chapter 15, note 4 and notes for chapter 19, note 6). By taking the values 387 and 474 for the radii of the inscribed sphere and the circle inscribed in the octahedron-square, Kepler implies that the radius of the circumscribed sphere (in other words, the least distance of Venus) is 670, which is the value according to the polyhedral hypothesis (and Kepler's calculation) when the moon's orbit is included in the earth's sphere (p. 162, line 11, column 4).

5. KGW 7, p. 435.

CHAPTER XVIII

1. Quotation from Horace, *Epistles*, I, i, 32.

2. Shortly after his first visit to Tycho Brahe, Kepler explains to Herwart that one of the principal aims of his visit was to obtain more accurate values of the eccentricities, in order to confirm the polyhedral hypothesis (KGW 14, p. 128). Kepler's chief difficulty at this time was that he did not know the archetypal causes of the eccentricities and their differences.

3. Kepler here paraphrases the text of Copernicus.

4. Bernhard Walther, patron of Regiomontanus, was a distinguished observer. His observations were first published in 1544.

5. The whole preface, of which only an extract is given here, may be found in L. Prowe, *Nicolaus Coppernicus*, Berlin, 1883-1884, vol. 2, pp. 387-396.

6. The third, fourth and fifth minutes are the corresponding sexagesimal fractions of a degree. For example, in his *Prutenic tables*, Reinhold gives the distance

of the equinox from the first point of Aries at the date of birth of his patron, the Duke of Prussia, as part 26 scrup. 59 1ª 28 2ª 47 3ª, where 1ª, 2ª, 3ª denote scrupla sexagesima prima, secunda, tertia. *Prutenicae tabulae coelestium motuum*, Wittenberg, 1585, f. 43a.

CHAPTER XIX

1. *Ephemerides novae ab annos 1577 ad annum 1590*, Tübingen, 1580.

2. *Disputatio de eclipsibus solis et lunae*, Tübingen, 1596, p. 20. Maestlin writes: "If the observations which we have made with the radius astronomicus are to be taken as dependable, we often find that the distance of Venus when higher above the horizon from the Sun lying near the horizon is noticeably smaller than if on the same day the distance of the same planet were taken from the Sun when higher in the sky and more clear from vapors. Therefore the height of the Sun has appeared through the vapors to be greater than true. Hence that the Sun itself and other stars similarly can appear to be above the horizon when they are still below it, we do not hold to be impossible, but conclude to be certainly the case."

3. The principal reason for the difference in the case of Venus is Kepler's mistake in calculating the distance of the planet. (See notes for chapter 15, note 4.)

4. Using the correct distances, the opposite is in fact the case, so that the fit is better when the moon's orbit is included in the earth's sphere. (See Introduction, p. 26).

5. This is an extract from Maestlin's letter of 11 April 1596 (KGW 13, pp. 77-79), though edited by Maestlin for inclusion in the *Mysterium cosmographicum*, so that there are differences from the original letter.

6. We have already referred to this difference (see notes for chapter 15, note 4). Originally, Maestlin calculated the distances of Mercury on the basis of the *Prutenic tables*, but as he indicates here (though not in the original version of the letter, which was the one used by Kepler), he eventually decided to base his distances of Mercury on the observations of Ptolemy rather than the *Prutenic tables*, and this is what he did in the calculation of the values given in the *Mysterium cosmographicum*.

7. Rheticus had argued in the *Narratio prima* (See Rosen, *Three Copernican treatises*, p. 161) that, in the cases of Jupiter and Mercury, the apogee of the planet was about a quadrant from the apogee of the sun (see Maestlin's diagrams), so that the change in the distance of the earth from the sun (as it moves from apogee to perigee) produces no observable change in the eccentricity of the planet. In the cases of Mars and Venus, however, the centers of the deferents were suitably placed to reveal the change in the eccentricity. Copernicus found that the eccentricity of Mars and Venus had decreased on account of the approach of the center of the earth's orbit towards the sun. (*De revolutionibus*, Book 5, chapters 15, 16 and 22). Maestlin's objection to the extension by Rheticus of his argument concerning Jupiter to the case of Mercury is based on the fact that the apogee of Mercury is really considerably more than a quadrant from the apogee of the sun. Again, this is clear from Maestlin's diagrams.

8. Since the eccentricities in the time of Copernicus were clearly in doubt, Maestlin considered that the data did not exist that would permit a test of the polyhedral hypothesis for the contemporary positions of the planets. This explains why he restricted his own calculations of distances to the configurations in the time of Ptolemy.

9. This is an allusion to the remark of Rheticus (p. 184, line 24) that Copernicus had congratulated himself when he came to within 10′ of the true positions.

Commentary Notes

CHAPTER XX

1. Aristotle, *De caelo*, 291 a 33.

2. The periodic time of Venus is 7½ months, not 8½ months, as Kepler takes it to be.

3. The errors in this table, arising from Kepler's assumption of wrong values for the mean distances of Jupiter and Mercury, do not significantly distort the general pattern.

4. Except in the case of Mercury, the distances in this table are in agreement with the mean distances implicit in column 2 of Kepler's table (p. 162) and hence with those implicit in Maestlin's values of the greatest and least distances (Plate V). In the case of Mercury, as already noted (notes for chapter 15, note 4), there is a slight discrepancy between the mean values given by Maestlin and Kepler. In the present table, the mean distance of Mercury is evidently taken to be 36000, which is approximately the mean of the values 48850 and 23110 given by Maestlin in his letter of 11 April 1596. (See notes for chapter 17, note 3).

5. Quoting a poem of Giovanni Gioviano Pontano, Rheticus describes the sun as the governor of nature. (See Rosen, *Three Copernican treatises*, p. 143.)

6. At this time Kepler evidently believed that the intensity of light weakened in proportion to distance from the source (he speaks of light spreading out in a circle, not a sphere) and he concluded that the effect of the moving soul in the sun weakened in the same way. When he discovered the inverse-square law for the intensity of light (*Astronomiae pars optica*, chapter 1, prop. 9, KGW 2, p. 22), Kepler was able to retain the inverse-distance law for the moving force in the sun, because a force spreading out in the plane of the ecliptic sufficed to explain the motion of the planets.

7. This means that, if r_1, r_2 are the distances of the two planets (with $r_2 > r_1$) and T_1, T_2 are the corresponding periodic times, then $\dfrac{T_2 - T_1}{T_1} = 2 \dfrac{r_2 - r_1}{r_1}$. In this form, the relation reveals quite clearly the conceptual basis.

8. Here Kepler introduces an algebraic transformation of the previous formula, which now becomes $\dfrac{T_1 + \frac{1}{2}(T_2 - T_1)}{T_1} = \dfrac{r_2}{r_1}$, so that the ratio of the distances can be calculated directly, as in the example which Kepler describes.

9. Kepler's second method of calculation is in fact algebraically equivalent to the first. In his original letter to Maestlin, Kepler used distances taken from Maestlin's lectures and he used the transformed formula as the basis for his calculation (KGW 13, p. 38).

10. *Astronomia nova*, chapter 33. On Kepler's ideas concerning the souls of the spheres, see H. A. Wolfson, The Problem of the souls of the spheres from the Byzantine commentaries on Aristotle through the Arabs and St. Thomas to Kepler, *Dumbarton Oaks papers*, No. 16, Washington, 1962, pp. 65-93, especially pp. 90-93.

11. On Kepler's physical theory of planetary motion, see F. Krafft, Johannes Keplers Beitrag zur Himmelsphysik, in *Internationales Kepler-Symposium* (see Introduction, note 4), pp. 95-139.

12. But in the meantime Kepler had discovered the inverse-square law for the intensity of light. (See note 6 above).

13. In the *Astronomia nova*, chapter 39, Kepler recognized that the effects of the length of the path and the strength of the solar force would combine to give periodic times in proportion to the squares of the distances from the sun. As he explains in this note, the correction may be effected by replacing the arithmetic

mean $\frac{1}{2}(T_1 + T_2)$ in his previous formula $\frac{r_2}{r_1} = \frac{T_1 + \frac{1}{2}(T_2 - T_1)}{T_1}$, or equivalently

$\frac{r_2}{r_1} = \frac{\frac{1}{2}(T_1 + T_2)}{T_1}$ by the geometric mean $\sqrt{}(T_1 T_2)$, so that the formula becomes

$\frac{r_2^2}{r_1^2} = \frac{T_2}{T_1}$. Kepler discovered his third or harmonic law $\frac{r_2^{3/2}}{r_1^{3/2}} = \frac{T_2}{T_1}$ on 15 May 1618. He announced it in the *Harmonice mundi*, Book 5, chapter 3 and gave a physical explanation in the *Epitome astronomiae copernicanae* (KGW 7, p. 306). Two new factors are introduced—the resistance arising from the bulk of the planet and the capacity of the planet to assimilate the solar force—which combine with the effects of the length of the path and the weakening of the solar force, to produce the harmonic law. Kepler took the bulk or quantity of matter proportional to $\sqrt{r}$ and the volume (measuring, on the analogy of a water-mill, the capacity to assimilate the solar force) proportional to r. While there was some observational evidence for the second relation, Kepler had to rely on archetypal causes for the first (KGW 7, pp. 283-284). On Kepler's harmonic law, see O. Gingerich, The origins of Kepler's third law, *Vistas in astronomy*, 18 (1975), 600 and R. Haase, Marginalien zum 3. Keplerschen Gesetz, *Kepler Festschrift 1971*, Regensburg, 1971, pp. 159-165.

14. *Harmonice mundi*, Book 3, Digressio politica (KGW 6, p. 188). (Cf. notes to chapter 1, note 13.) See also A. Nitschke, Keplers Staats—und Rechtslehre, in *Internationales Kepler-Symposium* (see Introduction, note 4), pp. 409-424.

15. Napier's tables are tables of logarithms of natural sines and therefore needed to be used in conjunction with a table of sines.

CHAPTER XXI

1. The value 559 is obtained by adding half the eccentricity to the mean distance 500 (see p. 174, lines 15-25), taking the eccentricity as half the difference between the greatest and least distances of Mercury given in column 1 of the table on p. 162. The two mean distances 500 and 559 correspond roughly to the maximum and minimum radii of the large eccentric of Mercury. (See notes for chapter 17, note 3.)

2. The distances in the first column are in agreement with those given by Kepler in the second column of his table on p. 162, except in the case of Mercury, where he uses the values communicated by Maestlin in the letter of 11 April 1596 (see notes for chapter 20, note 4). In the second column, Kepler calculates the mean distances in accordance with his formula relating distances and periodic times (see notes for chapter 20, note 7). In the last column, Kepler compares the ratios of the distances of neighboring planets with those predicted by the polyhedral hypothesis, seeking to show the polyhedral hypothesis in the best light by using the mean distances (that is, neglecting the thickness of the spheres) whenever these provide a closer fit than the extreme distances. Starting with the mean distance of Saturn according to the formula relating distances and periodic times, this distance is reduced in the ratio 577 to 1000, the relation between the inscribed and circumscribed spheres of the cube, to obtain the value 5290, which is found to correspond approximately to the mean distance of Jupiter. While the method of comparison in the case of the superior planets is fairly clear, the treatment of the inferior planets seems confused, and there is in fact an error, evidently arising out of Maestlin's failure to comprehend Kepler's intention (KGW 13, p. 109). As Kepler explains to Maestlin (KGW 13, p. 117), the comparison of the distances of Venus and Mercury starts with the mean

distance of Mercury according to the formula relating distances and periodic times. Then this distance 429 is increased in the ratio 1000 to 577, the relation between the inscribed and circumscribed spheres of the octahedron, to obtain the value 741, which is found to correspond to the greatest distance of Venus according to the Copernican data.

3. The statements concerning the outer solids (and also the tetrahedron between Mars and Jupiter) are directly supported by the figures in the right hand column. The statements concerning the inner solids should probably be interpreted as follows. Although the figures show that these solids could lie between the extreme distances according to the Copernican data, the spaces between the earth and the two adjacent planets would be smaller according to the (more reliable) distances computed from the motions, but the reduction would be less than the earth's eccentricity, so that there would still be room for the solids between the mean distance of the earth and the least distance of Mars on the one hand, and between the mean distance of the earth and the greatest distance of Venus on the other. (Cf. Introduction, pp. 28-29 and Table III.)

4. *Harmonice mundi,* Book 5, chapter 4 (KWG 6, p. 309).

CHAPTER XXII

1. See Maestlin's letter of 27 February 1596 (KGW 13, pp. 54-55). The epicycle-on-eccentric representation is a geometrically equivalent transformation of the eccentric-on-eccentric representation illustrated in Maestlin's diagrams. As is evident from the diagram of the epicycle-on-eccentric representation (see Appendix, fig. 4), the thickness of the sphere needed to accommodate the epicycle (assumed real) is twice as great as the thickness that would suffice to allow the variation in the distance of the planet, and hence twice the thickness that would be required in the eccentric-on-eccentric representation. (Cf. Introduction, p. 20.)

2. Maestlin modified Kepler's text in this passage to remove an error that Kepler had overlooked. Explaining the point in a letter to Kepler, Maestlin (KGW 13, pp. 109-111) quotes Kepler's original, where the point C, in the Copernican representation, is wrongly identified with the center of the Ptolemaic equant. In fact, AC = three-quarters of the eccentricity of the equant.

3. Kepler's reasoning, in which the two causes of difference in the periodic times of separate planets — namely, the length of the path and the weakening of the solar force in proportion to the distance from the sun — are applied to the motion in a single orbit, may be interpreted as follows. Taking r to be the radius of the eccentric EFGH, the distance of the planet from the sun A when in apogee is $r + e$, where $e = AB$ (that is, half the eccentricity of the equant). In accordance with Kepler's formula, two separate planets, moving at distances r and $r + e$ from the sun respectively, would have periodic times T_1 and T_2 given by

$$\frac{r+e}{r} = \frac{T_1 + \frac{1}{2}(T_2 - T_1)}{T_1}, \text{ or more simply, } \frac{r+2e}{r} = \frac{T_2}{T_1}.$$

Since the mean angular velocities ω_1 and ω_2 would be inversely proportional to the periodic times, we may write $\frac{r+2e}{r} = \frac{\omega_1}{\omega_2}$. Now interpreting ω_1 and ω_2 as the angular velocities, about the sun A, of a single planet in its mean distance and apogee respectively, Kepler infers that the planet moves in its path EFGH as if it were moving uniformly in the equant IKLM. This conclusion is a generalization to the whole orbit of a result that has been established only in the neighborhood of the apsides.

4. By the whole eccentricity, Kepler means AC ($= 3/2AB$), the eccentricity of the deferent in the Copernican epicycle-on-eccentric representation. In this

representation (fig. 1), the planet P moves uniformly on the epicycle, while the epicycle center moves uniformly on the deferent, center C. The angular velocity in the epicycle (relative to the radius vector of the deferent) is equal to the angular velocity of the epicycle center moving in the deferent. In the geometrically equivalent Copernican eccentric-on-eccentric representation (fig. 2), the planet P moves uniformly on a large eccentric whose center moves uniformly on a small eccentric with center C and radius BC, so that the center of the eccentric is at B when the planet is in the apsides. In both representations, the true anomaly $v = a - 2e \sin a + e^2 \sin 2a$, where e = AB and the radii of the deferent (fig. 1) and eccentric (fig. 2) are taken as 1. It follows that, to a first approximation, the angular velocity of the planet about A when in apogee is $(1 - 2e)\,\omega$, where ω is the mean angular velocity. This is clearly equivalent to an angular velocity of ω about the point D, the center of the Ptolemaic equant, since AD = 2e. As in the case of the Ptolemaic representation, Kepler has only verified that his physical theory is consistent with the Copernican representation in the neighborhood of the apsides.

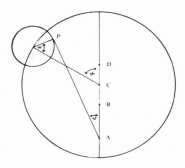

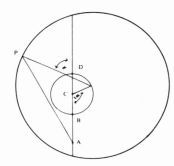

Fig. 1   Fig. 2

It should be noted that, in these diagrams, the letters A, B, C and D represent the similarly designated points in Kepler's own diagram (p. 216).

5. The term *nutus* employed here by Kepler, is the Latin equivalent of Aristotle's ῥοπή , treated extensively by Simplicius, for whom it meant the endeavor of a body to remain in its natural place, or to return to this place when displaced from it. By the sixteenth century, however, *nutus* came also to be identified with *impetus*. For example, in the commentary of Henri de Monanthueil on the *Questiones mechanicae* (Monantholius, *Aristotelis mechanica,* Paris, 1599, p. 108), the term is used in connection with any endeavor, whether a natural inclination or an impetus in the sense of Philoponus, or even a combination of the two. A *nutus* arising either from an external force or an internal volition is described by

Monantholius as *non naturalis,* while a *nutus* inherent in the nature of a body is described as *naturalis.* Here Kepler uses the term figuratively to mean *impetus* in the sense of a divinely inspired volition. See H. M. Nobis, Ropé und Nutus in Keplers Astronomie, *Kepler Festschrift 1971* (see Introduction, note 4), pp. 244-265.

6. Kepler's hypothesis of the inverse-distance relation for the solar force pushing a planet along its orbit was the first step in the path that eventually led him to the area law. See E. J. Aiton, Kepler's second law of planetary motion, *Isis,* 60 (1969), 75-90.

7. *Astronomia nova,* chapter 28. See C. Wilson, Kepler's derivation of the elliptical path, *Isis,* 59 (1968), 5-7.

8. *Harmonice mundi,* Book 4, chapter 7 (KGW 6, p. 264).

CHAPTER XXIII

1. The 'upper apsis' is the 'apogee'.

2. The terms 'head' and 'tail' refer respectively to the ascending and descending nodes and are derived from the mythological explanation of eclipses, found with variations in ancient India, China and Islam, according to which a dragon, with its head and tail twisted round the nodes, swallowed the sun and moon whenever the opportunity occurred. It was appropriate that the moon should have been placed initially at its greatest distance from the nodes, so that there would be no danger of an eclipse during the first night. For a description of the sources and variations of this mythological explanation of eclipses, see W. Hartner, *Oriens, occidens,* Hildesheim, 1968, pp. 268-286. The terms *caput Draconis* and *cauda Draconis* for the ascending and descending nodes (ἀναβιβάζων and καταβιβάζων σύνδεσμος) are defined in the *Prutenic tables* (see notes for chapter 18, note 6), f. 38b.

3. The Platonic Year (or World Year) is described in the *Timaeus,* 39D, as the interval which elapses before all the planets return simultaneously to their starting points. See also the commentary on this passage by Proclus, *Commentaire sur le Timée,* translated by A. J. Festugière, Paris, 1966-1968, vol. 4, pp. 118-122.

4. Copernicus, *De revolutionibus,* Book 1, chapter 10 and Pliny, *Historia naturalis,* ii, 1.

5. The intervals subsidiary to melodic intervals are the differences of melodic intervals (as the melodic intervals themselves are differences of consonances). Although not exactly melodic, these intervals — diesis, comma and limma — find application in melodic modulation. See *Harmonice mundi,* Book 3, chapter 4 (KGW 6, p. 132-133). See also M. Dickreiter (see notes for chapter 12 , note 15), p. 153.

6. This hymn is a paraphrase of Psalm 8 into which Kepler has worked a reference to the five Platonic solids. See F. Seck, Johannes Kepler als Dichter, in *Internationales Kepler-Symposium* (see Introduction, note 4), pp. 427-450, especially p. 431 and p. 443.

APPENDIX

PTOLEMAIC AND COPERNICAN GEOMETRICAL
REPRESENTATIONS MENTIONED OR USED
BY KEPLER AND MAESTLIN

Ptolemy found that a simple eccentric sufficed to represent the apparent motion of the sun about the earth. For the representation of the motions of the superior planets he introduced the device known as the equant. Copernicus rejected the equant as inconsistent with the principles of astronomy and found that the motions of all the planets except Mercury could be represented by two geometrically equivalent constructions, which may be described as eccentric-on-eccentric and epicycle-on-eccentric, respectively. Mercury required a more complicated combination of circles. Maestlin based his calculation of the distances of the planets from the sun on Copernicus's Mercury theory and eccentric-on-eccentric representations for the other planets. Kepler took his planetary distances directly from Copernicus and sought a physical basis for the Ptolemaic equant.

The Eccentric

Let E (fig. 3) be the center of the earth's orbit and C a point on the line of apsides of the planet, such that ED = 2e, where 2e is the eccentricity as defined by Ptolemy; that is, the eccentricity of the center of uniform motion. Then the eccentric, with center D, is taken to be the path of the planet.

The Equant

In the case of the superior planets Ptolemy found that the planet moved not on the eccentric with center D (fig. 2) but on an equal eccentric (the deferent) with center C, where EC = CD = e. Thus the eccentricity of the center of equal distances C is half the eccentricity of the center of equal angular motion D, so that the eccentricity may be said to be bisected. In this representation, the eccentric circle with center D is known as the equant circle and its center as the equant point. Both the circle and its center are often referred to simply as the equant.

Eccentric-on-eccentric

This representation is called by Copernicus *eccentri eccentrus, eccentricus eccentrici* and *eccentreccentricus.*

Let E (fig. 3) be the center of the earth's orbit and C a point on the line of apsides of the planet such that EC = ϵ where ϵ is three-quarters of the eccentricity of the planetary orbit considered as a simple eccentric (i.e., three-quarters of the eccentricity of the equant in the Ptolemaic theory). With center C and radius ⅓ ϵ construct the small eccentric FG. Then with center F and radius a construct the large eccentric LM.

Suppose that initially the planet is at L. As the planet moves uniformly on the large eccentric, the center F moves along the small eccentric in the same sense and with twice the angular velocity. It follows that, when the planet is in apogee at L, the center of the large eccentric is at F and EL = a + ⅔ ϵ ; when the planet is in perigee at M, the center of the large eccentric is again at F and EM = a − ⅔ ϵ . But when the planet is in the mean distances, the center of the large eccentric is at G. The path of the planet is nearly circular. (Cf. p. 252, fig 2.)

In this representation, one-quarter of the eccentricity is assigned to the small eccentric and three-quarters to the large eccentric. By this distribution of the eccentricity, Copernicus was able to approximate the Ptolemaic theory of the inequality of the planet's motion without having to depart from the principle of uniform circular motion by postulating an equant.

Appendix

Epicycle-on-eccentric-deferent

This representation is called by Copernicus *eccentrepicyclus*. Let E (fig. 4) be the center of the earth's orbit and C a point on the line of apsides of the planet such that EC = ε , where ε , as before, is three-quarters of the eccentricity of the planetary orbit condsidered as a simple eccentric. With center C and radius a construct the deferent AB, and with center A and radius ⅓ ε construct the epicycle FG. Then, as the center A of the epicycle moves uniformly on the deferent, the planet, initially at F, moves uniformly, relatively to the rotating radius CA of the deferent, with the same angular velocity. The dotted curve shows the path of the planet. The planet is in apogee at F, when EF = a + ⅔ ε , and in perigee at H, when EH = a − ⅔ ε .

This representation is geometrically equivalent to the eccentric-on-eccentric, the small eccentric having been exchanged for an epicycle.

The Mercury Theory

The Mercury theory is described by Copernicus in *De revolutionibus*, Book 5, chapter 25 and the derivation of the numerical parameters from the observations is given in chapter 27.

Let E (fig. 5) be the center of the earth's orbit ATB and let AB be the line of apsides of the planet. Then with center C, a point on AB, describe the small eccentric FG, and with center G, describe the large eccentric HI. Also, with center I, describe the small epicycle KL. Suppose now that the center G of the large eccentric describes the small eccentric twice in a year, while I completes a revolution of the large eccentric in Mercury's sidereal period of 88 days. Suppose also that the diameter LK of the epicycle always points to the center of the large eccentric. Then the planet completes two oscillations on the epicycle diameter in the course of a year, so that, when the earth is at A or B (on the line of apsides) the center of the large eccentric is at G and the planet at K, and when the earth is 90° from A or B, the center of the large eccentric is at F and the planet at L. The introduction of the oscillation on the epicycle diameter does not violate the principle of uniform circular motion, since the oscillation can be regarded as compounded of uniform circular motions.

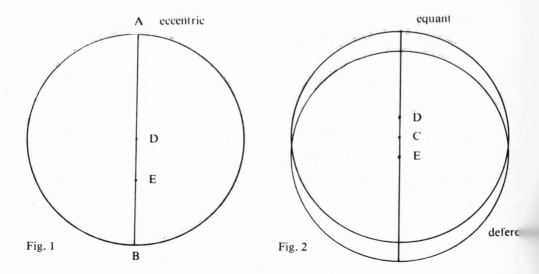

Fig. 1   Fig. 2

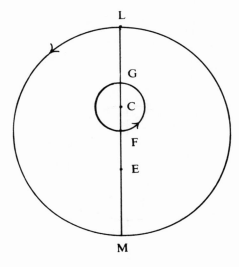

Fig. 3

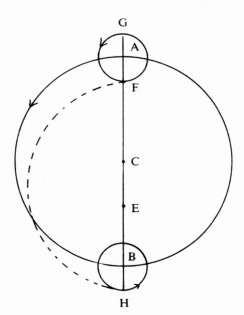

Fig. 4

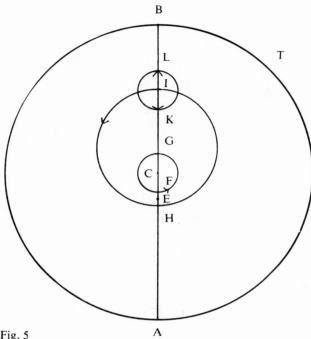

Fig. 5

SELECT BIBLIOGRAPHY

Abbreviations

K G W = Johannes Kepler, *Gesammelte Werke*, Munich, 1937 —.
K F R = E. Preuss (ed.) *Kepler Festschrift 1971*, Regensburg, 1971.
K L C = *Johannes Kepler, Werk und Leistung*, Linz, 1971.
K S W = F. Krafft, K. Meyer, B. Sticker (eds.) *Internationales Kepler-*
Symposium, Weil der Stadt 1971, Hildesheim, 1973.

Works by Kepler

Joannis Kepleri astronomi opera omnia, edited by C. Frisch, with commentary in
Latin, Frankfurt-Erlangen, 1858-1871.
Johannes Kepler, *Gesammelte Werke,* edited by Walther von Dyck, Max Caspar,
Franz Hammer and Martha List, with commentary in German, Munich,
1937 —.
...*Das Weltgeheimnis,* translated with commentary by Max Caspar, Augsburg,
1923 and Munich-Berlin, 1936.
...*Neue Astronomie,* translated with commentary by Max Caspar, Munich-
Berlin, 1929.
...*Weltharmonik,* translated with commentary by Max Caspar, Munich-Berlin,
1939, reprinted Darmstadt, 1967.
...*Conversation with Galileo's sidereal messenger,* translated with commentary
by E. Rosen, New York, 1965.
...*The six-cornered snowflake,* translated by C. Hardie, Oxford, 1966.
...*L'Étrenne ou la neige sexangulaire,* translated by R. Halleux, Paris, 1975.
...*Somnium,* translated with commentary by E. Rosen, Madison-London, 1967.
...*Selbstzeugnisse,* edited with introduction by Franz Hammer, translated by
Esther Hammer, Stuttgart-Bad Cannstatt, 1971.
Johannes Kepler in seinen Briefen, selections in translation, edited by Max
Caspar and Walther von Dyck, Munich-Berlin, 1930.

Bibliographical Works

Max Caspar, *Bibliographia Kepleriana,* Munich 1936; second edition, revised by
Martha List, Munich, 1968.
Martha List, Bibliographia Kepleriana, 1967-1975, in *Vistas in astronomy,* 18
(1975), 955-1012.

Other Works

E. J. Aiton, Kepler's second law of planetary motion, *Isis,* 60 (1969), 75-90.
...Johannes Kepler and the Astronomy without hypotheses, *Japanese Studies in
the history of science,* 14 (1975), 49-71.
...Johannes Kepler in the light of recent research, *History of science,* 14 (1976),
77-100.
...Johannes Kepler and the 'Mysterium Cosmographicum,' *Sudhoffs Archiv,* 61
(1977), 173-194.
Aristotle, *De caelo* (Loeb Classical Library).
...*Posterior analytics* (Loeb Classical Library).
H. Balmer, Keplers Beziehungen zu Jost Bürgi und anderen Schweizern, in KLC.
Carola Baumgardt, *Johannes Kepler: life and letters,* New York, 1951.
A. Beer and P. Beer (eds.), *Kepler. Four hundred years.* A special volume of
Vistas in astronomy (vol. 18) containing the Proceedings of Conferences held
in honor of Johannes Kepler, 1975.

Bibliography

Tycho Brahe, *Opera omnia,* edited by J. L. E. Dreyer, Copenhagen, 1913-1929.
Ruth Breitsohl-Klepser, *Heiliger ist mir die Wahrheit Johannes Kepler,* edited by Martha List, Stuttgart, 1976.
G. Buchdahl, Methodological aspects of Kepler's theory of refraction, in KSW.
K. H. Burmeister, *Georg Joachim Rheticus,* Wiesbaden, 1967-1968.
M. Caspar, *Kepler,* translated by C. Doris Hellman, London and New York, 1959.
I. B. Cohen, Dynamics, the key to the new science of the seventeenth century, *Acta historiae rerum naturalium necnon technicarum,* Special Issue No. 3 (Prague, 1967), pp. 83-100.
...Kepler's century, *Vistas in astronomy,* 18 (1975), 3-36.
N. Copernicus, *De revolutionibus orbium coelestium,* facsimile of Kepler's copy, with introduction by Johannes Müller, New York and London, 1965.
...*On the revolutions of the heavenly spheres,* translated with introduction and notes by A. M. Duncan, London, Vancouver and New York, 1976.
...*On the revolutions,* translated by Edward Rosen, Warsaw and London, 1978.
Nicolai Cusae Cardinalis Øpera, Paris, 1514; reprinted, Frankfurt am Main, 1962.
M. Dickreiter, Dur und Moll in Keplers Musiktheorie, in KLC.
...*Der Musiktheoretiker Johannes Kepler,* Bern and Munich, 1973.
P. Duhem, *Le système du monde,* Paris, 1913-1959.
M. Ficino, *Théologie platonicienne,* translated by R. Marcel, Paris, 1964-1970.
Walther Gerlach and Martha List, *Johannes Kepler, Dokumente zu Leben und Werk,* Munich, 1971.
O. Gingerich, Johannes Kepler and the new astronomy, *Quarterly Journal of the Royal Astronomical Society,* 13 (1972), 346-373.
...The origins of Kepler's third law, *Vistas in astronomy,* 18 (1975), 595-601.
...Kepler's treatment of redundant observations, in KSW.
A. Grafton, Michael Maestlin's account of Copernican planetary theory, *Proceedings of the American Philosophical Society,* 117 (1973), 523-550.
R. Haase, Marginalien zum 3. Keplerschen Gesetz, in KFR.
F. Hammer, Die Astrologie des Johannes Kepler, *Sudhoffs Archiv,* 55 (1971), 113-135.
W. Hartner, *Oriens, occidens,* Hildesheim, 1968.
C. Doris Hellman, *The comet of 1577: its place in the history of astronomy,* New York, 1971.
R. Hooykaas, *Humanisme, science et réforme: Pierre de la Ramée,* Leiden, 1968.
J. Hübner, *Die Theologie Johannes Keplers zwischen Orthodoxie und Naturwissenschaft,* Tübingen, 1975.
H. Hugonnard-Roche, E. Rosen and J. P. Verdet, *Introductions à l'astronomie de Copernic,* Paris, 1975.
Johannes Kepler — Werk und Leistung, Katalog der Ausstellung Linz 19 Juni bis 29 August 1971, Linz, 1971.
Kepler und Tübingen, Tübingen Kataloge Nummer 13, published by the Kulturamt der Stadt Tübingen, 1971.
U. Klein, Johannes Keplers Bemühungen um die Harmonieschriften des Ptolemaios und Porphyrios, in KLC.
A. Koestler, *The Sleepwalkers,* London, 1968.
A. Koyré, *From the closed world to the infinite universe,* Baltimore, 1957.
...*The astronomical revolution,* translated by R. E. W. Maddison, Paris, London and New York, 1973.
F. Krafft, K. Meyer and B. Sticker (eds.), *Internationales Kepler-Symposium Weil der Stadt 1971* (= *Arbor scientiarum,* Reihe A, Band 1), Hildesheim, 1973.

Bibliography

F. Krafft, Physikalische Realität oder mathematische Hypothese? *Philosophia naturalis,* 14 (1973), 243-275.

...Johannes Keplers Beitrag zur Himmelsphysik, in KSW.

:M. Maestlin, *Ephemerides novae ab annos 1577 ad annum 1590,* Tübingen, 1580.

...*Disputatio de eclipsibus solis et lunae,* Tübingen, 1596.

D. Mahnke, *Unendliche Sphäre und Allmittelpunkt,* Halle 1937; reprinted Stuttgart-Bad Cannstatt, 1966.

J. Mittelstrass, *Die Rettung der Phänomene. Ursprung und Geschichte eines antiken Forschungsprinzips,* Berlin, 1962.

...*Neuzeit und Aufklärung,* Berlin, 1970.

...Methodological elements of Keplerian astronomy, *Studies in history and philosophy of science,* 3 (1972), 203-232.

K. P. Moesgaard, The 1717 Egyptian years and the Copernican theory of precession, *Centaurus,* 13 (1968), 120-138.

H. Monantholius, *Aristotelis mechanica Graeca, emendate, Latina facta, & commentariis illustrata,* Paris, 1599.

A. Nitschke, Keplers Staats- und Rechtslehre, in KSW.

H. M. Nobis, Ropé und nutus in Keplers Astronomie, in KFR.

Florian de Ocampo, *De la Corónica de España,* continued by Ambrosio de Morales, Alcalá de Henares, 1574-1586.

W. Petri, Die betrachtende Kreatur in trinitarischen Kosmos, in KFR.

Plato, *Republic* (Loeb Classical Library).

...*Timaeus* (Loeb Classical Library).

Timaeus Platonis, sive de universitate, interpretibus M. Tullio Cicerone & Chalcidio, una cum eius docta explanatione, Paris, 1563.

Plutarch, *De facie quae in orbe lunae apparet* (*Moralia,* vol. 12, Loeb Classical Library).

F. D. Prager, Kepler als Erfinder, in KSW.

E. Preuss (ed.), *Kepler Festschrift 1971* (= *Acta Albertina Ratisbonensia,* Band 32), Regensburg, 1971.

Proclus, *Commentaire sur le Timée,* translated by A. J. Festugière, Paris, 1966-1968.

L. Prowe, *Nicolaus Coppernicus,* Berlin, 1883-1884.

Ptolemy, *Tetrabiblos* (Loeb Classical Library).

...*Handbuch der Astronomie,* translation of the *Almagest* by K. Manitius, reprinted Leipzig, 1963.

J. R. Ravetz, *Astronomy and cosmology in the achievement of Nicolaus Copernicus,* Warsaw, 1965.

E. Reinhold, *Prutenicae tabulae coelestium motuum,* Wittenberg, 1585.

E. Rosen, *Three Copernican treatises,* New York, 1971.

...Kepler and the Lutheran attitude to the Copernican hypothesis, in KLC.

J. L. Russell, Kepler and scientific method, *Vistas in astronomy,* 18 (1975), 733-745.

G. Simon, Kepler's astrology: the direction of a reform, *Vistas in astronomy,* 18 (1975), 439-448.

Simplicius, *In Aristotelis de caelo commentaria* (= *Commentaria in Aristotelem Graeca,* vol. 7), edited by J. L. Heiberg, Berlin, 1894.

B. Sticker, Johannes Kepler — homo iste, in KSW.

N. Swerdlow, The derivation of the first draft of Copernicus's planetary theory: a translation of the Commentariolus with commentary, *Proceedings of the American Philosophical Society,* 117 (1973), 423-512.

D. P. Walker, Kepler's celestial music, *Journal of the Warburg and Courtauld Institutes,* 30 (1967), 228-250.

Bibliography

R. S. Wesiman, Kepler's theory of hypothesis and the realist dilemma, in KSW.

...The Melanchthon circle, Rheticus and the Wittenberg interpretation of the Copernican theory, *Isis,* 66 (1975), 165-193.

...The comet and the cosmos: Kepler, Mästlin and the Copernican hypothesis, in J. Dobrzycki (ed.), *The reception of Copernicus' heliocentric theory,* Dordrecht, 1972.

C. Wilson, Kepler's derivation of the elliptical path, *Isis,* 59 (1968), 5-25.

...How did Kepler discover his first two laws? *Scientific American,* 226 (1972), 93-106.

H. A. Wolfson, The problem of the souls of the spheres from the Byzantine commentaries on Aristotle through the Arabs and St. Thomas to Kepler, *Dumbarton Oaks papers,* No. 16, Washington, 1962, pp. 65-93.

E. Zinner, *Leben und Wirken des Johannes Müller von Königsberg genannt Regiomontanus,* Munich, 1938.

INDEX

A

accidental proof, 75
Aiton, E.J., 7, 9, 30, 31, 253, 259
Alfonso, 22, 75
Alfonsine tables, 83, 235, 236
Ambrosio de Morales, 233, 261
America, 51
Anaxagoras, 169
anima movens, 18, 22, 23, 27, 199, 203. *See also* solar force
a posteriori derivation, 17, 77, 97
a priori reasons (or archetypal causes), 8, 9, 17-27, 59, 79, 99, 139, 141, 175, 187, 225, 236-40, 247; confirmed by experience, 24; for eccentricities, 187, 247; for harmonies, 141; hypothetical nature of, 24, 237, 238, 243; for number, dimensions and arrangements of planetary orbits, 9, 18, 21, 236; for order of polyhedra, 24, 238; for peculiarity of Mercury, 25, 26, 175, 247; for planetary distances, 20; source in geometrical relations, 19
Aristarchus, 207
Aristotle, 21, 31, 77, 95, 109, 125, 127, 129, 169, 197, 234-36, 240, 250, 259; *per se* rule, 235
artifices, 13. *See also* practitioners
aspects, astrological, 25, 135, 137, 145, 147, 195, 240-42; efficacy of, 25, 145, 195, 241, 242; origin of, 147, 241; properties of, 137, 241; relation to musical harmonies, 25, 135, 137, 241, 242
astrology, 24, 25, 171, 195, 239; influences of the heavens, 171, 195; natural powers of the planets, 24, 115-19, 195, 239
atmospheric refraction, 195, 248
Averroes, 169

B

Bacon, R., 15
Balmer, H., 243, 259
Baumgardt, C., 9, 259
Birkenmajer, L.A., 14
Bodin, J., 236
Brahe, Tycho, 10, 20-27, 39, 45, 59, 61, 77, 87, 91, 103, 163, 187, 195, 207, 217, 221, 237, 247, 248, 260
Buchdahl, G., 31, 260
Burmeister, K.H., 260

C

Campanus de Novara, 149-51, 238, 243
Cardanus, 241
Caspar, M., 13, 259, 260
Charles V, 57, 61, 233
Cicero, 23, 45, 69, 93, 193, 234, 237
Cohen, I.B., 10, 247, 260
colures, solsticial and equinoctial, 83, 236
comets, 10, 87
Copernicus, N., 17-30, 49, 75-107, 155-59, 161, 165-99, 215-23, 235-38, 245-48, 253-57, 260
Copernican system, 7-29, 49, 59-69, 75, 85-93, 157-59, 177-79, 197, 209-11, 227-31, 233-38, 255-58; distances of planets in, 8, 19-28, 157-59, 177-79, 197, 209-11; explanatory power of, 17, 235; motion of the earth in, 17, 235; perfect numbers in, 10; physical truth of, 235; reconciliation of, with the Bible, 9, 19-22, 75, 85, 235; representation of orbits in, 20, 255-58; unknown to Pythagoras, 238; Wittenberg interpretation of, 235
corpus, 14, 92, 93
creation, idea of, 17-24, 49, 53, 63, 93-99, 125, 209, 237
Crombie, A.C., 15
Crusius, M., 20, 30
Cusanus, 19, 23, 93, 237, 260

D

Dasypodius, C., 147, 243
demonstro and *demonstratio,* 14
Dickreiter, M., 240, 242, 243, 260
Diogenes Laertius, 169, 246
diurnal motion, 85, 91, 236
Dreyer, J.L.E., 237
Duhem, P., 236, 260
Duke of Württemberg, 19, 22, 233, 238; dedication of Plate III to, 22, 238
Duncan, A.M., 7, 30, 260
Düring, I., 241

E

earth, position of, 24, 238
eclipses, mythological explanation of, 253
eccentric, 79-89, 175, 237, 244, 247, 255

263

Index

M

Maestlin, M., 17-29, 59, 63, 69, 79, 81, 161, 165, 175, 181-93, 215, 223, 235-38, 244-50, 255, 261; atmospheric refraction observed by, 189, 248; on comet of 1577, 236; Copernican system accepted by, 17, 22, 236; distances of planets calculated by, 19, 20, 25, 161, 183, 244, 245; letter on Mercury, 20, 25, 193, 250; reports on Kepler's work by, 19-21, 235

magnetism, 91, 171

Mahnke, D., 237, 261

Mercury, 25, 26, 173, 175, 191-95, 243, 247, 250, 256-58; archetypal reason for peculiarity of, 25, 26, 247; astrological influence of, 191, 195; and the circle in the octahedron-square, 25, 173, 175, 243, 247; Copernican theory of, 256-58; large eccentricity of, 175, 247; letter of Maestlin concerning, 20, 25, 193, 250

Mittelstrass, J., 31, 261

Moesgaard, K.P., 236, 261

Monantholius, H., 252, 253, 261

moon, similar in nature to the earth, 24, 26, 165

motions of the planets, 18, 20, 27-30, 41, 63, 75, 83, 85, 97, 159-63, 169, 179, 181, 187, 197-213, 223, 225, 234, 235, 249, 250; correspondence with harmonic ratios, 187; number, extent and times of retrogressions, 75, 235; periodic times of, 197, 199, 205, 207, 223, 225, 249; physical cause of, 18, 27, 28, 29, 30, 199-203, 234; in relation to the distances, 20, 28, 63, 197-213, 249, 250; in relation to the zodiac, 159-63

mundus, 14

musical harmony, 19, 25, 27, 131-47, 225 239-42, 251-53; archetypal cause of, 141, 145; classification of concords, 133, 135, 143, 145, 240; consonances, 251, 252; differences of melodic intervals, 225, 252; melodic intervals, 141, 225, 242, 253; relation with astrological aspects, 25, 242; tempered intervals, 242; terminology, 240, 242; tonality, 240

N

Neoplatonism, 15

Newton, I., 9

Nicholas of Cusa. *See* Cusanus

ninth sphere, 83, 91, 236

Nitschke, A., 236, 250, 261

Nobis, H.M., 253, 261

numbers, 10, 19, 24, 49, 65, 71, 73, 109, 119, 121, 139, 234; counted (*numerati*), 24, 73, 234; counting (*numerantes*), 71, 139, 234; nobility (perfection) of, 10, 65, 71, 73, 119, 121; perfection (nobility) of number three, 109; source of nobility, 24, 49

nutus, 250, 253

O

orbis, 14, 15

order and arrangement of polyhedra, 24, 107-15, 123-29, 238; *a priori* reasons for, 238; empirical evidence for, 238

P

per se rule, 77, 235

Peurbach, G., 175, 185

Petri, W., 234, 240, 261

Philoponus, 252

planets, 7-9, 18, 24, 63, 115-19, 237, 239; astrological properties of, 24, 115-19, 239; imaginary, 63-65, 234; number, order and size of orbits of, 8, 9, 63; position of orbits in relation to the sun, 237. *See also* laws of planetary motion; motions of the planets

Plato, 8, 23, 30, 31, 43, 61, 63, 93, 97, 221, 234, 236, 238, 239, 261; construction of the world-soul, 234

Platonic Year, 221-23, 253

Pliny, 253

Plutarch, 169, 238, 246, 261

polyhedra, the five regular, 9, 17-28, 49, 61-71, 97-121, 143, 149-53, 167, 173-79, 197, 234, 238-40, 253; association with elements, 99, 238; circumscribed and inscribed spheres of, 21, 23, 240; classification of, 24, 105, 113, 115, 121, 149-53; properties of, 26, 101-21, 149-53, 239; relation with musical harmonies, 143

polyhedral hypothesis, 19-31, 69, 71, 97, 99, 107-15, 123-29, 141, 155-63, 169, 177-93, 209-15, 234-38, 243-53; astronomical proof of, 155; compared with distances derived from motions, 28, 209, 250, 251; disagreement with the motions, 209-15; discrepancies concerning individual planets, 189-93; errors in calculation, 25, 26, 31, 244-48; order and arrangement of polyhedra in, 24, 107-15, 123-29, 238; relation with musical harmonies, 25, 169; tests of, 19-28, 99, 157-63, 177, 179, 189, 191, 211-15, 243-48

265

Z

THE JANUS LIBRARY